Lecture Notes in Computer Science

Commenced Publication in 1973
Founding and Former Series Editors:
Gerhard Goos, Juris Hartmanis, and Jan van Leeuwen

T0237822

Alexandre Petrenko   Margus Veanes
Jan Tretmans   Wolfgang Grieskamp (Eds.)

Testing of Software and Communicating Systems

19th IFIP TC6/WG6.1 International Conference, TestCom 2007
7th International Workshop, FATES 2007
Tallinn, Estonia, June 26-29, 2007
Proceedings

 Springer

Volume Editors

Alexandre Petrenko
CRIM
Montreal, Canada
E-mail: petrenko@crim.ca

Margus Veanes
Microsoft Research
Redmond, WA, USA
E-mail: margus@microsoft.com

Jan Tretmans
Embedded Systems Institute
Eindhoven, The Netherlands
E-mail: jan.tretmans@esi.nl

Wolfgang Grieskamp
Microsoft Research
Redmond, WA, USA
E-mail: wrwg@microsoft.com

Library of Congress Control Number: 2007928493

CR Subject Classification (1998): D.2.5, D.2-3, C.2, F.3, K.6

LNCS Sublibrary: SL 5 – Computer Communication Networks and
Telecommunications

ISSN   0302-9743
ISBN-10   3-540-73065-6 Springer Berlin Heidelberg New York
ISBN-13   978-3-540-73065-1 Springer Berlin Heidelberg New York

Springer is a part of Springer Science+Business Media

springer.com

© Springer-Verlag Berlin Heidelberg 2007
Printed in Germany

Typesetting: Camera-ready by author, data conversion by Scientific Publishing Services, Chennai, India
Printed on acid-free paper   SPIN: 12076826   06/3180   5 4 3 2 1 0

Preface

Testing is one of the most important techniques for validating the correctness of communicating and software systems. Triggered by the quest for improved quality of systems, and by the high and ever-increasing costs of making and executing the necessary tests, testing has evolved during the last two decades from an ad-hoc and under-exposed area of systems development to an important and active area, where a lot of research and development, in academia as well as in industry, is taking place. This evolvement is also reflected in an increasing number of testing conferences and workshops being regularly organized. Two of these, TestCom the 19th IFIP TC6/WG6.1 International Conference on Testing of Communicating Systems – and Fates the seventh International Workshop on Formal Approaches to Testing of Software, were jointly held in Tallinn, Estonia, June 27–29, 2007. In addition, FORTE, the 27th IFIP International Conference on Formal Methods for Networked and Distributed Systems, was also held in Tallinn during these dates, together thus forming a large event on testing, validation, and specification of software, communicating, and distributed systems.

The objective of TESTCOM/FATES 2007 was to offer a forum for researchers from academia as well as industry, developers, and testers to present, discuss, and learn about new approaches, theories, methods, and tools in the field of testing of software and communicating systems.

TESTCOM - Testing of Communicating Systems is an IFIP-sponsored series of international conferences, previously also called the International Workshop on Protocol Test Systems (IWPTS) or International Workshop on Testing of Communicating Systems (IWTCS). It is devoted to testing of communicating systems, including testing of communication protocols, services, distributed platforms, and middleware. The previous events were held in Vancouver, Canada (1988); Berlin, Germany (1989); McLean, USA (1990); Leidschendam, The Netherlands (1991); Montreal, Canada (1992); Pau, France (1993); Tokyo, Japan (1994); Evry, France (1995); Darmstadt, Germany (1996); Cheju Island, Korea (1997); Tomsk, Russia (1998); Budapest, Hungary (1999); Ottawa, Canada (2000); Berlin, Germany (2002); Sophia Antipolis, France (2003); Oxford, UK (2004); Montreal, Canada (2005); and New York, USA (2006).

FATES - Formal Approaches to Testing of Software is a series of workshops devoted to the use of formal methods in software testing. Previous events were held in Aalborg, Denmark (2001); Brno, Czech Republic (2002); Montreal, Canada (2003); Linz, Austria (2004); Edinburgh, UK (2005); and Seattle, USA (2006).

This volume contains the proceedings of TESTCOM/FATES 2007, the joint conference of TESTCOM and FATES. Out of 61 submitted papers, the Program Committee selected 24 papers for presentation at the conference. Together with the invited presentation by Antti Huima from Conformiq Software Ltd., Finland, they form the contents of these proceedings. The conference itself, in addition, contained another invited presentation, jointly with Forte, by Susanne Graf from Verimag (France), and presentations of work-in-progress papers, position papers, short experience reports, and tool demonstrations, which were separately published. A tutorial day preceded the main conference.

We would like to thank the numerous people who contributed to the success of TESTCOM/FATES 2007: the Steering Committee of IFIP for the TEST-COM conference, the Program Committee and the additional reviewers for their support in selecting and composing the conference program, and the authors and the invited speakers for their contributions without which, of course, these proceedings would not exist. We thank Conformiq Software Ltd., CRIM, and Microsoft Research for their financial support and Springer for their support in publishing these proceedings. We acknowledge the use of EasyChair for the conference management and wish to thank its developers. Last, but not least, we thank the Institute of Cybernetics at Tallinn University of Technology and the Department of Computer Science of TUT, in particular, Juhan Ernits, Monika Perkmann, Jaagup Irve, Ando Saabas, Kristi Uustalu, and Tarmo Uustalu, for all matters regarding the local organization and for making TestCom/Fates 2007 run smoothly.

April 2007

Alexandre Petrenko
Margus Veanes
Jan Tretmans
Wolfgang Grieskamp

Conference Organization

Program Chairs

Alexandre Petrenko (CRIM, Canada)
Margus Veanes (Microsoft Research, USA)
Jan Tretmans (Embedded Systems Institute, The Netherlands)
Wolfgang Grieskamp (Microsoft Research, USA)

Steering Committee

John Derrick (University of Sheffield, UK), Chair
Ana R. Cavalli (INT, France)
Roland Groz (LSR-IMAG, France)
Alexandre Petrenko (CRIM, Canada)

Program Committee

Bernhard K. Aichernig (Graz University of Technology, Austria)
Paul Baker (Motorola, UK)
Antonia Bertolino (ISTI-CNR, Italy)
Gregor v. Bochmann (University of Ottawa, Canada)
Juris Borzovs (LVU, Latvia)
Rachel Cardell-Oliver (The University of Western Australia, Australia)
Richard Castanet (LABRI, France)
Ana R. Cavalli (INT, France)
John Derrick (University of Sheffield, UK)
Sarolta Dibuz (Ericsson, Hungary)
Khaled El-Fakih (American University of Sharjah, UAE)
Marie-Claude Gaudel (University of Paris-Sud, France)
Jens Grabowski (University of Gottingen, Germany)
Roland Groz (LSR-IMAG, France)
Rob Hierons (Brunel University, UK)
Teruo Higashino (Osaka University, Japan)
Dieter Hogrefe (University of Gottingen, Germany)
Antti Huima (Conformiq Software Ltd., Finland)
Thierry Jeron (IRISA Rennes, France)
Ferhat Khendek (Concordia University, Canada)
Myungchul Kim (ICU, Korea)
Victor Kuliamin (ISP RAS, Russia)
Hartmut Konig (BTU Cottbus, Germany)
David Lee (Ohio State University, USA)

Bruno Legeard (Leirios, France)
Alexander Letichevsky (Institute of Cybernetics, Ukraine)
Giulio Maggiore (Telecom Italia Mobile, Italy)
Brian Nielsen (University of Aalborg, Denmark)
Manuel Núñez (UC de Madrid, Spain)
Ian Oliver (Nokia Research, Finland)
Doron Peled (University of Bar-Ilan, Israel)
Alexander Pretschner (ETH Zurich, Switzerland)
Harry Robinson (Google, USA)
Vlad Rusu (IRISA Rennes, France)
Ina Schieferdecker (Fraunhofer FOKUS, Germany)
Kenji Suzuki (University of Electro-Communications, Japan)
Andreas Ulrich (Siemens, Germany)
Hasan Ural (University of Ottawa, Canada)
Mark Utting (University of Waikato, New Zealand)
M Umit Uyar (City University of New York, USA)
Juri Vain (Tallinn University of Technology, Estonia)
Carsten Weise (Ericsson, Germany)
Burkhart Wolff (ETH Zurich, Switzerland)
Jianping Wu (Tsinghua University, China)
Nina Yevtushenko (Tomsk State University, Russia)
Zheng Zhang (Microsoft Research, China)

Local Organization

Juhan Ernits, Monika Perkmann, Jaagup Irve, Ando Saabas, Kristi Uustalu,
Tarmo Uustalu (Tallinn University of Technology, Estonia)

External Reviewers

Benharef Abdel
Jongmoon Baik
Lydie du Bousquet
Henrik Brosenne
Patryk Chamuczynski
Camille Constant
Guglielmo De Angelis
Jeremy Dubreil
Maxim Gromov
Toru Hasegawa
Hatem Hamdi
Jukka Honkola
Akira Idoue
Guy-Vincent Jourdan
Sungwon Kang

Fang-Chun Kuo
Mounir Lallali
Shuhao Li
Luis Llana
Stephane Maag
Wissam Mallouli
Mercedes G. Merayo
Marius Mikucionis
Helmut Neukirchen
Tomohiko Ogishi
Jean-Marie Orset
Patrizio Pelliccione
Larry Quo
Ismael Rodriguez
Soonuk Seol

Table of Contents

Implementing Conformiq Qtronic
(Invited Talk)

Antti Huima

Conformiq Software Ltd
antti.huima@conformiq.com

Abstract. Conformiq Qtronic[1] is a commercial tool for model driven testing. It derives tests automatically from behavioral system models. These are black-box tests [1] by nature, which means that they depend on the model and the interfaces of the system under test, but not on the internal structure (e.g. source code) of the implementation.

In this essay, which accompanies my invited talk, I survey the nature of Conformiq Qtronic, the main implementation challenges that we have encountered and how we have approached them.

Problem Statement

Conformiq Qtronic is an ongoing attempt to provide an industrially applicable solution to the following technical problem: Create an automatic method that provided an object (a model) M that describes the external behavior of an open system (begin interpreted via well-defined semantics) constructs a strategy to test real-life black-box systems (implementations) I in order to find out if they have the same external behavior as M.

(Now the fact that we have created a company with the mission to solve this technical problem hints that we believe it to be a good idea to *employ* automatic methods of this kind in actual software development processes—which is not *a priori* self-evident. In this essay, however, I will not touch these commercial and methodological issues, but will concentrate on the technical problem only.)

Two notes are due, and the first concerns the *test harness*. It is namely usually so that the model M describes a system with different interfaces and behavior than the real implementation that is being tested (IUT, implementation under test). Virtually always, the model M is somehow described on a higher level of abstraction than the real implementation. As a concrete example, SIP (Session Initiation Protocol) [2] is a packet-oriented protocol that is run over a transport (e.g. UDP [3]) and that has an ASCII encoding: every SIP packet is encoded as a series of ASCII characters. However, some possible model M for testing an SIP implementation could define the logical flow of packets with their main contents but not describe, for instance, their actual encoding. How can this model M be used to test a real SIP implementation that requires ASCII encoded packets over UDP? The answer is that the "abstract" tests generated automatically from

---

[1] www.conformiq.com

A. Petrenko et al. (Eds.): TestCom/FATES 2007, LNCS 4581, pp. 1–12, 2007.
© IFIP- International Federation for Information Processing 2007

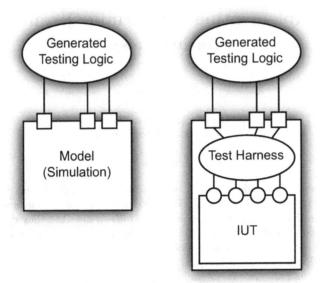

Fig. 1. The relation between a model and a corresponding implementation under test. The test harness bridges the higher-level model with the concrete executable implementation. The logic of the test harness itself is not explicitly tested by the generated testing logic.

M are run through a *test harness* that provides the "low-level details" such as encoding and decoding data and managing the UDP packets.

What does this mean in terms of testing? The automatic tests derived from M are not testing the codec layer of the SIP implementation *per se*, because the codec was not specified in M. Instead, the tests focus on the "high-level" logic above the codec layers. See Fig. 1.

This is a very commonly occurring pattern. Yet testing the codecs could also benefit from model-based testing. In practical deployments this would be usually carried out by using a different model and different testing set-up that would focus *solely* on the testing of the codecs. Regardless, we have found that succesful deployments of model-based testing usually focus on the "higher level" issues.

The second note concerns the models M in the context of Conformiq Qtronic. For us, the models are—on the internal level—multithreaded Scheme [4] programs that can "talk" with an external, unspecified (open) environment, and the semantics for these programs are given via small-step virtual machine operational semantics [5,6], defined in the corresponding language specification [7].

Design of Conformiq Qtronic

The development of Conformiq Qtronic has always been driven by two major design goals, which are *industrial applicability* and *openess*. In other words, we have strived to create a tool that can be applied to real-world problems by "normal" software engineers in the industry, and that can be integrated easily into heterogenous software development tool chains.

The most visible consequence is that Conformiq Qtronic allows the user to use in modeling many constructs that are known to create difficulties for model checking algorithms. Namely, Conformiq Qtronic supports infinite data types (e.g. rational numbers and strings), fully dynamic data with garbage collection, classes with static and dynamic polymorphism, concurrency, and timings. Furthermore, the user is not asked to provide abstractions [8, 9] or slicings [10, 11] of the model, as the design philosophy has been that Conformiq Qtronic should relieve the user of such technicalities.

CHALLENGE:
Legal models can have arbitrarily complex, infinite state spaces

It follows immediately that there exist models that are too difficult for Conformiq Qtronic to handle. Our solution to this problem is that we do not try to solve it, because to provide the tool's *raison d'être*, it is sufficent that there exist *enough* contexts where Conformiq Qtronic can create substantial value.

Another consequence is that many of the known methods for model checking and test generation that assume finite-state specifications [12,13,14,15,16,17,18, 19,20], or even only finitely branching specifications, are of limited value for us. I do not claim that research into model driven testing from finite-state models would be useless in general, especially as there exist case studies that prove otherwise [21]. However, our experience is that the number of industrial systems that have natural finite-state specifications is limited. There exists certainly a larger body of systems that have somewhat natural finite-state *abstractions*, but as I already mentioned, we do not want to force the user to do abstraction manually.

Modeling

The main artifact the user of Conformiq Qtronic must manage is the model, and we believe it is very important that the model can be described in a *powerful* language that is *easy to adopt*.

CHALLENGE:
Providing industrial-strength specification and modeling language

In the out-of-the-box version of Conformiq Qtronic, the behavioral models are combinations of UML statecharts [22, 23] and blocks of an object-oriented programming language that is basically a superset of Java [24] and C# [25]. We have dubbed this semi-graphical language QML (Qtronic Modeling Language). The use of UML statecharts is optional so that models can be described in pure textual notation also.

QML [26] extends Java by adding value-type records (which exist already in C#), true static polymorphism (i.e. templates or generics), discretionary type inference [27,28,29], and a system for free-form macros (see Fig. 2). To avoid any misunderstanding it must be mentioned that QML does *not* include the standard libraries that come with Java or C#.

```
_rec -> _port ::= _port.send(_rec, -1)
```

Fig. 2. A free-form macro that gives a new syntax for sending data through ports without timeouts. Note that there is no built-in meaning for the arrow literal, and when matched, _port and _rec can correspond to arbitrary expressions.

A QML program must specify an open system, i.e. a system with one or more message passing interfaces that are open to the "environment" (please see Fig. 1 again). These interfaces correspond to the testing interfaces of the real system under test. For example, if a QML program starts by sending out a message X, then a conforming system under test must also send out the message X when it starts. Thus, in a very concrete manner, a QML program is an *abstract reference implementation* of the system it specifies. As a matter of fact, a great way to test Conformiq Qtronic itself is to run tests derived from a model against a simulated execution of the very same model (expect no failures!), or mutants [30,31,32] of it.

In order to clarify this further, we do *not* rely on any existing Java or C# compilers, but have a full custom-built translator for QML. (I elaborate this later in this paper.)

Online and Offline Testing

Conformiq Qtronic offers two complementary ways for deploying the derived tests: *online testing (on-the-fly)* and *offline test generation*. Online testing means in our context that Conformiq Qtronic is connected "directly" with the system under test via a DLL (dynamically linked library) plug-in interface. In this mode, the selection and execution of test steps and the validation of the system under test's behavior all take place in parallel. In contrast, *offline test generation* decouples test case design from the execution of tests. In the offline mode, Conformiq Qtronic creates a library of test cases that are exported via an open plug-in interface and that can be deployed later, independent of the Conformiq Qtronic tool.

Conformiq Qtronic supports the testing of systems against *nondeterministic system models*, i.e. again models that allow for multiple different observable behaviors *even against a deterministic testing strategy*—but only in the online mode.

CHALLENGE:

Supporting nondeterministic system models

At the present, the offline test case generator assumes a deterministic system model. One of the reasons for this is that the test cases corresponding to a nondeterministic model resemble trees as at the test generation time the choices that the system under test will make are not yet known. The branching factor of such trees is difficult to contain, especially in the case of very wide nondeterministic branches (e.g. the system under test chooses a random integer). In contrast, the online algorithm can adapt to the already observed responses of the system under test, and choose the next test steps completely dynamically. [21, 33]

CHALLENGE:

Timed testing in real time

One important source of nondeterminism is time [33, 34, 35], especially because the testing setup itself typically creates communication latencies. For example, if the SUT sends out a timeout message after 10 seconds, it can be that the testing harness actually sees the message only after 10.1 seconds due to some slowness in the testing environment. In the same way the inputs to the SUT can get delayed. This is so important that the user interface for Conformiq Qtronic provides a widget for setting the maximum allowed communication latency. This widget actually controls the time bound of a bidirectional queue object that the tool adds implicitly "in front of" the provided system model.

Multilanguage Support

Even though QML is the default modeling language provided with Conformiq Qtronic, the tool supports also other languages. This multi-language support is implemented internally by translating all user-level models into an intermediate process notation. This notation, which we call CQλ, is actually a variant of Scheme [4], a lexically scoped dialect of the LISP language family [36]. For an example, see Fig. 3.

CHALLENGE:

Compiling models into an intermediate language

So, all QML models are translated eventually into LISP programs. Our pipeline for doing this consists of the following components: (1) a model and metamodel [37] loading front-end based on the ECore specification [38] and XMI [39]; (2) an in-memory model repository [40, 41]; (3) a parsing framework for textual languages that supports fully ambiguous grammars [42, 43]; and (4) a graph-rewriting [44, 45, 46] based framework for model transformation [40].

First, a QML model consisting of UML statecharts and textual program blocks is loaded into the model repository via the model loading front-end. At the same time, two metamodels are loaded: one for QML and one for CQλ. Originally, the textual program blocks appear in the repository as opaque strings. The next step is to parse them and to replace the strings with the corresponding syntax trees (for which there is support in the QML metamodel). Macro expansion, type checking and type inference happen at this stage. Then the graph rewriter is invoked with a specific *rule set*, which is iterated until a fixpoint; this causes the model to be gradually transformed from an instance of the QML metamodel to an instance of the CQλ metamodel. Eventually, the resulting CQλ model is linearized into a textual CQλ program and a fixed CQλ library pasted in. [47, 48] A simple example of a rewriting rule is shown in Fig. 4.

```
(define-input-port input)
(define-output-port output)
(define main
  (lambda ()
    (let* ((msg (ref (handshake input #f) 1))
           (_ (handshake (tuple output msg) #f)))
      (main))))
```

Fig. 3. A minimalistic CQλ program that defines an "echo" system: every message sent to the port input must be echoed back immediately through the port output

```
replace "Timer trigger" {
} where {
    Transition t;
    TimeoutTrigger trigger;
    t.trigger == trigger;
    t.trigger_cql == nil;
} with {
    t.trigger_cql := '(tuple ,CQL_Symbol("__after__") ,trigger.timeout);
};
```

Fig. 4. A simple graph rewriting rule that generates the CQλ counterpart for a timeout trigger in an UML statechart

It is important to be able to map the translated CQλ program back to the original user-level model in order to support traceability. There is specific support for this in the CQλ language: program blocks can be linked both lexically as well as dynamically to the user-level model.

Test Generation

So how does Conformiq Qtronic generate tests? The core algorithm is an enumerator for simulated executions of the given model. What makes this difficult is that this enumerator needs to be able to simulate an *open* model, i.e. a model that communicates an environment that has not been specified. Basically the enumerator assumes that a fully nondeterministic environment has been linked with the open model. This creates infinitely wide nondeterministic branches in the state space, because the hypothetical environment could send e.g. any integer whatsover to the system (Conformiq Qtronic supports as a datatype the full set $\mathbb{Z}$). Another set of nondeterministic branches is caused by the internal choices in the model itself, and these can be also infinitely wide (the *model* generates a free integer value). So the trick is how to handle a state space with infinitely many states *and* an infinite branching factor. This we do with *symbolic execution* and I get back to this shortly.

CHALLENGE:
Supporting known testing heuristics

In order to be able to support different testing heuristics, such as transition or state coverage or boundary value analysis [1], Conformiq Qtronic has a built-in capability to include *coverage checkpoints* in the intermediate CQλ-level models. Similar constructs have been called also e.g. coverage items in the literature. [49] A coverage checkpoint is marked in a LISP model by a call to the built-in **checkpoint** procedure. This means that the generation of coverage checkpoints can be fully controlled in the model transformation stage. Indeed, all the various user-level testing heuristics such as transition coverage or boundary value analysis have been implemented in the model transformator via this generic checkpoint facility.

Let us give the set of all possible traces, i.e. sequences of messages, the name $\mathbb{T}$, the set of all coverage checkpoints the name $\mathbb{C}$, and denote the set of booleans by $\mathbb{B}$. The state space enumerator can be seen as an oracle that implements the following functions:

$$\mathsf{valid} : \mathbb{T} \to \mathbb{B}$$

$$\mathsf{coverage} : \mathbb{T} \to 2^{\mathbb{C}}$$

$$\mathsf{plan} : \mathbb{T} \times 2^{\mathbb{C}} \to \mathbb{T}$$

The function valid tells whether a given trace is something that the model could produce or not, so it embodies a (bounded) model checker. The next function coverage calculates the set of checkpoints that *must* have been passed on every execution of the model that produces the given trace. Finally, plan calculates an extension (suffix) for a valid trace that can be produced by the model, attempting to find such an extension that it would cause a checkpoint that is not included in the given set to be passed.

Given these oracles, a simplified version of the online mode of Conformiq Qtronic can be described by the following algorithm. Initialize a variable C—which will contain checkpoints—with the empty set. Initialize another variable t—which will contain a trace—with the empty trace. Then repeat *ad infinitum*: If $\mathsf{valid}(t) = \mathrm{false}$, signal 'FAIL' and stop testing. Otherwise, update C to $C \cup \mathsf{coverage}(t)$. Then calculate $t' = \mathsf{plan}(t, C)$. If the next event in t' is an input to the system under test, wait until the time of the event and then send it. Otherwise just wait for some time. In any case, if a message is received from the SUT during waiting, update t accordingly; if not, update t with the message potentially sent, and in any case with the current wallclock reading.

Offline script generation is even easier. Because the model must be deterministic modulo inputs, anything returned by plan works as a test case. The basic idea is to find a set T of traces such that $|T|$ is small and $\bigcup_{t \in T} \mathsf{coverage}(t)$ is large.

In practice, the oracles valid, coverage and others are built around a *symbolic executor* for CQλ. Symbolic execution is well known and has been applied for test generation and program verification. [50, 51, 52] Usually implementing it requires some form of *constraint solving* [53,54], and so Conformiq Qtronic also sports a constraint solver under the hood. Maybe interestingly, the data domain for our solver is the least D such that

$$\mathbb{Q} \cup \mathbb{B} \cup \mathbb{S} \cup D^0 \cup D^1 \cup \cdots = D$$

where $\mathbb{S}$ denotes the infinite set of *symbols* (opaque, enumerated values). In words, any constraint variable in a constraint problem within Conformiq Qtronic can *a priori* assume a numeric, boolean or symbol value, or a tuple of an arbitrary size containing such values and other tuples recursively. In particular, we present strings (e.g. strings of Unicode [55] characters) as tuples of integers, and value records (`structs`) as tuples containing the values of the fields. This weakly typed structure of the constraint solver reflects the dynamic typing in the CQλ language.

CHALLENGE:

Implementing constraint solver over infinite domains

The constraint solver for Conformiq Qtronic has been developed in-house in C++ because it is tightly integrated with the symbolic executor itself. For example, our symbolic executor has a garbage collector [56] for the LISP heap, and when *symbolic* data gets garbage collected (references to constraint variables), the solver attempts to eliminate the variables from the constraint system by bucket elimination [57]. Handling infinite data domains provides a big challenge, because many of the state-of-the-art methods for solving difficult constraint problems assume finite-domain problems, and we do active research on this area on daily basis.

Scalability

At least in the form implemented in Conformiq Qtronic, model-based testing is a computationally intensive task.

CHALLENGE:

Practical time and space complexity

In practice, both time and memory space are important resource factors for us. Certainly, we work continuously to incrementally improve the time and memory characteristics of Conformiq Qtronic, but we employ also some categorical solutions.

The performance of Conformiq Qtronic (as for most other software applications) becomes unacceptable when it runs out of physical memory and begins swap trashing. To prevent this, Conformiq Qtronic swaps proactively most of the runtime objects—like parts of the symbolic state space—on hard disk. Our architectural solution for this is based on a variant of *reference counting* [56].

One challenge in the near future is to scale Conformiq Qtronic from single-workstation application to *ad hoc* grids [58] or clusters. Simply, we want to provide the option to run the core algorithms in parallel on all free CPUs in an internal network, such as all the idling Windows and Linux workstations within an office.

Post Scriptum

According to John A. Wheeler,

> We live on an island surrounded by a sea of ignorance. As our island of knowledge grows, so does the shore of our ignorance.

How true is this of model-based testing also! When we started the Conformiq Qtronic journey in early 2003—one hundred years after Wheeler's famous quote—we had a few fundamental questions that we had to solve. Today, our questions have changed in nature but only increased in number! Here are some of those which we have encountered:

—Is it possible to automatically explain the "logic behind" computer-generated test cases to a human engineer?

—Can computer-generated test cases be grouped and catalogued intelligently?

—Is it possible to generate offline test scripts that handle infinite-valued non-determinism without resorting to full constraint solving during test script execution?

—What forms of precomputation (e.g. forms of abstract interpretation [59, 28, 60]) can be used to reduce the runtime computational burden of online testing?

—Are the better modeling formalisms for system modeling for model-based testing than general modeling languages?

—How can well-researched finite-domain constraint solving techniques (like nonchronological backtracking or conflict set learning) be used to the full extent in the context of infinite-domain problems?

Acknowledgements. Conformiq Qtronic research and development has been partially supported by TEKES[2], and ITEA[3], an EUREKA[4] cluster. Special thanks are due to the co-chairs of the conference for their kind invitation.

References

[1] Craig, R.D., Jaskiel, S.P.: Systematic Software Testing. Artech House Publishers (2002)

[2] Rosenberg, J., Schulzrinne, H., Camarillo, G., Johnston, A., Peterson, J., Sparks, R., Handley, M., Schooler, E.: SIP: Session initiation protocol. Request for Comments 3261, The Internet Society (2002)

---

[2] www.tekes.fi
[3] www.itea-office.org
[4] www.eureka.be

[3] Postel, J.: User datagram protocol. Request for Comments 768, The Internet Society (1980)

[4] Abelson, H., Dybvig, R.K., Haynes, C.T., Rozas, G.J., Iv, N.I.A., Friedman, D.P., Kohlbecker, E., Steele, J.G.L., Bartley, D.H., Halstead, R., Oxley, D., Sussman, G.J., Brooks, G., Hanson, C., Pitman, K.M., Wand, M.: Revised report on the algorithmic language scheme. Higher Order Symbol. Comput. 11(1), 7–105 (1998)

[5] Gunter, C.A.: Semantics of Programming Languages. MIT Press, Cambridge (1992) ISBN 0-262-07143-6

[6] Plotkin, G.D.: A Structural Approach to Operational Semantics. Technical Report DAIMI FN-19, University of Aarhus (1981)

[7] Huima, A. (ed.): CQλ specification. Technical report, Conformiq Software (2003) Available upon request.

[8] Clarke, E.M., Grumberg, O., Long, D.E.: Model checking and abstraction. ACM Trans. Program. Lang. Syst. 16(5), 1512–1542 (1994)

[9] Ammann, P., Black, P.: Abstracting formal specifications to generate software tests via model checking. In: Proceedings of the 18th Digital Avionics Systems Conference (DASC99), vol. 2, 10.A.6, IEEE, New York (1999)

[10] Reps, T., Turnidge, T.: Program specialization via program slicing. In: Danvy, O., Glueck, R., Thiemann, P. (eds.) Proceedings of the Dagstuhl Seminar on Partial Evaluation, Schloss Dagstuhl, Wadern, Germany, pp. 409–429. Springer, New York (1996)

[11] Weiser, M.: Program slicing. In: ICSE '81: Proceedings of the 5th international conference on Software engineering, Piscataway, NJ, USA, pp. 439–449. IEEE Press, New York (1981)

[12] Clarke Jr., E.M., Grumberg, O., Peled, D.A.: Model Checking. MIT Press, Cambridge (2000) ISBN 0-262-03270-8

[13] Luo, G., Petrenko, A., Bochmann, G.V.: Selecting test sequences for partially-specified nondeterministic finite state machines. Technical Report IRO-864 (1993)

[14] Lee, D., Yannakakis, M.: Principles and methods of testing finite state machines - A survey. In: Proceedings of the IEEE,vol. 84, pp. 1090–1126 (1996)

[15] Pyhälä, T., Heljanko, K.: Specification coverage aided test selection. In: Lilius, J., Balarin, F., Machado, R.J. (eds.) Proceeding of the 3rd International Conference on Application of Concurrency to System Design (ACSD'2003), Guimaraes, Portugal, pp. 187–195. IEEE Computer Society, Washington (2003)

[16] Tretmans, J.: A formal approach to conformance testing. In: Proc. 6th International Workshop on Protocols Test Systems. Number C-19 in IFIP Transactions, pp. 257–276 (1994)

[17] Luo, G., von Bochmann, G., Petrenko, A.: Test selection based on communicating nondeterministic finite state machines using a generalized wp-method. IEEE Transactions on Software Engineering SE-20(2), 149–162 (1994)

[18] Feijs, L., Goga, N., Mauw, S.: Probabilities in the TorX test derivation algorithm. In: Proc. SAM'2000, SDL Forum Society (2000)

[19] Petrenko, A., Yevtushenko, N., Huo, J.L.: Testing transition systems with input and output tester. In: TestCom 2003, Springer, Heidelberg (2003)

[20] Tretmans, J.: Test generation with inputs, outputs and repetitive quiescence. Software—Concepts and Tools 17(3), 103–120 (1996)

[21] Veanes, M., Campbell, C., Schulte, W., Tillmann, N.: Online testing with model programs. In: ESEC/FSE-13: Proceedings of the 10th European software engineering conference held jointly with 13th ACM SIGSOFT international symposium on Foundations of software engineering, pp. 273–282. ACM Press, New York, NY, USA (2005)

[22] Object Management Group: Unified Modeling Language: Superstructure. Technical Report formal/2007-02-05 (2007)

[23] Selic, B.: UML 2: a model-driven development tool. IBM Syst. J. 45(3), 607–620 (2006)

[24] Gosling, J., Joy, B., Steele, G., Bracha, G.: The Java Language Specification, 3rd edn. Prentice-Hall, Englewood Cliffs (2005)

[25] Michaelis, M.: Essential C# 2.0. Addison-Wesley, London (2006)

[26] Conformiq Software: Conformiq Qtronic User Manual. Conformiq Software, Publicly available as part of product download (2007)

[27] Milner, R.: A theory of type polymorphism in programming. Journal of Computer and System Science 17(3), 348–375 (1978)

[28] Nielson, F., Nielson, H.R., Hankin, C.: Principles of Program Analysis. Springer, Heidelberg (1999) ISBN 3-540-65410-0

[29] Pierce, B.C.: Types and Programming Languages. MIT Press, Cambridge (2002)

[30] Budd, T.A., DeMillo, R.A., Lipton, R.J., Sayward, F.G.: Theoretical and empirical studies on using program mutation to test the functional correctness of programs. In: POPL '80: Proceedings of the 7th ACM SIGPLAN-SIGACT symposium on Principles of programming languages, pp. 220–233. ACM Press, New York, USA (1980)

[31] Offutt, A.J., Lee, S.: An empirical evaluation of weak mutation. IEEE Transactions on Software Engineering 20(5), 337–344 (1994)

[32] Zhu, H., Hall, P., May, J.: Software unit test coverage and adequacy. ACM Computing Surveys 29(4), 366–427 (1997)

[33] Larsen, K.G., Mikucionis, M., Nielsen, B.: Online testing of real-time systems using UPPAAL. In: Grabowski, J., Nielsen, B. (eds.) FATES 2004. LNCS, vol. 3395, pp. 79–94. Springer, Heidelberg (2005)

[34] Bohnenkamp, H., Belinfante, A.: Timed testing with TorX. In: Fitzgerald, J.A., Hayes, I.J., Tarlecki, A. (eds.) FM 2005. LNCS, vol. 3582, pp. 173–188. Springer, Heidelberg (2005)

[35] Briones, L., Brinksma, E.: Testing real-time multi input-output systems. In: Lau, K.-K., Banach, R. (eds.) ICFEM 2005. LNCS, vol. 3785, pp. 264–279. Springer, Heidelberg (2005)

[36] Steele Jr., G.L., Gabriel, R.P.: The evolution of Lisp. ACM SIGPLAN Notices 28(3), 231–270 (1993)

[37] Object Management Group: Meta Object Facility (MOF) Core Specification. Technical Report formal/06-01-01 (2006)

[38] Budinsky, F., Steinberg, D., Merks, E., Ellersick, R., Grose, T.J.: Eclipse Modeling Framework, 1st edn. Addison-Wesley, London (2003)

[39] Object Management Group: MOF 2.0/XMI Mapping Specification. Technical Report formal/05-09-01 (2005)

[40] Mellor, S.J., Scott, K., Uhl, A., Weise, D.: MDA Distilled. Addison-Wesley, London (2004)

[41] Kleppe, A., Warmer, J., Bast, W.: MDA Explained. Addison-Wesley, London (2003)

[42] Aho, A.V., Johnson, S.C., Ullman, J.D.: Deterministic parsing of ambiguous grammars. Commun. ACM 18(8), 441–452 (1975)

[43] Aycock, J., Horspool, R.N.: Faster generalized LR parsing. In: Jähnichen, S. (ed.) CC 1999 and ETAPS 1999. LNCS, vol. 1575, pp. 32–46. Springer, Heidelberg (1999)

[44] Rozenberg, G. (ed.): Handbook of Graph Grammars and Computing by Graph Transformation. World Scientific, vol. 1 (1997)

[45] Engelfriet, J., Rozenberg, G.: Node replacement graph grammars, vol. 44, pp. 1–94

[46] Drewes, F., Kreowski, H.J., Habel, A.: Hyperedge replacement graph grammars, vol. 44, pp. 95–162

[47] Nupponen, K.: The design and implementation of a graph rewrite engine for model transformations. Master's thesis, Helsinki University of Technology (2005)

[48] Vainikainen, T.: Applying graph rewriting to model transformations. Master's thesis, Helsinki University of Technology (2005)

[49] Blom, J., Hessel, A., Jonsson, B., Pettersson, P.: Specifying and generating test cases using observer automata. In: Grabowski, J., Nielsen, B. (eds.) FATES 2004. LNCS, vol. 3395, pp. 125–139. Springer, Heidelberg (2005)

[50] Gotlieb, A., Botella, B., Rueher, M.: Automatic test data generation using constraint solving techniques. In: ISSTA '98: Proceedings of the 1998 ACM SIGSOFT international symposium on Software testing and analysis, pp. 53–62. ACM Press, New York, USA (1998)

[51] Khurshid, S., Pasareanu, C.S.: Generalized symbolic execution for model checking and testing. In: Garavel, H., Hatcliff, J. (eds.) ETAPS 2003 and TACAS 2003. LNCS, vol. 2619, pp. 553–568. Springer, Heidelberg (2003)

[52] Lee, G., Morris, J., Parker, K., Bundell, G.A., Lam, P.: Using symbolic execution to guide test generation: Research articles. Softw. Test. Verif. Reliab. 15(1), 41–61 (2005)

[53] Dechter, R.: Constraint Processing. Morgan Kaufmann Publishers, San Francisco (2003)

[54] Apt, K.R.: Principles of Constraint Programming. Cambridge University Press, Cambridge (2003)

[55] The Unicode Consortium: The Unicode Standard, Version 5.0. 5th edn. Addison-Wesley Professional (2006)

[56] Jones, R., Lins, R.D.: Garbage Collection: Algorithms for Automatic Dynamic Memory Management. Wiley, Chichester (1996)

[57] Dechter, R.: Bucket elimination: A unifying framework for reasoning. Artificial Intelligence 113(1-2), 41–85 (1999)

[58] Cunha, J.C., Rana, O.F. (eds.): Grid Computing: Software Environments and Tools, 1st edn. Springer, Heidelberg (2005)

[59] Cousot, P.: Abstract interpretation. ACM Computing Surveys 28(2), 324–328 (1996)

[60] Bozga, M., Fernandez, J.C., Ghirvu, L.: Using static analysis to improve automatic test generation. In: Tools and Algorithms for Construction and Analysis of Systems, pp. 235–250 (2000)

New Approach for EFSM-Based Passive Testing of Web Services

Abdelghani Benharref, Rachida Dssouli, Mohamed Adel Serhani,
Abdeslam En-Nouaary, and Roch Glitho

Concordia University
1455 de Maisonneuve West Bd, Montreal, Quebec
H3G 1M8, Canada
{abdel,m_serhan,ennouaar}@ece.concordia.ca,
{dssouli,glitho}@ciise.concordia.ca

Abstract. Fault management, including fault detection and location, is an important task in management of Web Services. Fault detection can be performed through testing, which can be active or passive. Based on passive observation of interactions between a Web Service and its client, a passive tester tries to detect possible misbehaviors in requests and/or responses. Passive observation is performed in two steps: passive homing and fault detection. In FSM-based observers, the homing consists of state recognition. However, it consists of state recognition and variables initialization in EFSM-based observers. In this paper, we present a novel approach to speed up homing of EFSM-based observers designed for observation of Web Services. Our approach is based on combining observed events and backward walks in the EFSM model to recognize states and appropriately initialize variables. We present different algorithms and illustrate the procedure through an example where faults would not be detected unless backward walks are considered.

Keywords: EFSM-based passive testing, Web Services testing.

1 Introduction

Web services, a rapidly emerging technology, offer a set of mechanisms for program-to-program interactions over the Internet [1]. Managing Web Services is critical because they are being actually used in a wide range of applications. Fault management including fault detection is an important issue in this management.

Active testing and passive testing have been used for fault detection. An active tester applies test cases to the Web Service Under Test (WSUT) and checks its responses. In passive testing, messages received (requests) and sent (responses) by the Web Service Under Observation (WSUO) are observed, and the correct functioning is checked against the WSUT's model. The observation is done by entities known as observers.

Passive testing can complement active testing because it helps detecting faults that have not been detected before deployment. Furthermore, in many cases, it is a better alternative to active testing when the system is already deployed in its final operating

A. Petrenko et al. (Eds.): TestCom/FATES 2007, LNCS 4581, pp. 13–27, 2007.

environment. It enables fault detection without subjecting the system to test cases. Test cases consume resources and may even imply taking the WSUT off-line.

Passive testing is conducted in two steps: passive homing (or state recognition) and fault detection. During the first phase, the observer tries to figure out the state where the WSUO is moving actually. This phase is necessary when the observation starts a while after the interaction has started and previous traces are not available.

Few models have been used for model-based observers but most of the published work on passive testing are on control part of systems and are based on Finite State Machine (FSM) model ([2], [3], [4]). Although this model is appropriate for control parts of WSUO, it does not support data flow. Extended FSM (EFSM) is more appropriate for the handling of variables.

The homing procedure in an EFSM-based observer consists of recognizing the actual state of the WSUO in addition to assigning appropriate values to different variables. In the few published papers on EFSM-based passive testing, the homing procedure is either ignored or it depends on the upcoming observed request/responses. In the first case ([5], [6], [7]), the observer must get all the traces to be able to initiate the fault detection process. In the second case ([8], [9]), the observer waits for exchanged messages before moving forward in the homing procedure. Ignoring the homing phase is a very restrictive assumption. Waiting for exchanged messages to continue on the homing procedure may delay the fault detection. Moreover, if there is a significant time gap between requests and responses, the observer spends most of its time waiting.

In this paper, we present a novel approach for homing online EFSM-based observers. Unlike offline observers, online observers analyze the observed traces in real time and report faults as soon as they appear. The observer performs forward walks whenever a new event (request or response) is observed. It performs backward walks in the EFSM model of the WSUO in absence of observed events. The information gathered from the backward and forward walks help speeding up the homing procedure.

The remaining sections of this paper are organized as follows: in the next section, we present related work for passive observation based on FSM and EFSM models. Section 3 presents our new approach using backward and forward walks to speed up the homing procedure and discusses different algorithms illustrated through an example. Section 4 concludes the paper and gives an insight for future works.

2 Related Work

Active testing refers to the process of applying a set of requests to the WSUT and verifying its reactions. In this configuration [10], the tester has complete control over the requests and uses selected test sequences to reveal possible faults in the WSUT. Even though it is performed before deployment, active testing is not practical for management once a Web Service is operating in its final environment. Under normal conditions, the tester has no control over requests and/or responses. Passive observation is a potential alternative in this case.

Fig. 1 shows an observer monitoring interactions between the WSUO and its client during normal operations without disturbing it. Disturbing in this case means no injection of requests messages for testing purposes. If the observed responses are different from what's expected, the WSUO is then declared faulty.

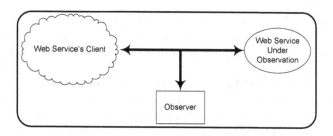

Fig. 1. Passive Testing Architecture

Formal methods, especially Finite State Machine (FSM) have been used in passive observation. Lee et al. in [2] propose algorithms for passive homing and fault detection for FSM-based observers for network management. These algorithms are extended in ([11]; [12]) to deal with fault location in networks and in [13] for avionics telecommunication. They have also been applied to GSM-MAP protocol in [4] and to TCP in [14]. Miller et al. extended the algorithms using Communicating FSM (CFSM) in [3]. While these algorithms work fine, they provide no support for dataflow which requires consideration of EFSM models.

EFSM is an extension of FSM by the following:

- Interactions have certain parameters, which are typed.
- The machine has a certain number of local variables, which are typed.
- Each transition is associated with an enabling predicate. The predicate can be any expression that evaluates to a Boolean (TRUE or FALSE). It depends on parameters of the received input and/or current values of local variables.
- Whenever a transition is fired, local variables can be updated accordingly and parameters of the output are computed.

Formally, an EFSM is described by a tuple $M = (S, S_0, I, O, T, V, \delta)$ ([10]) where:

- S is a set of states,
- $S_0 \in S$ is the initial state,
- I is a finite set of inputs,
- O is a finite set of outputs,
- T is a finite set of transitions
- V is a finite set of variables
- $\delta: S \times (I \cup O) \rightarrow S$ is a transition relation

In an EFSM, each transition of T is represented as t: I|P|A|O where:

- t: label/ID of the transition,
- S_s: starting state of the transition,

- I: the input that triggers the transition,
- P: the enabling predicate (data conditions),
- A: variables assignments
- O: the output produced by the transition
- S_e: ending state of the transition

Using EFSM models allows the detection of input/output faults in addition to faults related to data flow. The latter faults include calling of wrong function, wrong specification of data type, wrong initial values of variables, and referencing undefined or wrong variables.

In the literature on EFSM-based observers, the homing procedure is either ignored or depends fully on observed messages. In the first case, the authors in ([5]; [15]; [7]; [16]) suppose that the observation will start sharply with the interaction between the WSUO and its client. A passive observer based on this assumption will not be able to detect faults if it does not get whole traces. In the second case ([8]; [9]), the observer must wait for exchange of messages before moving forward in the homing procedure.

Since we are interested in online observation of Web Services, ignoring the homing procedure is not an option. We suppose that an EFSM-based online observer can initiate its observation at any time without having access to previously exchanged requests/responses. In the work presented in ([8]; [9]), the observer uses the exchanged messages for state and variables homing. This approach is efficient when the time gap between requests and responses is too short so the observer will be processing traces most of its time. If this time gap is relatively high, the observer spends a significant amount of time waiting for events while valuable information can be gathered by analyzing the EFSM model of the WSUO. The example presented in section 3.6 shows an example where the approaches presented in ([8]; [9]) fail to detect a fault that would have been detected if the observer was performing appropriate analysis of the EFSM machine of the WSUO.

3 EFSM-Based Observation: Forward and Backward Walks

In client-server communication as in Web Services, it is reasonable to assume that there will be a delay between requests and responses. In fact, the client takes time to formulate and send its request. Once a response is received, the client takes again some time to process the response and decide what to do with it. Moreover, the Web Service requires time to process a request, generate, and send its response.

To speed up the homing procedure, the observer should make a concise use of the information contained within the EFSM model in addition to the information carried by observed events. The homing algorithm can perform backward walks in the EFSM model to guess what transitions the WSUO fired before getting into its actual state. By analyzing the set of predicates and variable definitions on these transitions, the observer can reduce the set of possible states and/or the set of possible values of variables. Performing both backward and forward walks provides a set of possible execution trees: the forward process adds execution sequences to the root of trees, and the backward process adds execution sequences to the leaf states of trees.

During the homing procedure, the observer manipulates the following entities:

- Set of Possible States (**SPS**): this is the set of possible states with regards to what has been observed and processed up to now. At the beginning, all states are possible.
- Tree of Possible Previous States for state s (**TPPS(s)**): this tree contains the possible paths that could lead to each state s in the SPS. During the homing procedure, there is a TPPS for each state in the SPS.
- Set of Possible Variable Values for variable v (**SPVV(v)**): this is the set of all possible values that variable v can have with regards to what has been received and processed before. It consists of a list of specific values or ranges. At the beginning, all values in the definition's domain of variable v are possible.
- Set of Known Variables (**SKV**): the set of known variables. A variable is said to be known if it is assigned a specific value. In this case, SPVV(v) contains one element, i.e. $|SPVV(v)| = 1$.
- Set of Unknown Variables (**SUV**): the set of variables not yet known.

The next three sub-sections present in detail the processes of analyzing observed requests and responses and performing backward walks within an EFSM-based observer using both backward and forward walks.

3.1 The Homing Controller Algorithm

While the observer is going through the homing procedure (Algorithm 1), it has 3 possible options:

1. process a request that has been received by the WSUO (line 11),
2. process a response that has been sent by the WSUO (line 17), or
3. perform a one-step backward walk (line 20). In this case, the algorithm considers the event e that triggers the loop as empty.

Processing observed events has priority and the backward walk is performed if and only if there are no observed events waiting for processing. This procedure is repeated until:

- a fault is detected (unexpected input/output which results in an empty set of possible states and/or contradictory values of variables), or
- the set of possible states has one item ($|SPS| = 1$) **and** the set of unknown variables is empty ($SUV = \emptyset$).

The complexity of Algorithm 1 depends on the number of events required to successfully achieve the homing procedure (line 4) and the complexity of processInput (line 11), processOutput (line 17), and performBackWalk (line 20). Lets denote the number of cycles to achieve the homing by n, and the complexities of processInput, processOutput and performBackWalk by O(BF_PI), O(BF_PO), O(BF_BW) respectively. The complexity O(H) of the homing algorithm is given in Equation 1 and will be developed through the following sections when individual complexities will be computed.

$$O(H) = n.O(BF\_PI) + n.O(BF\_PO) + n.O(BF\_BW) \qquad (1)$$

$SPS := S$ // At startup, all states are possible

$SUV := V$ // At startup, all variables are unknown

Expected_Event ← "Any"

Data: event e

4 **repeat**

    e ← observed event

    switch *(e)* **do**

        case *(e is an input)*

            if *(Expected_Event == "Output")* **then**

            | **return** "Fault: Output expected not Input"

            else

11   processInput(e);   // Complexity: **O(BF_PI)**

                Expected_Event ← "Output";

        case *(e is an output)*

            if *(Expected_Event == "Input")* **then**

            | **return** "Fault: Input expected not Output"

            else

17   processOutput(e) ;   // Complexity: **O(BF_PO)**

                Expected_Event ← "Input";

        otherwise

20   performBackWalk ;   // Complexity: **O(BF_BW)**

until $(|SPS| == 1) AND (|SUV| == 0)$

Algorithm 1. Homing controller

3.2 Processing Observed Requests

When the observer witnesses an input, if the observer was expecting an output, a fault ("Output expected rather than Input") is generated. Otherwise, it removes all the states in the set of possible states that don't accept the input, and the states that accept the input but the predicate of the corresponding transition is evaluated to FALSE. For each of the remaining possible transitions, the input parameters are assigned (if applicable) to appropriate state variables. Then, the predicate condition is decomposed into elementary expressions (operands of AND/OR/XOR combinations). For each state variable, the set of possible values/ranges is updated using the elementary conditions. If this set contains a unique value, this latter is assigned to the corresponding variable; this variable is then removed from the set of unknown variables and added to the set of known variables. The transition's assignments part is processed, then updating the sets of known/unknown variables accordingly (Algorithm 2). The observer expects now the next observable event to be an output.

Input: event e

Data: boolean possibleState

1 **foreach** *(S ∈ SPS)* **do**
    possibleState = false

3   **foreach** *Transition t so that ((t.S_s == S) AND (t.I == e) AND (t.P ≠ FALSE))* **do**
      possibleState = true

      assign appropriate variables the values of the parameters of e

6   update $SPVV$, SKV, and SUV

7   decompose the predicate into elementary conditions

8   **foreach** *(elementary condition)* **do**
9   update the $SPVV$, SKV, and SUV

      if *(contradictory values/ranges)* **then**
        return "Contradictory values/ranges"

  if *(possibleState == false)* **then**
    remove S from SPS

    if *(SPS == ∅)* **then**
      return "Fault detected before homing is complete"

Algorithm 2. Processing observed requests

The complexity of Algorithm 2 is affected by the maximum number of states in the SPS (line 1), maximum number of transitions at each state in the SPS (line 3), and the complexity of updating the SPVV, SKV, and SUV. In fact, we can assume that a predicate will have very few elementary conditions, then decomposing the predicate (line 7) and using the elementary conditions to update the variables (line 8) does not affect the complexity of the whole algorithm. If the number of variables is V, the complexity of updating the SPVV, SKV, and SUV is in the order of $O(V)$ since the procedure should go through all the variables. The complexity of Algorithm 2 is depicted in Equation 2 where S_{max} is the maximum number of states in the SPS, and T_{max} is the maximum number of transitions that a state in the SPS can have.

$$O(BF\_PI)=O(S_{max}.T_{max}.V) \qquad (2)$$

3.3 Processing Observed Responses

In case the event is a response (output), if the observer was expecting an input, a fault ("Input expected rather than Output") is generated. Otherwise, the observer removes all the states in the set of possible states that don't have transitions that produce the output. If a state has two (or more) possible transitions, the TPPS is cloned as many as possible (number of possible transitions) so that each clone represents a possible transition. The assignment part of the transition is processed and variables are updated. The set of possible states holds the ending states of all the possible transitions. In the context of SOAP communication between a Web Service and its client, the response (message) holds basically one parameter. Whenever an output message is observed, a variable becomes known, or at least a new condition on variable values is augmented unless the message carries no parameter or the variable is already known. The observer expects now the next observable event to be an input.

Let's now determine the complexity of Algorithm 3. If we denote the maximum number of nodes (i.e states) in a TPPS tree by P_{max}, cloning a TPPS tree (line 5) is in the order of $O(P_{max})$. Moreover, the complexity of removing a TPPS tree (lines 14 and 19) is also in the order of $O(P_{max})$. Lines 8 and 9 do not affect the complexity since the number of assignments in a transition is somehow low compared, for instance, to P_{max}. The complexity of Algorithm 3 then can be written as:

$$O(BF\_PO) = O(S_{max}.T_{max}.(P_{max} + V)) \qquad (3)$$

3.4 Performing Backward Walk

While the observer is waiting for a new event (either request or response), it can perform a 1-step backward walk in the EFSM model to guess the path that could bring the Web Service to its actual state. From each state in the set of possible states, the observer builds a tree of probable-previously visited states and fired transitions. Every time a transition could lead to the actual state or one of its possible previous states, the variables constraints in the enabling condition is added as a set of successive elementary conditions connected with logical operators OR and AND: constraints of two successive transitions are connected with AND, while constraints on two transitions ending at the same state are connected with OR.

Data: event e

Data: boolean possibleState

1 **foreach** *(S ∈ SPS)* **do**

    possibleState = false

3   **foreach** *Transition t so that ((t.S_s == S) AND (t.I == e) AND*

    (t.P ≠ FALSE) AND (t **can produce** *e))* **do**

      possibleState = true

5   clone the corresponding *TPPS*

      t.S_e becomes the root of the cloned *TPPS*

      S becomes its child

8   process the transition's assignment part

9   assign appropriate variables values of the parameter (if any) of e

10   update the *SPVV*, *SKV* and *SUV*

      if *(contradictory values/ranges)* **then**

        ⌊ **return** "Contradictory values/ranges"

13   remove S from the *SPS*

14   remove the original *TPPS* ;   /* no longer useful, cloned (and

    updated) trees will be used */

  if *(possibleState == false)* **then**

16   remove S from *SPS*

   if *(SPS == ∅)* **then**

    ⌊ **return** "Fault detected before homing is complete"

19   remove the corresponding *TPPS*

Algorithm 3. Processing observed responses

Algorithm 4 has three embedded loops. The first loop (line 1) is bounded by the number of TPPS trees; that is, the number of states in the SPS (S_{max}). The second loop (line 2) goes through all leaf states of a TPPS, which is at the worst case P_{max}. The third loop (line 3) explores all the states in the EFSM that can lead to a particular state in a TPPS. Lets denote the number of states in an EFSM by S_{EFSM}. Propagating a constraint through the root of a TPPS (line 4) is in the order of $O(P_{max}.V)$ since the procedure has to process all states and update the SPVV at each state. The complexity of Algorithm 4 can be written as:

$$P(BF\_BW) = O(S_{max}.S_{EFSM}.P^2_{max}.V) \qquad (4)$$

Input: EFSM

1 foreach *(TPPS)* do

2 | foreach *(Leaf state S of TPPS)* do

3 | | foreach *(state S⌐ in the EFSM that leads to S)* do

4 | | | Propagate the constraints of the corresponding transition

   | | | toward the root of *TPPS*;

   | | | /* Lets consider that propagation cannot go beyound

   | | | state S_p */

   | | | if *(S_p is the root of TPPS)* then

   | | | | /* this path is possible→ consider it in the

   | | | | *TPPS*   */

   | | | | add S⌐ as child of S;

7 | | | update *SPVV*, *SKV*, and *SUV*;

   | | | if *(contradictory values/ranges)* then

   | | | | return "Contradictory values/ranges";

Algorithm 4. Performing Backward walk

From Equation **1**, Equation 2, Equation **3**, and Equation **4**, the overall complexity for homing an observer using Algorithm 1 can be developed as follows:

$$O(H) = O(n.S_{max}.T_{max}.V + n.S_{max}.T_{max}.(P_{max} + V) +$$
$$n.S_{max}.S_{EFSM}.P^2_{max}.V)$$
$$= n.S_{max}.T_{max}.(P_{max} + V) + n.S_{max}.S_{EFSM}.P^2_{max}.V)$$

3.5 Discussion

Although backward walks-based observers require a little bit more resources than an observer without backward walks, this overload is acceptable. First of all, backward walks are performed whenever there is no trace to analyze so the observer does not use additional processing time. It just uses the slots initially allocated to trace analysis. Second, limiting the backward to a unique step at a time reduces the duration of cycles of Algorithm 4 and does not delay processing of eventual available traces.

As for convergence of Algorithm 1, it is not possible to decide if the observer will converge or not. This is the case for both brands of observers: with backward and without backward. This limitation is out of the scope of the homing approach used but fully tied to the fact that the observer has no control on exchanged events. The Web

Service and its client can continuously exchange messages that do not bring useful information to reduce the SPS and the SPVV.

However, the backward approach can be compared to the approach without backward, for the same WSUO and observed traces, as follows:

Property 1: if an observer without backward walks converges, an observer with backward walks converges too.

Property 2: if an observer without backward walks requires n cycles to converge, and an observer with backward walks requires m cycles to converge, then $m \leq n$.

The next sub-section presents a proof of property 2 which can be considered also as proof for property 1.

Proof
The homing algorithm converges when the SPS has one element and the SUV is empty. The SUV is empty when, for each variable v in V, SPVV(v) contains a unique element.

As discussed above, analysis of traces adds states as roots of TPPS and backward walks adds states as leaves of TPPS. Whenever a trace can generate two different execution paths, the corresponding TPPS is cloned. This will build TPPS trees where the root has a unique child. In such trees, all constraints propagation from backward walks will propagate using AND operator between the root and its child. This propagation tries to reduce the SPVV; in the worst case the SPVV is neither reduced nor extended.

In Fig. 2, at a cycle i, a TPPS has S_i as root, S_j is its child, and $SPVV_i(v)$ is the set of possible values of variable v at S_i as computed from a previously observed trace. Suppose that during cycle $i+1$, the backward walk adds two leaves to S_j: S_l and S_k. In Fig. 2, the labels on transitions represent the SPVV that result from the predicate of the transitions.

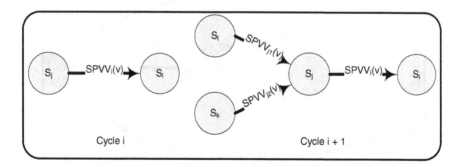

Fig. 2. SPVV and constraints propagation

Propagation of constraints from S_l and S_k to S_j and then to S_i modifies SPVV(v) as follows: $SPVV_{i+1}(v) = SPVV_i(v) \cap ((SPVV_l(v) \cup SPVV_k(v)))$. There are three cases:

1. $SPVV_i(v) \subseteq (SPVV_l(v) \cup SPVV_k(v))$: in this case, the $SPVV_{i+1}(v)$ is equal to $SPVV_i(v)$. The backward walks do not bring useful information to reduce

SPVV$_i$(v). If subsequent backward walks do the same, the number of required cycles for homing remains unchanged: *m=n*.

2. SPVV$_{i+1}$(v) = ∅: this indicates that the variable, at S$_i$ after cycle i+1 can not have any value from its definition domain. The observer detects a fault immediately without waiting for the next observed event which results in *m* strictly less than *n* (*m<n*).

3. SPVV$_{i+1}$(v) ⊂ SPVV$_i$(v): in this case, the SPVV(v) is reduced. If following backward walks, associated to trace analysis, reduce further the SPVV(v), the homing with backward is likely to require less than *n* cycles (*m<n*) or at most *n* cycles (*m=n*).

The following example illustrates the first case where backward walks reduce the number of required cycles (*m<n*) and allows detection of faults that can not be detected without backward walks. The execution of the homing procedure is detailed hereafter in a step by step scenario.

3.6 Example

Let's consider the portion of an EFSM of a Web Service illustrated in Fig. 3 where variables u, x, y, and z are integers. Events I1(15), O(13), and I2(0) are observed respectively. Each transition is represented as t:I|P|A|O where t is the label of the transition, I its input, P its predicate, A is the set of assignments, and O is the output. A predicate of a transition is evaluated to TRUE/FALSE if its condition is true/false, otherwise it is said INCONCLUSIVE if the predicate can not be evaluated. The latter case occurs if some of the variables in the predicate are not yet known.

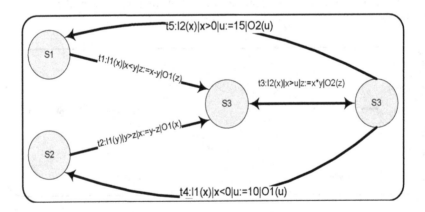

Fig. 3. EFSM Example

Observation without backward walks
After observing I1(15), transitions t1, t2, and t4 can be fired but not t3 or t5. However, since the input parameter is bigger than 0, the predicate of t4 is evaluated to FALSE.

Only transitions t1 and t2 should be considered since the variables y and z are, up to now, unknown and the predicates are evaluated to INCONCLUSIVE. This reduces the set of possible states to S1 and S2. If t1 is executed then x := 15, y > 15, and z := 15 − y, if t2 is executed then y := 15, z < 15, x := 15 − z.

When O1(13) is observed, the value of the output parameter (13) indicates that transition t2 has been executed. Later on, when event I2(0) is observed, since the variable u is unknown, the predicate (x > u) is evaluated to INCONCLUSIVE, which enables the transition. So, the sequence I1(15), O1(13), I2(0) executes properly.

However, the sequence I1(15), O1(13), I2(0) is a faulty sequence and the fault would be detected if backward walks have been considered as discussed in the next section.

Observation with backward walks

The delay after each event (I1, O1, and I2) gives the observer opportunities to perform backward walks. The observer executes the following operations: processInput(I1(15)), performBackWalk, processOutput(O1(13)), performBackWalk, processInput(I2(0)).

As illustrated in Table 1, after executing the first three operations, SPS contains S3. In TPPS, S2 is the child of S3. To get to S2, the only previous transition is t4 which assigns 10 to variable u. From this point forward, the homing procedure is completed since SPS has one state and SUV is empty. Later on when receiving I2(0), transition t3 can not be fired since its predicate (x>u) is evaluated to FALSE. The observer notifies the WSO that a fault just occurred.

Table 1. Content of SPS, SPVV, SKV, SUV, TPPS

	I1(15)	Backward walk	O1(13)
SPS	S1, S2	S1, S2	S3
SPVV	t1 : x:=15, y> 15, z:= x-y or t2 : y:=15, z <15, x:=y-z	t4 : u:=10 or t5 : u:=15	x:=13, y:=15, z:=2, u:=10
SKV	t1 : x, or t2 : y	u, x or u, y	x, y, z, u
SUV	t1 : y, z, u or t2 : x, z, u	y, z or x, z	Ø
TPPS	Figure 4.a	Figure 4.b	

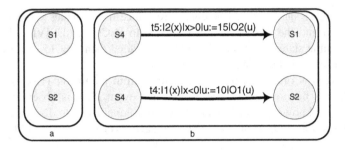

Fig. 4. TPPS

4 Conclusion

Fault detection is a basic operation in management of Web Services. It is conducted through testing which can be passive or active. An active tester applies selected test cases to the WSUT and checks the responses. Unlike active testers, a passive tester observes, passively, the interactions between the WSUO and its client. Based on this observation, correctness of requests and/or responses is verified.

FSM models have been used for passive testing for network management. However, it does not support data flows, an important aspect in Web Services XML-messaging. EFSM has the ability to specify both control and data flow parts of Web Services. When designing EFSM-based observers, the homing procedure has to assign appropriate values for different variables.

In this paper, we presented a novel approach for homing EFSM-based observers. This approach is based on observed events and on backward walks in the EFSM model of the WSUO. Whenever a trace is observed, it's immediately processed by the observer. Otherwise, the observer analyzes the possible paths that could bring the WSUO to its actual state. Analyzing the set of constraints on different paths could reduce the set of possible values variables can have at a specific state.

We are currently implementing observers based on the algorithms presented above and Web Services that will be used to evaluate the detection capabilities of such observers.

References

[1] W3C, World Wide Consortium (2006) at http://www.w3.org
[2] Lee, D., Netravali, A.N., Sabnani, K.K., Sugla, B., John, A.: Passive testing and applications to network management. In: International Conference on Network Protocols, Atlanta, GA, USA, pp. 113–122. IEEE Computer Society, Washington (1997)
[3] Miller, R.E.: Passive testing of networks using a CFSM specification. In: International Performance, Computing and Communications Conference, Tempe/Phoenix, AZ, USA, pp. 111–116. IEEE, New York (1998)
[4] Tabourier, M., Cavalli, A., Ionescu, M.: A GSM-MAP protocol experiment using passive testing. In: Brauer, W. (ed.) Formal Methods. World Congress on Formal Methods in the Development of Computing Systems. LNCS, vol. 1, pp. 915–934. Springer, Heidelberg (1999)

[5] Cavalli, A., Gervy, C., Prokopenko, S.: New approaches for passive testing using an Extended Finite State Machine specification. Information and Software Technology 45, 837–852 (2003)

[6] Alcalde, B., Cavalli, A., Chen, D., Khuu, D., Lee, D.: Network protocol system passive testing for fault management: a backward checking approach. In: Formal Techniques for Networked and Distributed Systems (FM), Madrid, Spain. LNCS, pp. 150–166. Springer, Heidelberg (2004)

[7] Ladani, B.T., Alcalde, B., Cavalli, A.: Passive testing - a constrained invariant checking approach. In: 17th International Conference on Testing of communicating systems (TestCom), Montreal, Que, Canada. LNCS, pp. 9–22. Springer, Heidelberg (2005)

[8] Lee, D., Dongluo, C., Ruibing, H., Miller, R.E., Jianping, W., Xia, Y.: A formal approach for passive testing of protocol data portions. In: 10th International Conference on Network Protocols, Paris, France, pp. 122–131. IEEE Computer Society, Washington (2002)

[9] Lee, D., Dongluo, C., Ruibing, H., Miller, R.E., Jianping, W., Xia, Y.: Network protocol system monitoring-a formal approach with passive testing. IEEE/ACM Transactions on Networking 14, 424–437 (2006)

[10] Dssouli, R., Saleh, K., Aboulhamid, E., En-Nouaary, A., Bourhfir, C.: Test development for communication protocols: towards automation. Computer Networks 31, 1835–1872 (1999)

[11] Miller, R.E., Arisha, K.A.: On fault location in networks by passive testing. In: International Performance, Computing, and Communications Conference, Phoenix, AZ, USA, pp. 281–287. IEEE, New York (2000)

[12] Miller, R.E., Arisha, K.A.: Fault identification in networks by passive testing. In: 34th Annual Simulation Symposium, Seattle, WA, USA, pp. 277–284. IEEE Computer Society, Washington (2001)

[13] Arisha, K.A.: Fault management in avionics telecommunication using passive testing. In: 20th Digital Avionics Systems Conference (DASC), Daytona Beach, FL, USA, vol. 1, pp. 1–7. IEEE, New York (2001)

[14] Dongluo, C., Jianping, W., HuiCheng, C.: Passive testing on TCP. In: International Conference on Communication Technology (ICCT). Beijing, China: Beijing Univ. Posts & Telecommun, pp. 182–186 (2003)

[15] Arnedo, J.A., Cavalli, A., Nunez, M.: Fast testing of critical properties through passive testing. In: 15th IFIP International Conference on Testing of Communicating Systems, Sophia Antipolis, France. LNCS, pp. 295–310. Springer, Heidelberg (2003)

[16] Bayse, E., Cavalli, A., Nunez, M., Zaidi, F.: A passive testing approach based on invariants: application to the WAP. Computer Networks 48, 247–266 (2005)

Automation of Avionic Systems Testing*

David Cebrián, Valentín Valero, and Fernando Cuartero

Albacete Computer Science Research Institute
University of Castilla-La Mancha, Avda. España s/n, Albacete (Spain)
http://www.dsi.uclm.es

Abstract. In this paper we present an automatic testing process to validate Avionic Systems. To do that, we have developed a tool that interprets scripts written in Automated Test Language and translate them to user codes written in C language. To carry out this work, the syntax of scripts has been defined by a context free grammar. This testing process is based on the execution of a pre-defined set of test cases. Currently, these test sets are obtained from Test Description Document and they are introduced in the system in C code manually. Therefore, automation of this process would reduce the time used for the testing, as a great quantity of tests are realized and a great quantity of errors are made when tests are made by hand.

Keywords: Testing, Avionics systems, Real time systems, grammar testing.

1 Introduction

In the development of Avionic Systems, testing and validation, have become a very important part into the software development process, due to the critical real environment where they are going to work. For this reason, testing is usually a very time consuming task. Test automation facilities are desirable in order to reduce the time required for this task.

Actually, for the development of Avionic Systems Software, and in general for the development of complex critical systems, the use of formal techniques for evaluating the capabilities provided by the system and the expected ones becomes very important. Depending on the system to be designed, very different specification formalisms can be used [2,4,12,15].

Real-time and embedded systems are nowadays so complex that to completely specify their behavior is a very difficult task. In particular, these systems are very heterogeneous and include a big amount of components with different natures (sensors, busses, displays, keyboards, storage devices, etc.).

For this reason, software testing, and mainly in this kind of systems, has become a very important part into the software development process [15], as

---

* Supported by the Spanish government (cofinanced by FEDER founds) with the project TIN2006-15578-C02-02, and the JCCLM regional project PAC06-0008-6995.

A. Petrenko et al. (Eds.): TestCom/FATES 2007, LNCS 4581, pp. 28–40, 2007.

systems are more and more complex and not detected failures can have fatal consequences.

A failure in a software system can occur for several reasons, now we just mention some of them:

- Specification deficiencies:
  - Incomplete description of functionality.
  - Inconsistent description of functionality.
- Design errors:
  - Misinterpretation of specification.
  - Erroneous control logic.
  - Insufficient error handling.
- Coding errors:
  - Non-initialized data.
  - Usage of wrong variables.

In the development of Avionic systems, testing and validation is a decisive job due to the critical real environment where they are going to work. The effects of an avionic system malfunction would be catastrophic in many cases, so intense efforts to avoid failures are always taken and heavy tests are performed on the equipments. That is why testing is usually a very time consuming human effort and so much time is dedicated to achieve the necessary qualification.

In the literature about avionic software testing, we can see in [14] a description of the main aspects that must be considered to test this kind of systems. Another related work is [7], where model-based simulation testing is used to test avionic software systems.

Then, the main purpose in these works is to be able to qualify before getting the system totally operative in the Avionic systems in order to improve the whole process. For that purpose, test automation facilities are desirable in order to reduce the time required for qualification. Thus, in this paper our goal is to describe a tool that we have developed in order to automate a part of the testing process for Helicopter embedded software.

This tool interprets scripts written in Automated Test Language and translates them to user codes written in C. This Automated Test Language is closer to human language and it permits to describe test cases easily. This test automation tool has been applied to helicopter software testing, in a real corporation (Eurocopter company).

The particular language that we use has been defined by the software testing group of the helicopter company, and it is specific for this purpose. There are, of course, some standard notations that could be used to accomplish this task too, like TTCN-3 [3].

To interpret these scripts written in Automated Test Language, a context-free grammar has been used. Nowadays, grammars are omnipresent in software development. They are used for the definition of syntax, exchange formats and others. Several important kinds of software rely on grammars, e.g., compilers, debuggers, profilers, slicing tools, pretty printers, (re-) documentation tools, language reference manuals, browsers, software analysis tools, code preprocessing tools, and software modification tools [6,9,16].

A grammar defines a formal language and provides a device for generating sentences. From a perspective of software engineering, a grammar may be considered as both a specification (defining a language) and a program (serving as a parser generators input). In practice, ensuring that a grammar specifies an intended language which can be considered as user requirements is indeed a validation problem [5].

Testing is a standard way to validate specifications or programs (formal analysis is another). Grammar testing covers various technical and pragmatic aspects such as coverage notions [10], test set generation [11], correctness and completeness claims for grammars or parsers, and integration of testing and grammar transformations.

In this paper we focus our attention on test set generation. Test data generation requires a variety of techniques [8], for example, to minimize test cases, to accomplish negative test cases, etc.

The rest of the paper is structured as follows. In the next Section a description of a system that accepts user codes generated by the tool will be described. Next, the tool operation is shown. In Section 4 a case study is presented. Finally, our conclusions and future work are presented.

2   System Overview

In this section, we describe the specific system that we consider for our Avionic System Testing environment. This System accepts user codes generated by the implemented tool. The System is a combination of a real-time platform designed to execute Avionic Equipment Tests and a Unix workstation which is used as a user interface to drive the test.

The Avionic Equipment of the considered helicopter basically consists of a core computer that integrates, among others, the functions concerning control and display subsystem, navigation subsystem and communication subsystem. These subsystems are connected via redundant busses to improve reliability.

Then, the Real-Time System platform contains the simulated equipment of the helicopter and it is mainly composed by I/O cards for the different busses of the helicopter (MILBUS 1553, ARINC 429, RS485, etc.) and an avionic database containing specific information about the helicopter to be tested (Fig. 1).

And finally, the System Unix Workstation shall provide the capability to control the operation of the tests and run the simulations. By means of it, we can manage the simulations, specifying the concrete datas that are to be used, we can also inject some types of errors to test the system reactions, we can prepare the scripts for the tests, and of course, we can monitor the system, to view and record the results of the simulation.

Another feature of interest of this system is that of scenarios, which allow the testers to establish the context in which tests are to be made. Then, a scenario is a specific test directly linked to an upper context that defines the set of objects which should be operational during the test like codes, data items to be displayed in dashboards, data items to be modified in dashboards and

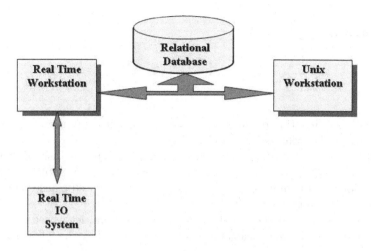

Fig. 1. Real-Time platform

configuration of simulated equipment. The descriptions of the system and of the scenarios are stored in the database.

2.1 Testing Environment Constraints

Due to the nature of the system environment there are important constraints related to how user code should be generated. User code will be always executed in the same way: there is a period indicated in the test information which serves as a basis to cyclic execution of pieces of code. Each cycle will have the duration of the specified period, and each piece of code is forced to be executed within one cycle. Generated user code has to fit this rule and special care has to be taken in controlling that no piece of code extends the cycle duration, since this would cause a general system failure.

Thus user code must be divided according to this restriction, so a control code is introduced to select the concrete piece of code that must be executed on each cycle. The easiest solution to fulfil this as a set of switches, each one including a persistent counter that will indicate in each cycle what case clause to execute:

```
static int Counter = 1;

int TestRun ( )
{
  switch ( Counter )
  {
    case 1 :
      <execution piece 0>
      Counter = 2;
      break ;
```

```
case 2 :
  <execution piece 1>
  Counter = 3;
  ...
 }
}
```

Thus, the modification of the value of the persistent counter will allow navigating between different execution pieces.

This hard coding constraint makes the most challenging task of the code generator to establish a set of mechanisms that will allow translating a sequence of instructions to an equivalent code made of a set of switch clauses connected and controlled by auxiliary variables.

Then, once we have described the testing environment, let us see the format that system user codes have (each test). They are defined by the following items:

- A name - (32 bytes length).
- A period - integer expressed in milliseconds.
- An Interface: a text file description describing the exchange of information between the user code and the system.
- A specific main module which will be automatically called by the system according to the period.
- Some user modules.

Furthermore, a test is composed of three hierarchical levels: procedures, tasks and steps. This division obeys the grammar definition that we are considering, in which each test is divided in this way. Then, each of these three levels may be run in a kind of control loop for some specific set of values, which may be defined directly in the test description or in a text file which can be modified from an execution to another.

3   The Tool

The system accepts C code to specify the tests. With the purpose of making easier the specification of these tests, a tool called Code Generator, has been built. This tool accepts as input user scripts that make easier the tests specification. The function of this software is to interpret these scripts written in Automated Test Language and translate them to system user codes written in C.

These user scripts (written in Automated Test Language) are generated from a document called, test description document. This document is written by expert testers and it is written in natural language. So to carry out these tests, they have to translate the test description document to Automated Test Language (Fig. 2). This task is made manually but in the future the test description document will be written in a high level language and this task will be automated. An example of Automated Test Language is shown in Fig. 3.

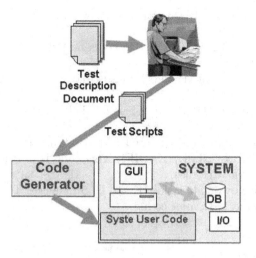

Fig. 2. Testing process

```
Example_Script 1000
Procedure P1
    Task T1
        Step 1
            GET Variable INSERT_BOTH
            IF WAIT_VALUE VALUE_AVAILABLE TRUE 0
            THEN
                ADDOUTPUT "IRS1 available"
            ELSE
                ADDOUTPUT "IRS1 not available"
            ENDIF
Procedure P2
    Task T2
        Step 1
            GET Variable INSERT_BOTH
```

Fig. 3. Example script 1000

The tool must read the scripts specified by the user and translate them into C code. C functions will be created and grouped into different files in order to increase modularity. The file structure will follow the scheme shown in Fig. 4:

Some execution levels are considered for system user code execution control. Each level will call the level immediately below. In the example script 1000 (Fig. 3), the resulting user code will be executed in different nesting levels:

- **Level 0**: code in INIT, RUN (not nested switch) and END phase in Main User Code file.
- **Level 1**: Example_Script 1000 related code: nested switch in RUN phase in Main User Code file.
- **Level 2**: P1 and P2 related code. This is the code in *ProcedureName_*Run.cc files.
- **Level 3**: T1 and T2 related code.This code is located in *ProcedureName_TaskName*.cc files.
- **Level 4**: Step 1 and Step 1 related code (but not the nested code they contain: IF/ THEN/ ELSE/ ENDIF). This code is also located in *ProcedureName_TaskName*.cc files.
- **Level 5 and below**: code inside IF and ELSE clauses. This code is located in *ProcedureName_TaskName*.cc files.

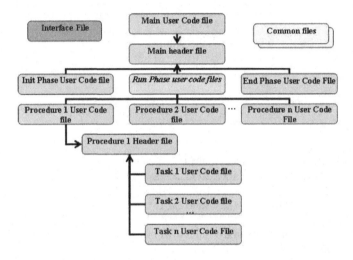

Fig. 4. Generated file structure

3.1   General Strategies

This tool has been implemented in a platform independent way. This approach allows the system to be compiled for several platforms with no changes in the source code.

The formal syntax of the scripts has been defined by a non-ambiguous context-free grammar, so each valid input will have only one possible derivation tree. The complete grammar is omitted, because it is very large, it has 72 production rules, and it is unimportant for our purposes.

The tool must read the scripts and translate them into C code. For this reason, a LALR(1) interpreter has been built, which can deal with many context-free grammars by using small parsing tables. The parsing process of these interpreters is divided into several levels:

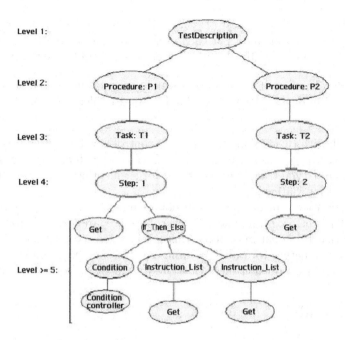

Fig. 5. AST associated with Example Script 1000

- **Lexical level:** This is the simplest and lowest level of the interpreter. In
  this level the parser reads the input character by character and translates
  these sets of characters into words.
- **Syntactic Level:** Once the input has been divided into tokens, the processing is much easier. This level checks the correctness of a given input
  according to the specified grammar. Then, once the words of the input have
  been identified (which is done by the lexical level), this level just tests if
  there is a derivation tree in the grammar, which leads to the given input.
  As this level generates the derivation trees of the given grammar, it is very
  important that this grammar has no ambiguity. Each given input will have,
  if it has any, only one possible derivation tree.
- **Semantic Level:** A grammar must define all the correct sequences of the
  language. But there are conditions which might be really difficult to represent in the definition of a language. To avoid this, grammars which define a
  superset of the correct sequences accepted are used and some tests to check
  that the accepted sequences fulfil these constraints are added. In this step
  we check that some features (like the declaration of variables) are consistent. These features have to be checked over the whole test. For this reason,
  a structure capturing a logical representation of the scripts has been introduced. Specifically, we have used ASTs (Abstract Syntax Trees). Figure 5
  shows an example of AST, in this case that one associated with the Example
  Script 1000 (Fig. 3). In this example, we can see how the defined tree has
  a structure according to the level division in the scripts. These nodes which

keep the structure of the test (TestDescription, Procedure, Task, Step) could be called structural trees. For each of the available instructions in the test (If-then-else, GET, ADDOUTPUT) there is an appropriate tree too.

For the development of the project, lex and yacc tools have been used. Particularly, flex 2.5 implementation of lex and bison 2.1 implementation of yacc have been chosen.

- **Flex**[13]: with this tool programs whose control flow is directed by instances of regular expressions in the input stream can be written. It is well suited for editor-script type transformations and for segmenting input in preparation for a parsing routine.
- **Bison**[1]: Yacc provides a general tool for imposing structure on the input to a computer program. The yacc user must prepare a specification for the syntax that the input must fulfil: this specification includes rules describing the input structure (productions from our context free grammar), but also code to be invoked when these rules are recognized (semantic actions).

4   Case Study

We now use another example (Antenna_Selection) to describe the tool operation (Fig. 6). In practice, the scripts are obtained manually by using the test description document. Notice that we do not need to know how these scripts are obtained (which technique is used in particular to generate them), neither to fully understand their mission because these tasks are carried out by testing engineers of helicopter company.

Code Generator takes this script as input and translates it to system user code. The system user code thus obtained consists of some files, the structure of which is shown in Fig. 7.

The contents of these files are:

- **Antenna_Selection.x:** This is the interface file. It contains imports and exports of variables from the database.
- **CommonFunctions.h:** This file contains a set of common functions used by other files.
- **Antenna_Selection.cc:** This is the main file and controls the test operation. Below, its source code is shown in Fig. 8:
- **Antenna_Selection_Init.cc:** It contains the code for the inicialization of the test.
- **Antenna_Selection_End.cc:** It contains the code for the conclusion of the test.
- **P1_Antena_Run.cc, P2_Antena_Run.cc and P3_Antena_Run.cc:** They contain the related code with procedures P1, P2 and P3.
- **P1_Antena_T1.cc, P2_Antena_T1.cc, P3_Antena_T1.cc, P1_Antena_T2.cc, P1_Antena_T3.cc:** These files contain the related code with the diferent tasks.

```
Antenna_Selection 200               ENDIF
PROCEDURE P1_Antena                 TASK T2
  TASK T1                             STEP 0
    STEP 0                              GET External Internal
      SET External 5                TASK T3
      GET External1 Internal1         STEP 0
      SET External1c 6.2                GET External1 Internal1
      ADDOUTPUT "HELLO"                 SLEEP 10
    STEP 1                              GET External2 Internal2
      GET External2 Internal2     PROCEDURE P2_Antena
      FREEZE Proof1                 TASK T1
      UNFREEZE Proof2                 STEP 0
      FREEZE Proof3                     SLEEP 10
    STEP 2                              SLEEP 12
      GET External2 Internal2           GET External3b Internal3b
      ADDOUTPUT hey                     SLEEP 15
      INSERT_LABEL hello label1         GET External3c Internal3c
      REMOVE_LABEL hello label1         GET External3d Internal3d
      IF (6==6) THEN              PROCEDURE P3_Antena
        INSERT_LABEL hello label1    TASK T1
      ELSE                             STEP 0
      ADDOUTPUT hey                      ADDOUTPUT "hello"
      INFORMATION "proof"
```

Fig. 6. Example script

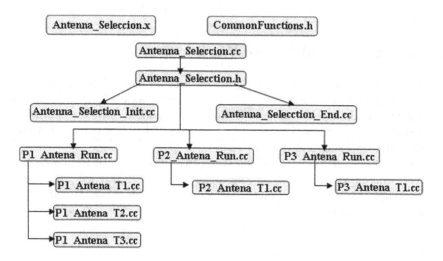

Fig. 7. Generated files

These files are accepted by the test system as input and they are used to drive the test. We can observe, in Fig. 8, that the generated code obeys the constraints imposed by the testing environment as it is divided into some pieces of code by

```
#include Antenna_Selection.h"
                                       case 1:
void CODE (unsigned long inCodeState)      P2_Antena_Run();
{                                          if ( Level==1 )
  switch(inCodeState)                        Step[Level]
  {                                              =Step[Level]+1;
    case CO_INIT:                        case 2:
      Level = 1;                             P3_Antena_Run();
      for(int i=0; i<4; i++)                 if ( Level==1 )
        Step[i]=0;                             Step[Level]
      Antenna_Selection_Init();                  =Step[Level]+1;
      AnaisEndOfPhase ();                    break;
      break;                             }
    case CO_RUN:                         break;
      switch ( Step[1] )                 case CO_END:
      {                                    Antenna_Selection_End();
        case 0:                            fclose(Fout);
          P1_Antena_Run();                 AnaisEndOfPhase ();
          if ( Level==1 )                  break;
            Step[Level]            }
                =Step[Level]+1;     }
```

Fig. 8. Antenna_Selection.cc

means of a set of switchs. Each piece of code will require a time that will not be greater than the cycle duration (this is controlled by the *inCodeState* variable).

5   Conclusions and Future Work

In this paper, a first step for helicopter software automated testing has been shown. We cannot provide real experimental results because the tests are carried out by the helicopter company testing group, and we have no information comparing the time required for the testing process when it was made manually and currently, by using our tool.

As future work our intention is to extend the automation of testing to other aspects of this process. The whole concept of Automated Test will allow generating a complete test set based on the test description document. It will concern a complete environment including:

1. Automatic database frame import.
2. Automatic scenario definition.
3. Automatic code generation.

The principle is to define a high-level language with a friendly syntax which will be used by test teams to describe their tests. A simple way may be to directly use this high-level language as part of the test description document and to be able to extract it automatically. The main advantage that we can obtain with this automated testing is to save a lot of time in testing execution and humman effort to achieve the necessary qualification.

For this purpose, some constraints must be fulfilled:

1. A friendly syntax and format easily generated from the specifications. This description must be accessible to a test team guy even if he is not an expert in programming languages.
2. Common format agreed by all test teams. The idea is to use a simple support which may be provided by test guys using standard editors like: text editors, Microsoft Office editors, etc.
3. Comprehensive language, not directly mapped on a specific software compiler. The idea is that test team guys must be able to describe a test procedure, even if they are not experts in software programming language.
4. The test description language must cover at least all the functionalities actually covered by the specifics ones already existing.

Our intention for the immediate future is to increase the automation level of testing environment, by including scenarios in the Code Generation tool, and even it would be important to replace the Test Description Document by another document which can be automatically interpreted by a tool.

Acknowledgement. We would like to thank to the anonymous referees for their suggestions and corrections that have contributed to improve this paper significantly.

References

1. Stallman, R., Donnelly, C.: Bison. The YACC-compatible Parser Generator (1995)
2. Cavalli, A.R., Favreau, J.P., Phalippou, M.: Formal methods for conformance testing: Results and perspectives. In: Proceedings of the IFIP TC6/WG6.1 Sixth International Workshop on Protocol Test systems VI, pp. 3–17, Amsterdam, The Netherlands, The Netherlands, North-Holland Publishing Co (1994)
3. Willcock, C.: Introduction to TTCN-3 (2002)
4. Geilen, M.C.W.: Formal Techniques for Verification of Complex Real-Time Systems. M.C.W. Geilen (2002)
5. Li, C.L.H., Jin, M., Gao, Z.: Test criteria for context-free grammars. Computer Software and Applications Conference. COMPSAC 2004. In: Proceedings of the 28th Annual International, pp. 300–305 (2004)
6. Kort, R.L.J., Verhoef, C.: The grammar deployment kit. Electronic Notes in Theoretical Computer Science 65(3), 7 (2002)
7. Peleska, G.J.J., Brumm, K., Hartmann, T.: Advancement in automated simulation and testing technology for safety-critical avionic systems. Aerospace Testing 2006 (2006)
8. Klint, P., Lämmel, R., Verhoef, C.: Toward an engineering discipline for grammarware. ACM Trans. Softw. Eng. Methodol. 14(3), 331–380 (2005)
9. Lämmel, R.: Grammar testing. In: Hussmann, H. (ed.) ETAPS 2001 and FASE 2001. LNCS, vol. 2029, pp. 201–216. Springer, Heidelberg (2001)

10. Lämmel, R., Schulte, W.: Controllable combinatorial coverage in grammar-based testing. TestCom 2006, pp. 19–38 (2006)
11. Maurer, P.M.: Generating test data with enhanced context-free grammars. IEEE Software 7(4), 50–55 (1990)
12. Núñez, M., Pelayo, F.L., Rodríguez, I.: A formal methodology to test complex embedded systems: Application to interactive driving system. In: IFIP TC10 Working Conf.: International Embedded Systems Symposium, IESS'05, pp. 125–136. Springer, Heidelberg (2005)
13. Paxson, V.: Flex, version 2.5. A fast scanner generator (1995)
14. Peleska, J.: Test automation for avionic systems and space technology (extended abstract) (1996)
15. Peleska, J.: Formal methods for test automation - hard real-time testing of controllers for the airbus aircraft family. Integrated Design and Process technology, IDPT-2002 (2002)
16. Wu, H.: Grammar-driven generation of domain-specific language tools. In: OOPSLA '06: Companion to the 21st ACM SIGPLAN conference on Object-oriented programming systems, languages, and applications, pp. 772–773, ACM Press, New York, NY, USA (2006)

Automatic Test Generation from Interprocedural Specifications

Camille Constant[1], Bertrand Jeannet[2], and Thierry Jéron[1,*]

[1] IRISA/INRIA, Campus de Beaulieu, Rennes, France
{constant, jeron}@irisa.fr
[2] INRIA Rhône-Alpes, Saint Ismier, France
Bertrand.Jeannet@inrialpes.fr

Abstract. This paper adresses the generation of test cases for testing the conformance of a reactive black-box implementation with respect to its specification. We aim at extending the principles and algorithms of model-based testing for recursive interprocedural specifications that can be modeled by Push-Down Systems (PDS). Such specifications may be more compact than non-recursive ones and are more expressive.

The generated test cases are selected according to a test purpose, a (set of) scenario of interest that one wants to observe during test execution. The test generation method we propose in this paper is based on program transformations and a coreachability analysis, which allows to decide whether and how the test purpose can still be satisfied. However, despite the possibility to perform an exact analysis, the inability of test cases to inspect their own stack prevents it from using fully the coreachability information. We discuss this partial observation problem, its consequences, and how to minimize its impact.

1 Introduction

Testing is the most used validation technique to assess the correctness of reactive systems. Among the aspects of software that can be tested, e.g. functionality, performance, timing, robustness, etc, we focus here on conformance testing and specialize it to reactive systems [1]. Conformance testing compares the observable behaviour of an actual black-box implementation of the system with the observable behaviour described by a formal specification, according to a conformance relation. It is an instance of *model-based testing* where specifications, implementations, and the conformance relation between them are formalised. Test cases are automatically derived from specifications, and the verdicts resulting from test execution on an implementation are proved to be consistent with respect to the conformance relation. Moreover, in addition to checking the conformance of the implementation, the goal of the test case is also to guide the test execution towards the satisfaction of a test purpose, typically a set of scenarii of interest. The *test selection* problem consists in finding a strategy that maximizes the likelihood for the test case to realize the test purpose.

* This work was partly supported by France Telecom R&D, contract 46132862.

A. Petrenko et al. (Eds.): TestCom/FATES 2007, LNCS 4581, pp. 41–57, 2007.

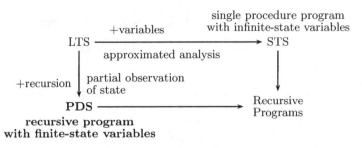

Fig. 1. Test selection on various models

This problem has been previously addressed in the case where the specifications, the test cases and the test purposes are modeled with finite Labelled Transition Systems (LTS) [2,3]. It was more recently addressed in the case where the same objects are modeled with Symbolic Transition Systems (STS), which extend LTS with infinite datatypes and can model non-recursive imperative programs [4]. The aim of this paper is to address the test selection problem in the case where the specification is modeled as a Push-Down System (PDS), which extends LTS with a stack over a finite alphabet and can model recursive programs manipulating finite datatypes. Such specifications may be more compact that non-recursive ones and are more expressive than single procedure programs. Fig. 1 summarizes the different models.

Outline and Contributions. We first illustrate our test selection methodology on an example in Sect. 2. Then, we recall in Sect. 3 the testing theory framework we use. Next, we present our contribution which is twofold:

- First, we describe in Sect. 4 a test generation method, that takes as inputs a recursive specification and a non-recursive test purpose and that returns a recursive test case. This method is based on program transformations. Technical choices are guided by theoretical properties of the underlying PDS and LTS models, but the generation is defined in terms of programming language concepts.
- Second, we present in Sect. 5 a selection algorithm which takes as input the previously generated test case and specializes it. This algorithm is based on a (co)reachability analysis of the test case. We formalize a partial observation problem, due to the inability of test cases to inspect their own stacks. We compare its consequences on the generated test cases with the impact of using a non-exact, over-approximated coreachability analysis as done for test selection based on symbolic STS models [4]. We also propose an improvement of the selection algorithm which minimizes the negative impact of this partial observation aspect on test selection.

2   Introductive Example

We illustrate in this section the concepts we will develop and our testing methodology on a running example, before formalizing it in the next sections.

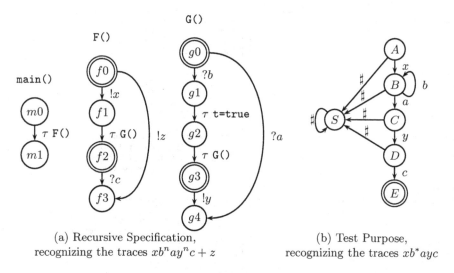

(a) Recursive Specification,
recognizing the traces $xb^n ay^n c + z$

(b) Test Purpose,
recognizing the traces xb^*ayc

Fig. 2. Control-flow Graphs of Specification and Test Purpose

Specification. In our testing theory, the *IUT* is considered as a black box reactive system, and its observations are the messages exchanged with its environment. The specification we consider as an example is the small recursive program of Fig. 4. Its control flow graph is given on Fig. 2(a). Double circles denote the observation points (see Sect. 4). Inputs and outputs are distinguished by the symbols ? and ! (inputs and outputs alphabets are disjoints). The behavior of this specification program is the following: the main function calls the function $F()$, which either emits the output $!z$ and returns, or emits the output $!x$, calls the function $G()$, then receives the input c and returns to its caller. According to its first input, the function $G()$ has two different behaviors: if the input is $?a$, it returns to the caller, whereas if the input is $?b$, the function $G()$ is called again (recursively) and after it returns, the output $!y$ is emitted.

What is really important is the traces recognized (or generated) by the specification, which is here $Traces(S) = \{!z\} \cup \{!x \cdot (?b)^n \cdot ?a \cdot (!y)^n \cdot c? \mid n \geq 0\}$, a context free language. The (non-)conformance of an *IUT* w.r.t. a specification S will be only based on $Traces(S)$. Intuitively, an *IUT* will be defined as *conformant* to S if after any execution trace which is a prefix of $Traces(S)$, it emits only outputs that S can emit as well. Note that the global variable t has no influence on $Traces(S)$, its usefulness will be explained later (see Sect. 5).

For instance, the *IUT* besides (where [] stands for the non-deterministic choice operator) is not conformant to S. After the execution trace $!x$, it may emit $!y$, whereas S specifies that no output may be emitted at this point: one or more $?b$ and then one $?a$ should be received first). Note that on all states, *IUT* is ready to receive any input on the parameter p.

```
void main(){
    emit(x) [] receive(p);
    emit(y) [] receive(p);
}
```

Fig. 3. Non-conformant *IUT*

```
enum out_t { x,y,z };
enum inp_t { a,b,c };


bool t = false;
void main(){
  f();
}

void f()
{
  f0: emit(p) when (p == x || p == z){


        if (p==z) goto f3;
     }
  f1: g();
  f2: receive(p) when (p == c) {};


  f3:
}

void g()
{
  g0: receive(p) when (p == a || p == b){
        if (p == a) goto g4;
     };



  g1: t = true;
  g2: g();
  g3: emit(p) when (p == y) {}


  g4:
}
```

```
enum out_t { a,b,c };
enum inp_t { x,y,z };
enum verdict_t { none, fail, pass, inconc };
enum verdict_t verdict = none;
bool t = false;
void main(){
  f();
}

void f()
{
  f0: receive (p) when true {
        if (p != x && p != z)
          { verdict = fail; abort(); }
        if (p == z) goto f3;
     };
  f1: g();
  f2: emit(p) when (p == c) {}
      []
      receive(p) when true
        { verdict = fail; abort() };
  f3:
}

void g()
{
  g0: emit(p) when (p == a || p == b){
        if (p == a) goto g4;
     }
     []
     receive(p) when true
        { verdict = fail; abort() };
  g1: t = true;
  g2: g();
  g3: receive(p) when true {
        if (p != y){ verdict = fail; abort(); }
     }
  g4:
}
```

Fig. 4. Specification corresponding to Fig. 2

Fig. 5. Canonical tester associated to the specification of Fig. 4

Canonical Tester. A tester, called *canonical tester* can be generated from S very straightforwardly, according to the (yet intuitive) definition of conformance. The name *canonical tester* stems from the fact that it can detect any non-conformant execution of the implementation. It is actually the most general tester, from which any sound test case can be derived. The program transformation consists in mirroring inputs into outputs and *vice-versa*, and to emit a failure verdict when the transformed program receives an unexpected input at an observation point. Fig. 5 gives the canonical tester Can(S) associated to S. A new type and a global variable **verdict** have been introduced for storing the verdict. Can(S) stimulates the *IUT* by sending to it input messages, and checks that the outputs of the *IUT*, which correspond to its own inputs, are conformant w.r.t. S.

If this canonical tester is run in parallel with the non-conformant program of Fig 3, and if the conformant program chooses to emit !x and then !y, the tester will perform the execution $m_0 \xrightarrow{\tau} f_0 \xrightarrow{?x} f_1 \xrightarrow{\tau} g_0$ and will reach location g_0, where it will receive an unexpected $?y$ input and will abort.

Notice that not only the (canonical) tester, but also the example IUT accept any input at an observation point. These are usual asumptions in the ioco theory. For the tester, the reason is that it should check any output from the IUT for conformance. For the IUT, this allows to prevent deadlocks.

Test purpose. For large specifications, the canonical tester is too general. It tests the IUT in a completely random way. One is often more interested in guiding the execution of the IUT so as to realize a specific scenario that may reveal an error, and to stop the test execution successfully when the scenario has been completed without conformance error.

In this context, a test purpose is a (set of) scenario one wants to observe during a *conformant* test execution. The test purpose depicted as an automaton on Fig. 2 specifies that one is interested in detecting conformance errors occuring along the traces in $Traces_E(TP) = xb^*ayc$. The symbol $\sharp$ means "all other elements in the alphabet" and the double circle denotes the final state E. This test purpose indicates that we want to test the case where the IUT emits $!x$ at control point f_0 and performs one recursive call of G from G.

The aim of test selection is to transform the canonical tester so that it is more likely to produce the execution trace xb^*ayc until completion when executed in parallel with the IUT. If a conformance error occurs, the tester aborts immediately with a `fail` verdict. For instance, the first time the tester enters in function `G()`, it should first emit a $!b$, because a matching $?y$ should be later received to realize the scenario. Moreover, the second time it enters (recursively) in function `G()`, it should emit an $!a$, because only one $?y$ message should be observed before $!c$.

On the other hand, if an IUT starts its execution by emitting one $!z$ (which is conformant to S), the scenario cannot be completed. The tester should detect such a case and abort gracefully with an `inconclusive` verdict.

Selected test case. Fig. 8 depicts the test case we obtain with the method we will develop in the paper. Compared to the canonical tester of Fig. 5, we have first inserted at each observation point a call to the function `TP()` (after having checked the absence of conformance error at this point). The function `TP()` defined on Fig. 6 takes as input the last message exchanged and implements the automaton of Fig. 2(b). If the final state is reached, it emits the `pass` verdict.

There are two other modifications to the canonical tester. At control point f_0, when a $?z$ is received, the `inconclusive` verdict is emitted. Last, at control point g_0, the condition for emitting a message has been enforced: $!a$ is emitted iff the variable `t` is true. Indeed, `t` allows to distinguish if `G()` is called for the first time from f_1, in which case `t` is false, or if it is called recursively from g_2, in which case `t` is true. Hence, the knowledge of the value of `t` allows the test case to realize exactly the scenario defined by the test purpose (once $?x$ has been received from the IUT).

```
enum pc_t { A,B,C,D,E,S };
enum pc_t pc = A;
void TP(enum msg_t p)
{
    if    (pc == A && p == x ) pc = B;
    elsif (pc == B && p == b)  pc = B;
    elsif (pc == B && p == a)  pc = C;
    elsif (pc == C && p == y)  pc = D;
    elsif (pc == D && p == c ){
        pc = E;
        verdict = pass;
        abort();
    }
    else pc = S;
}
```

Fig. 6. Test Purpose corresp. to Fig. 2.(b)

```
// Type and global variables Declarations
// ...
void main(){
  m0: f();
  m1:
}

void f()
{
  f0: receive (p) when true {
  f0r:   if (p != x && p != z)
             { verdict = fail; abort(); }
         TP(p);


       };
  f1: g();
  f2: emit(p) when (p == c) {
  f2e:   TP(p)
         }
       []
       receive(p) when true
  f2r:   { verdict = fail; abort(); };
  f3:
}

void g()
{
  g0: emit(p) when (p == a || p == b){

  g0e:   TP(p);
         if (p == a) goto g4;
         }
       []
       receive(p) when true
  g0r:   { verdict = fail; abort(); };
  g1: t = true;
  g2: g();
  g3: receive(p) when true {
  g3r:   if (p != y)
             { verdict = fail; abort(); }
         TP(p);
         }
    g4:
}
```

```
// Type and global variables Declarations
// ...
void main(){
  m0: f();
  m1:
}

void f()
{
  f0: receive (p) when true {
  f0r:   if (p != x && p != z)
             { verdict = fail; abort(); }
         TP(p);
         if (p == z)
             { verdict = inconc; abort(); }
       };
  f1: g();
  f2: emit(p) when (p == c) {
  f2e:   TP(p)
         }
       []
       receive(p) when true
  f2r:   { verdict = fail; abort(); };
  f3:
}

void g()
{
  g0: emit(p) when ((p == a && t == true)
                 || (p == b && t == false)){
  g0e:   TP(p);
         if (p == a) goto g4;
         }
       []
       receive(p) when true
  g0r:   { verdict = fail; abort(); };
  g1: t = true;
  g2: g();
  g3: receive(p) when true {
  g3r:   if (p != y)
             { verdict = fail; abort(); }
         TP(p);
         }
    g4:
}
```

Fig. 7. Product   **Fig. 8.** Test Case after selection

The next sections describe the theoretical fundations of this test selection scheme sketched on the running example. After having presented the testing theory on LTS models in Sect. 3, we adapt it on recursive specifications in Sect 4.

3   Testing Theory

The testing theory we consider is based on the notions of *specification, implementation*, and *conformance relation* between them [2] and on the model of *Labelled Transition Systems* (LTS). An LTS is defined by a tuple $M = (Q, Q_0, \Lambda, \rightarrow)$ where Q is a set of states, Q_0 is the set of initial states, $\Lambda = \Lambda_v \cup \{\tau\}$ is an alphabet of visible (Λ_v) and internal ($\{\tau\}$) actions and $\rightarrow \subseteq Q \times \Lambda \times Q$ is a set of labelled transitions. The notation $p \xrightarrow{a} q$ stands for $(p, a, q) \in \rightarrow$, and $p \xrightarrow{a}$ for $\exists q : p \xrightarrow{a} q$. An execution is a sequence $q_0 \xrightarrow{a_0} q_1 \xrightarrow{a_1} \ldots q_{n+1}$ with $q_0 \in Q_0$. $Traces(M) \subseteq \Lambda_v^*$ denotes the projection of the set of executions of M onto visible actions. For a subset $X \subseteq Q$ of states, $Traces_X(M)$ denotes the projection of the set of executions of M ending in a state $q' \in X$ onto visible actions. It is also named the set of traces *accepted* by X. The set of prefixes (resp. strict prefixes) of a set of traces Y is denoted by $pref_{\leq}(Y)$ (resp. $pref_{<}(Y)$). M is called *deterministic* if Q_0 has a single element q_0, if $p \xrightarrow{a} q \wedge p \xrightarrow{a} q' \implies q = q'$ and if $p \xrightarrow{\tau} q \implies \neg(\exists \alpha \in \Lambda_v : p \xrightarrow{\alpha})$. Note that this definition is not the standard definition of a deterministic LTS because transitions with internal actions are possible. M is *complete* for $A \subseteq \Lambda$, if $\forall a \in A, \forall p \in Q, p \xrightarrow{a}$.

The specification is a deterministic LTS $S = (Q^S, Q_0^S, \Lambda, \rightarrow_S)$, and the Implementation Under Test (IUT) is assumed to be an LTS $IUT = (Q^{IUT}, Q_0^{IUT}, \Lambda, \rightarrow_{IUT})$ which is unknown except for its alphabet, which is assumed to be the same as that of the specification. Moreover, it is assumed that the IUT is *input-complete*, which reflects the hypothesis that the IUT cannot refuse an input from its environment.

In this context, a *test case* for the specification S is a deterministic LTS $TC = (Q^{TC}, Q_0^{TC}, \Lambda, \rightarrow_{TC})$ which is able to interact with an implementation and to emit verdicts:

- its alphabet is the mirror of that of S ($\Lambda_?^{TC} = \Lambda_!^S$ and $\Lambda_!^{TC} = \Lambda_?^S$)
- it is input-complete (outputs of IUT are not refused) except in verdict states;
- it is equipped with 3 disjoint subsets of sink, *verdict* states Pass, Fail, Inconc $\subseteq Q^{TC}$. Intuitively, Fail means rejection, Pass that some targetted behavior has been realized (this will be clarified later), and Inconc that a targetted behavior cannot be realized any more.

The conformance relation defines which implementations are considered correct w.r.t. the specification. We will consider the following conformance relation:

Definition 1 (Conformance relation). *Let $S = (Q^S, Q_0^S, \Lambda, \rightarrow_S)$ and $IUT = (Q^{IUT}, Q_0^{IUT}, \Lambda, \rightarrow_{IUT})$ be two LTS with same alphabet. A trace σ of IUT conforms to S, denoted by σ conf S, iff*

$$pref_{\leq}(\sigma) \cap [Traces(S) \cdot \Lambda_! \setminus Traces(S)] = \emptyset$$

IUT conforms to S, denoted by IUT conf S, iff all its traces are conformant:
$Traces(IUT) \cap [Traces(S) \cdot \Lambda_! \setminus Traces(S)] = \emptyset$.

Intuitively, IUT conf S if after each trace of S, IUT may emit only outputs that S can emit as well, while its inputs are unconstrained. Except for the notion of quiescence (absence of outputs), conf corresponds to the ioco relation of [2].

The set of traces $Traces(S) \cdot \Lambda_! \setminus Traces(S)$ is the set of minimal non-conformant traces. It is characterized by a test case called the canonical tester, which is obtained from the specification S by inversion of inputs and outputs, followed by an input-completion, where each unspecified input leads to Fail.

Definition 2 (Canonical Tester). Let $S = (Q^S, Q_0^S, \Lambda, \rightarrow_S)$ be the deterministic LTS of the specification. The canonical tester of S for conf is the deterministic LTS $\mathrm{Can}(S) = (Q^S \cup \mathsf{Fail}, Q_0^S, \Lambda^{\mathrm{Can}}, \rightarrow_{\mathrm{Can}})$ such that

- Fail $= \{q_{\mathsf{Fail}}\}$, with $q_{\mathsf{Fail}} \notin Q^S$ a new state;
- its alphabet is the mirror of that of S ($\Lambda_?^{\mathrm{Can}} = \Lambda_!^S$ and $\Lambda_!^{\mathrm{Can}} = \Lambda_?^S$)
- $\rightarrow_{\mathrm{Can}}$ is defined by the rules:

$$\frac{q, q' \in Q^S \quad q \xrightarrow{\alpha}_S q'}{q \xrightarrow{\alpha}_{\mathrm{Can}} q'} \qquad \frac{q \in Q^S \quad \alpha \in \Lambda_!^S = \Lambda_?^{\mathrm{Can}} \quad \neg(q \xrightarrow{\alpha}_S)}{q \xrightarrow{\alpha}_{\mathrm{Can}} q_{\mathsf{Fail}}}$$

We have $Traces(\mathrm{Can}(S)) = pref_{\leq}(Traces(S) \cdot \Lambda_!)$
$Traces_{\mathsf{Fail}}(\mathrm{Can}(S)) = Traces(S) \cdot \Lambda_! \setminus Traces(S)$

$\mathrm{Can}(S)$ is already a test case. However, it is typically too large and is not focused on any part of the system. It is more interesting in practice to test what happens in the course of a given scenario (or set thereof), and if no error has been detected, to end the test successfully when the scenario is completed.

Definition 3 (Test Purpose). A test purpose for a specification S is a deterministic LTS $TP = (Q^{TP}, Q_0^{TP}, \Lambda, \rightarrow_{TP})$ equipped with a subset $\mathsf{Accept} \subseteq Q^{TP}$ of accepting sink states. TP is complete except in these Accept states.

The completeness assumption allows not to constrain S in the product $S \times TP$ (unless the execution trace is accepted).

The test case should now not only detect conformance errors, but also try to satisfy the test purpose. For this, it has to take into account output choices of the specification (observable non-determinism) and to detect incorrect outputs of the IUT w.r.t. the test purpose.

The verdicts and their meanings are summarised as follows. The Fail verdict is emitted if the implementation does not conform to the specification, so iff TC observes an unspecified output after a trace of S. Pass means that the behavior wanted by the test purpose has been realized by the implementation. The verdict Pass is thus emitted iff TC observes a trace of S accepted by TP. The Fail and Pass verdicts are uniquely defined, so that they are emitted appropriately and as soon as possible, whereas the Inconc verdict is not uniquely defined. Indeed, Inconc, which means that the behavior wanted by TP cannot be realized any more, may be emitted *only if* the trace observed by TC belongs to S (it is conformant) but is refused by TP. The test execution can thus be interrupted, as Pass cannot be emitted any more. We have adopted this definition because checking whether a trace is refused is not always possible, either because it is undecidable, for instance with infinite-state symbolic model [4], or because of partial observation issues as discussed in Sect. 5. We refer to [4] for how *optimal* test cases are defined.

Test selection for LTS. We briefly recall how to generate an optimal test case from a specification S and a test purpose TP given as finite LTS [3]. One first builds the canonical tester $Can(S)$ using Def. 2. One then builds the product $P = Can(S) \times TP$ combining the information about conformance given by $Can(S)$ and the information about the wanted scenario given by TP. One defines the set Pass of verdict states as $\text{Pass} = Q^S \times \text{Accept}^{TP}$. Adding the Inconc verdict is done by observing that

$$pref_{\leq}(\textit{Traces}_{\text{Pass}}(P)) = \textit{Traces}_{coreach(\text{Pass})}(P)$$

where $coreach(\text{Pass}) = \{q \in Q^P \mid \exists q' \in \text{Pass} : q \to^* q'\}$ denotes the set of states that may reach a state in Pass. A valid test case TC is obtained from P by adding a new state Inconc and by modifying $\to_P$ as follows:

$$\frac{q \xrightarrow{\alpha}_P q' \quad q' \in coreach(\text{Pass})}{q \xrightarrow{\alpha}_{TC} q'} \qquad \frac{q \xrightarrow{\alpha}_P q' \quad \alpha \in \Lambda_?^{Can(S)} \quad q' \notin coreach(\text{Pass})}{q \xrightarrow{\alpha}_{TC} \text{Inconc}}$$

The first rule keeps only transitions maintaining the execution in $coreach(\text{Pass})$. In particular, it selects the appropriate outputs w.r.t. TP that should be sent to the IUT. As TC should remain input-complete, the second rule redirects the input transitions not selected by the first rule to the Inconc verdict, which is thus emitted as soon as the execution leaves the prefixes of accepted traces by a conformant input.

4 Modeling Recursive Specifications and Test Purposes

The previous section recalled our framework for model-based testing, based on the low-level semantics model of LTS. We already extended these principles and designed sound algorithms for infinite-state symbolic transition systems in [4]. Our aim here is to do the same for recursive specifications which can be compiled into (input/output) pushdown automata, PDS. Such specifications manipulate finite data but may have an infinite control due to the recursion, hence they are more expressive than finite LTS. In terms of traces, which is a relevant notion for the conformance relation, they generate context-free languages instead of regular languages. Moreover, even if there are cases where the recursion is bounded and the specification may be flattened into a LTS (by inlining), such a process may result in a huge LTS.

Expressions	*expr*	
Atomic Instructions	*atom*	$::=$ *var* = *expr* | if (*expr*) goto *label*
Interproc. Instructions	*callret*	$::=$ *proc*() | return
Communications	*com*	$::=$ emit(*p*) when *expr* {*block*}
		| receive(*p*) when *expr* {*block*}
		| *com* [] *com*
Instructions	*instr*	$::=$ *atom* | *callret* | *com*
Sequences	*block*	$::=$ ϵ | *instr*; *block*

Fig. 9. Language Syntax

Control point	$k \in K$
Global environment	$g \in GEnv = GVar \rightarrow Val$
Local environment	$l \in LEnv = LVar \rightarrow Val$
Configuration	$(g, \sigma) \in C = GEnv \times (K \times LEnv)^+$

Fig. 10. Language Semantic domains

Push-Down Systems. A labelled *Push-Down System* (PDS) is defined by a tuple $\mathcal{P} = (G, \Gamma, \Lambda, c_0, \hookrightarrow)$ where G is a finite set of locations, Γ is a finite stack alphabet, $c_0 \in G \times \Gamma^*$ is the inital configuration, $\Lambda = \Lambda_v \cup \{\tau\}$ is a finite set of visible (Λ_v) and internal ($\{\tau\}$) actions, and $\hookrightarrow \subseteq (G \times \Gamma) \times \Lambda \times (G \times \Gamma^*)$ is a finite set of labelled transitions. Such a labelled PDS $\mathcal{P}$ generates an infinite LTS $M = (Q^M, Q_0^M, \Lambda, \rightarrow_M)$ where $Q^M = G \times \Gamma^*$, $Q_0^M = \{c_0\}$, and $\rightarrow$ is defined by the rule:
$$(g, \gamma) \overset{\alpha}{\hookrightarrow} (g, \gamma') \wedge \omega \in \Gamma^* \implies (g, \omega \cdot \gamma) \overset{\alpha}{\rightarrow} (g, \omega \cdot \gamma')$$
The notions of deterministic and complete PDS are defined in terms of LTS.

A small programming language. The syntax and semantics of the small language we used in the example of Sect. 2 is inspired by BEBOP [5], an input language of the MOPED tool, which is a model-checker for linear-time temporal logic on pushdown systems [6].

BEBOP uses a classical imperative language syntax. We assume for the sake of simplicity that control structures have been transformed into test and branch intructions, and that parameter passing and returns for procedures are emulated by using dedicated global variables. This results in the syntax given in Fig. 9.

The features added to BEBOP are the communication instructions, and the non deterministic choice operator between them. Emission and reception instructions use a special global variable p which contains the message, and which may be used only in the condition and in the block associated to these instructions. We assume that emissions and receptions are not nested. The operator [] is the non-deterministic choice operator. It may be used only for communication instructions. The reason is that while we allow non-determinism, it should remain observable, so that to any trace of the program corresponds an unique execution.

Its semantics as a Push-Down System (PDS). We assume that the special variable p takes its values in the alphabet Λ. The semantics of this language is defined using the domains defined on Fig. 10. It is given as a labelled PDS $\mathcal{P} = (G, \Gamma, c_0, \Lambda, \hookrightarrow)$ where $G = GEnv$, $\Gamma = K \times LEnv$, $c_0 = (g_0, (k_0, l_0))$ is the initial configuration, and $\hookrightarrow$ is defined by the following inference rules, using the control flow graph associated to the program. We just sketch the standard inference rules and we refer to [6] for more details, as we focus more precisely on the semantics of the emission and reception instructions.

– An atomic instruction generates a rule of the form $\dfrac{k \xrightarrow{atom} k'}{(g, (k, l)) \overset{\tau}{\hookrightarrow} (g', (k', l'))}$
  with a condition on (g, l) in the case of a test and branch instruction.

– A procedure call generates a rule $\dfrac{k \xrightarrow{proc()} k'}{(g, (k, l)) \overset{\tau}{\hookrightarrow} (g, (k', l) \cdot (s_{proc}, l'_0))}$ where s_{proc} is the start point of the caller. Such a transition means that a new activation record is pushed onto the stack, with an initial local environment l'_0, which reflects the assumption that the variables are uninitialized. A procedure return generates a rule $\dfrac{k \xrightarrow{proc()} k' \quad e_{proc} \xrightarrow{return} \ldots}{(g, (k', l') \cdot (e_{proc}, l)) \overset{\tau}{\hookrightarrow} (g, (k', l'))}$ where the activation record is popped and the control goes back to the caller.

– An emission instruction generates a rule

$$\dfrac{k \xrightarrow{emit(p) \ when \ expr \ \{k':block\}} k'' \qquad \forall v \neq p : g'(v) = g(v)}{\begin{array}{l} (g, (k, l)) \overset{p}{\hookrightarrow} (g', (k', l)) \text{ if } [\![expr]\!](g', l) = \mathsf{true} \\ (g, (k, l)) \overset{\tau}{\hookrightarrow} (g', (k'', l)) \text{ if } [\![expr]\!](g', l) = \mathsf{false} \end{array}}$$

One first forgets the previous value of p when introducing g', in order to make it uninitialized, as its real scope is the condition and the block associated to the emission. Then, if the current environment (g, l) satisfies the condition, p is emitted and the control passes to the beginning of the block k'. Otherwise, the control passes to k''. Notice that an non-deterministic choice is performed here: the instruction may emit any message p which satisfies the condition.

The semantics of the reception is identical to the emission. Emissions and receptions need to be distinguished only w.r.t. the conformance relation.

All instructions generate internal transitions labelled by τ, except emission and reception instructions. The *observation points* of a program are defined as the control points at the beginning of communication instructions. They are the only control points from which a message may be exchanged. Such observation points may be separated by (sequences of) ordinary control points linked by internal τ-transitions. Notice that we do not use the term "observation point" in the sense given to it in the testing community, when refering to the testing architecture.

Interprocedural specification and its canonical tester. An interprocedural specification S (*c.f.* Fig. 4) is a program defined with the language of Fig. 9, which is deterministic, in the sense that the allowed non-determinism should be observable, so that to a trace corresponds a unique possible execution ending in an observation point. A choice can still exist between two emissions and/or receptions, but we cannot have a choice between two internal instructions (generating τ-transitions).

This deterministic assumption allows to build easily the canonical tester of S, which is an executable, hence deterministic observer of $Traces(S) \cdot \Lambda_?^S \setminus Traces(S)$. The canonical tester $Can(S)$ is obtained from S using the following program transformation at each observation point:

$$
\begin{array}{l}
\mathsf{emit}(p) \text{ when } expr_e \;\{ \\
\\
\quad block_e \\
\\
\} \\
\, [] \\
\mathsf{receive}(p) \text{ when } expr_r \;\{block_r\}
\end{array}
\quad \Rightarrow \quad
\begin{array}{l}
\mathsf{receive}(p) \text{ when } true \;\{ \\
\quad \mathsf{if}(\mathsf{not}\ expr_e)\{verdict = fail;\ abort()\ \} \\
\quad block_e \\
\} \\
\, [] \\
\mathsf{emit}(p) \text{ when } expr_r \;\{block_r\}
\end{array}
$$

This operation mimics the corresponding operation defined for LTS in Sect. 3. Here, it could be done on the PDS generated by the program, but we prefer to proceed directly by program transformations.

Test purpose. When performing test generations from LTS, the test purpose is an LTS that is taken into account by computing the product $Can(S) \times TP$ (*c.f.* Sect. 3). Now, $Can(S)$ is a PDS. It is known that the product of two PDS is not a PDS, hence we cannot specify test purposes using PDS if we do not want to manipulate more expressive computational models. However, as the product of a PDS with an LTS is still a PDS, we can consider test purposes defined by finite LTS. We can compute the synchronous product of $Can(S)$ with TP to add the Pass verdict to the canonical tester.

However, our goal is to proceed by program transformations. This excludes to work directly on the underlying LTS and PDS models. The solution consists:

- in implementing the LTS TP (which should satisfy Def. 3) by a procedure $\mathsf{TP(p)}$ that takes as input the last exchanged message and implements the LTS, *c.f.* Figs. 2(b) and 6;
- and in instrumenting $Can(S)$ by inserting calls to TP at observation points:

$$
\begin{array}{l}
\mathsf{receive}(p) \text{ when } true \;\{ \\
\quad \mathsf{if}(\mathsf{not}\ expr_r) \\
\quad\quad \{\ verdict = fail;\ abort()\ \} \\
\quad block_r \\
\} \\
\, [] \\
\mathsf{emit}(p) \text{ when } expr_e \;\{block_e\}
\end{array}
\quad \Rightarrow \quad
\begin{array}{l}
\mathsf{receive}(p) \text{ when } true\{ \\
\quad \mathsf{if}(\mathsf{not}\ expr_r) \\
\quad\quad \{\ verdict = fail;\ abort()\ \} \\
\quad TP(p);\ block_r \\
\} \\
\, [] \\
\mathsf{emit}(p) \text{ when } expr_e \;\{TP(p);\ block_e\}
\end{array}
$$

The call to TP is performed after having checked the conformance, because accepted traces are conformant. The procedure TP is in charge of emitting the Pass verdict. This transformed canonical tester will be denoted by P. Fig. 7 depicts the obtained program for our running example.

Location	Coreachable states from $\langle(-,\mathsf{pass},E),\omega\rangle$
m_0	$\langle(\mathsf{ff},-,A),\omega.m_0\rangle$ $\langle(\mathsf{tt},-,A),\omega.m_0\rangle$
m_1	$\langle(-,\mathsf{pass},E),\omega.m_1\rangle$
f_0	$\langle(\mathsf{ff},-,A),\omega.f_0\rangle$ $\langle(\mathsf{tt},-,A),\omega.f_0\rangle$
f_{0r}	$\langle(\mathsf{ff},-,A,x),\omega.f_{0r}\rangle$ $\langle(\mathsf{tt},-,A,x),\omega.f_{0r}\rangle$
f_1	$\langle(\mathsf{ff},-,B),\omega.f_1\rangle$ $\langle(\mathsf{tt},-,B),\omega.f_1\rangle$
f_2	$\langle(\mathsf{ff},-,D),\omega.f_2\rangle$ $\langle(\mathsf{tt},-,D),\omega.f_2\rangle$
f_{2e}	$\langle(\mathsf{ff},-,D,c),\omega.f_{2e}\rangle$ $\langle(\mathsf{tt},-,D,c),\omega.f_{2e}\rangle$
f_{2r}	$\perp$
f_3	$\langle(-,\mathsf{pass},E),\omega.f_3\rangle$
g_0	$\langle(\mathsf{ff},-,B),\omega.(f_1g_0+f_1g_2g_0)\rangle$ $\langle(\mathsf{tt},-,B),\omega.(f_1g_0+f_1g_2g_0)\rangle$
g_{0e}	$\langle(\mathsf{ff},-,B,a),\omega.f_1g_2g_{0e}\rangle$ $\langle(\mathsf{ff},-,B,b),\omega.f_1g_{0e}\rangle$ $\langle(\mathsf{tt},-,B,a),\omega.f_1g_2g_{0e}\rangle$ $\langle(\mathsf{tt},-,B,b),\omega.f_1g_{0e}\rangle$
g_{0r}	$\perp$
g_1	$\langle(\mathsf{ff},-,B),\omega.f_1g_1\rangle$ $\langle(\mathsf{tt},-,B),\omega.f_1g_1\rangle$
g_2	$\langle(\mathsf{ff},-,B),\omega.f_1g_2\rangle$ $\langle(\mathsf{tt},-,B),\omega.f_1g_2\rangle$
g_3	$\langle(\mathsf{ff},-,C),\omega.f_1g_3\rangle$ $\langle(\mathsf{tt},-,C),\omega.f_1g_3\rangle$
g_{3r}	$\langle(\mathsf{ff},-,C,y),\omega.f_1g_{3r}\rangle$ $\langle(\mathsf{tt},-,C,y),\omega.f_1g_{3r}\rangle$
g_4	$\langle(\mathsf{ff},-,C),\omega.f_1g_2g_4\rangle$ $\langle(\mathsf{ff},-,D),\omega.f_1g_4\rangle$ $\langle(\mathsf{tt},-,C),\omega.f_1g_2g_4\rangle$ $\langle(\mathsf{tt},-,D),\omega.f_1g_4\rangle$

(a) Coreachable states

Location	Reachable states from $\langle(\mathsf{ff},\mathsf{none},A),m_0\rangle$
f_{0r}	$\langle(\mathsf{ff},\mathsf{none},A,x),m_1f_{0r}\rangle$ $\langle(\mathsf{ff},\mathsf{none},A,z),m_1f_{0r}\rangle$
f_{2e}	$\langle(\mathsf{ff},\mathsf{none},C,c),m_1f_{2e}\rangle$ $\langle(\mathsf{tt},\mathsf{none},D,c),m_1f_{2e}\rangle$ $\langle(\mathsf{tt},\mathsf{none},S,c),m_1f_{2e}\rangle$
f_{2r}	$\langle(\mathsf{ff},\mathsf{none},C,-),m_1f_{2r}\rangle$ $\langle(\mathsf{tt},\mathsf{none},D,-),m_1f_{2r}\rangle$ $\langle(\mathsf{tt},\mathsf{none},S,-),m_1f_{2r}\rangle$
g_{0e}	$\langle(\mathsf{ff},\mathsf{none},B,a),m_1f_2g_{0e}\rangle$ $\langle(\mathsf{ff},\mathsf{none},B,b),m_1f_2g_{0e}\rangle$ $\langle(\mathsf{tt},\mathsf{none},B,a),m_1f_2g_3^+g_{0e}\rangle$ $\langle(\mathsf{tt},\mathsf{none},B,b),m_1f_2g_3^+g_{0e}\rangle$
g_{0r}	$\langle(\mathsf{ff},\mathsf{none},B,-),m_1f_2g_{0r}\rangle$ $\langle(\mathsf{tt},\mathsf{none},B,-),m_1f_2g_3^+g_{0r}\rangle$
g_{3r}	$\langle(\mathsf{tt},\mathsf{none},C,y),m_1f_2g_{3r}^+\rangle$ $\langle(\mathsf{tt},\mathsf{none},D,y),m_1f_2g_{3r}^+\rangle$ $\langle(\mathsf{tt},\mathsf{none},S,y),m_1f_2g_{3r}^+\rangle$

(b) Reachable states in observation points

Location	Intersection reachable and coreachable states
f_{0r}	$\langle(\mathsf{ff},\mathsf{none},A,x,),m_1f_{0r}\rangle$
f_{2e}	$\langle(\mathsf{ff},\mathsf{none},C,c),m_1f_{2e}\rangle$
f_{2r}	$\perp$
g_{0e}	$\langle(\mathsf{ff},\mathsf{none},B,a),m_1f_2g_3g_{0e}\rangle$ $\langle(\mathsf{tt},\mathsf{none},B,b),m_1f_2g_{0e}\rangle$
g_{0r}	$\perp$
g_{3r}	$\langle(\mathsf{tt},\mathsf{none},C,y),m_1f_2g_{3r}\rangle$

(c) Intersection between reachable and coreachable states

Fig. 11. Analysis of the program of Fig. 7. *The configurations are composed of the values of global variables* $(t,verdict,pc,p)$ *and the stack* $(-$ *means any value, and* $\omega=K^*$*). As* p *is "active" only at observation points, its value is not precised elsewhere.*

5   Test Selection on the Recursive Canonical Tester

Test selection is based on the same principle as for LTS, *c.f.* Sect. 3. In particular we will exploit the identity $pref_{\leq}(\mathit{Traces}_{\mathsf{Pass}}(P)) = \mathit{Traces}_{coreach(\mathsf{Pass})}(P)$ to recognize (conformant) traces that may be accepted in the future by the test purpose. However, the inability of test cases to inspect their own stack prevents it from using fully the coreachability information. We analyse this partial observation problem in this section.

Coreachability Analysis. In the PDS generated by the semantics of our programming language, a configuration is a pair $(g,\sigma) \in C$ of a global environment and a call-stack. The set of configurations corresponding to the Pass verdict is $\mathsf{Pass} = \{(g,\sigma) \mid g(verdict) = \mathsf{Pass}\}$. The wanted coreachable set is $coreach = \{c \in C \mid \exists c' \in \mathsf{Pass} : c \rightarrow^* c'\}$.

We will exploit nice theoretical properties of PDS for computing *coreach*. These properties justify the choice of PDS as the semantic model of our language, and the restriction to finite-state variables. Given a PDS $\mathcal{P} = (G, \Gamma, c_0, \Lambda, \hookrightarrow)$, a set of configurations $X \in \wp(G \times \Gamma^*) = G \to \wp(\Gamma^*)$ is *regular* if it associates to each global state a regular language. The first result is that the coreachability (resp. reachability) set of a PDS is regular if the final (resp. initial) set of configurations is regular [7]. The second result is that in this case, the coreachability (resp. reachability) set is computable with polynomial complexity [8,9]. The MOPED tool implements efficient symbolic algorithms to compute these sets, using a model of symbolic PDS where the transition relation $\hookrightarrow$ is represented with BDDs [6].

As the set Pass is regular, we can provide to MOPED the PDS generated by our recursive program and the set Pass, and we obtain the regular set of coreachable configurations. Coming back to our running example, the table of Fig. 11(a) indicates, for every location, the configurations from which we can reach the final configuration $\langle (-, \mathsf{pass}, E), \omega \rangle$. As there are no local variables in the example, the stacks contain only control points.

The problem of partial observation. The selection consists in adding tests in the program P, using coreachability information, for selecting the outputs to emit, and for detecting the inputs which make P leave the set of accepted traces. However, in an usual imperative language like ours, a program can only observe the top of the stack, whereas deciding whether the current configuration is coreachable or not may require the inspection of the full stack.

Let us define the observation function α :
$$C \to GEnv \times K \times LEnv$$
$$(g, \omega \cdot (k, l)) \mapsto (g, k, l)$$
extended to sets, and $\gamma = \alpha^{-1}$ the corresponding inverse function. (α, γ) forms a Galois connection. At some location k of the program, given a set of configurations X, and $X(c) = \{(c \in X \mid c = (g, \omega \cdot (k, l))\}$ its projection on location k, the program can only decide if the current valuation of variables (g, l) is included in $\alpha(X(c))$. This means that *in term of configurations*, one can only test inclusion in $\gamma \circ \alpha(X) \supseteq X$. In particular one may be in a case with

$$\gamma \circ \alpha(coreach(k)) \cap \gamma \circ \alpha(\overline{coreach}(k)) \neq \emptyset \qquad (1)$$

where one cannot decide, using only the observable part of the configuration, whether the configuration is coreachable or not. For instance, in Fig. 11(a), in location g_{0e}, $\alpha(coreach(g_{0e})) = (p \in \{a, b\} \wedge pc = B)$ and $\alpha(\overline{coreach}(g_{0e})) = $ tt.

Selection rules. Because of the partial obervation phenomenon, we have to be conservative in the selection. Let $cond_{co(k)}(g, l)$ be the logical characterization of $\alpha(coreach(k))$. We transform the program P as follows:

$$
\begin{array}{l|l}
\begin{array}{l}
\mathsf{receive}(p) \text{ when } true\,\{ \\
\quad \text{if } (\text{not } expr_r) \\
\qquad \{\ verdict = fail;\ abort()\ \} \\
\quad k_r: \\
\\
\qquad TP(p);\ block_r \\
\} \\
\\
\,[] \\
\mathsf{emit}(p) \text{ when } expr_e \\
\quad \{k_e:\ TP(p);\ block_e\}
\end{array}
& \Rightarrow &
\end{array}
$$

$$
\begin{array}{l}
\mathsf{receive}(p) \text{ when } true\,\{ \\
\quad \text{if } (\text{not } expr_r) \\
\qquad \{\ verdict = fail;\ abort()\ \} \\
\quad k_r:\ \text{if not } (cond_{co(k_r)}) \\
\qquad \{\ verdict = inconc;\ abort()\ \} \\
\qquad TP(p);\ block_r \\
\} \\
\\
\,[] \\
\mathsf{emit}(p) \text{ when } expr_e\, and\, cond_{co(k_e)} \\
\quad \{k_e:\ TP(p);\ block_r\}
\end{array}
$$

For receptions, at location k_r, after having checked the conformance, $\neg cond_{co(k)}$ is a *sufficient* condition to leave the prefixes of accepted traces ($\gamma(\neg cond_{co(k)}) \subseteq \overline{coreach}(k)$). So if it is satisfied we emit Inconc. For emissions, $cond_{co(k)}$ is a *necessary* condition to stay in prefixes of accepted traces ($\gamma(cond_{co(k)}) \supseteq coreach(k)$).

The obtained program is a sound test case. There is a strong similarity between this test selection algorithm and the test selection algorithm for symbolic infinite-state transition systems defined in [4]. Here partial observation may prevent us to perform an optimal selection. In [4], it is the impossibility to compute the exact coreachability set, and the need to resort to an overapproximation.

Improving selection with reachability information. One can improve the selection algorithm using reachability information. Let *reach* denote the set of reachable configurations of the program P. At a point k, we can exploit the knowledge that the current configuration is anyway included in $reach(k)$, and testing the inclusion in $\gamma \circ \alpha(reach(k) \cap coreach(k))$ instead of $\gamma \circ \alpha(coreach(k))$.[1] The problematic case identified by Eqn (1) becomes

$$
\gamma \circ \alpha(reach(k) \cap coreach(k)) \cap \gamma \circ \alpha(reach(k) \cap \overline{coreach}(k)) \neq \emptyset \qquad (2)
$$

It is clear that Eqn. (2) implies Eqn. (1) but that the converse if false.

Coming back to our example, Fig. 11(b) gives the reachability set of P projected on observation points, and Fig. 11(c) the intersection $reach(k) \cap coreach(k)$ for these points. If $cond_{co(k)}(g, l)$ denotes now a formula characterizing $\alpha(reach(k) \cap coreach(k))$, we now have $cond_{co(g_{0a})} = (pc = B) \wedge (t \wedge p = b \vee \neg t \wedge p = a)$ instead of just $(pc = B)$. One can check that Eqn. (2) is not true for $k = g_{0e}$, thus the selection is optimal at this point. Fig. 8 depicts the test case obtained by this improved selection algorithm. It should be noted that the presence of the variable t helps to perform an accurate selection at location g_0, because it allows to distinguish whether G() has been called from f_1 or from g_2. If we remove this variable, which does not change the semantics of S w.r.t. the conformance relation, one could not select optimally the output a or b to send to the IUT.

---

[1] As reachability and coreachability sets are regular, so is their intersection.

6   Concluding Remarks

The selection algorithm of Sect. 5 is currently under implementation by the extension of the MOPED tool. MOPED acts as a model-checker returning a Boolean answer, possibly with a counter-example. For our application, we need to get the sets of configurations computed by MOPED, to intersect reachability and coreachability sets, to project this intersection on the visible part, and to convert the result in terms of a programming language expression.

It is interesting to note the similarity of the two combinations: partial observation and exact analysis w.r.t. full observation and approximated analysis. In case of partial observation, the observation function α we introduced acts exactly as an abstraction (approximation) function. This means that one could apply our method to general recursive programs, on which the analysis would be in general approximated. The non-optimality of the selection would then be a consequence of the combination of partial observation and inexact analysis. One gets the diagram of Fig. 1.

Alternative methods. Our selection method described in Sect. 5 is based on (i) an exact analysis computing full configurations (instead of just visible parts of configurations), and (ii) on pure program transformations. These two choices could be revised. Concerning (i), one could use a less precise, classical interprocedural analysis method, which could still be exact for the observable part of the stack (for instance using the BEBOP tool [5]). However it would lead to a less precise selection scheme. In particular, intersecting the coreachabe set with reachable set would filter out less values. Concerning (ii), one could instrument the program so as to get more knowledge about the invisible part of the configuration. For instance, one could add a data-structure maintaining a stack of procedure return points, and using it when testing if one is still in a coreachable configuration. Although the resulting test case could not be any more transformed into a PDS, the analysis would still be performed on the same intermediate program P as in Sect. 5. Test case execution would be however slower, as testing for coreachability would involve more complex datatypes.

References

1. ISO/IEC 9646: Conformance Testing Methodology and Framework (1992)
2. Tretmans, J.: Test generation with inputs, outputs and repetitive quiescence. Software—Concepts and Tools, vol. 17(3) (1996)
3. Jard, C., Jéron, T.: TGV: theory, principles and algorithms. Int. Journal on Software Tools for Technology Transfer, vol. 6 (2004)
4. Jeannet, B., Jéron, T., Rusu, V., Zinovieva, E.: Symbolic test selection based on approximate analysis. In: Halbwachs, N., Zuck, L.D. (eds.) TACAS 2005. LNCS, vol. 3440, Springer, Heidelberg (2005)
5. Ball, T., Rajamani, S.: Bebop: A symbolic model checker for boolean programs. In: Havelund, K., Penix, J., Visser, W. (eds.) SPIN Model Checking and Software Verification. LNCS, vol. 1885, Springer, Heidelberg (2000)

6. Esparza, J., Schwoon, S.: A BDD-based model checker for recursive programs. In: Berry, G., Comon, H., Finkel, A. (eds.) CAV 2001. LNCS, vol. 2102, Springer, Heidelberg (2001)
7. Caucal, D.: On the regular structure of prefix rewriting. Theoretical Computer Science, vol. 106 (1992)
8. Finkel, A., Willems, B., Wolper, P.: A direct symbolic approach to model checking pushdown systems. Electronic Notes on Theoretical Computer Science, vol. 9 (1997)
9. Bouajjani, A., Esparza, J., Maler, O.: Reachability analysis of pushdown automata: Application to model checking. In: Mazurkiewicz, A., Winkowski, J. (eds.) CONCUR 1997. LNCS, vol. 1243, Springer, Heidelberg (1997)

A New Method for Interoperability Test Generation

Alexandra Desmoulin and César Viho

IRISA/Université de Rennes 1,
Campus de Beaulieu,
35042 Rennes Cedex, France
{alexandra.desmoulin,viho}@irisa.fr

Abstract. Interoperability testing aims at verifying the possibility for two or more components to communicate correctly while providing the foreseen services. In this paper, we describe a new method for generating interoperability test cases. This method is equivalent to classical methods in terms of non-interoperability detection. Contrary to classical approaches, this method avoids the well-known state-space explosion problem. It has been implemented in the CADP Toolbox and applied to a simplified version of the ISDN connection protocol. The obtained results confirm the real contribution of this method: test cases has been derived while classical approaches face the state-space explosion problem.

1 Introduction

Interoperability testing is used to verify that different protocol implementations communicate correctly while providing the services described in their respective specification. Contrary to conformance testing which is precisely characterized with testing architectures, formal definitions [1, 2] and tools for generating automatically tests [3, 4], interoperability is not formally defined. Some formal definitions [5, 6] and methods for generating interoperability tests [7, 8] exist, but there is no precise characterization of interoperability for the moment, and consequently no method based on formal definitions.

In this paper, we consider interoperability formal definitions of [9]. These definitions, called interoperability criteria, describe the conditions that two implementations must satisfy to be considered interoperable: providing the expected service while interacting correctly. Based on a proved equivalence between two of these criteria, a new method to generate automatically interoperability test cases is described. This method is equivalent to classical methods in terms of non-interoperability detection. But it avoids the construction of the specification interaction that may lead to the well-known state-explosion problem [6]. Moreover, we apply this method -implemented in the CADP Toolbox [10]- on a simplified version of the ISDN (Integrated Service Digital Network) connection protocol [11]. The obtained results show that the proposed method is a real contribution as we are able to derive interoperability test cases while classical methods were not applicable because of the state-space explosion problem.

A. Petrenko et al. (Eds.): TestCom/FATES 2007, LNCS 4581, pp. 58–73, 2007.

This paper is structured as follows. First, we present the formal background including interoperability formal definitions needed in Section 2. Then, Section 3 focuses on methods for generating interoperability test cases. We present first classical method and then our new method. Section 4 describes the results of the application of both methods on a simplified version of the ISDN connection protocol. Conclusion and future work are in Section 5.

2   Formal Background

In this section, we present the different notions that are used in the following. First, interoperability is defined in Section 2.1. Section 2.2 presents the formal model (IOLTS) and the related definitions used for interoperability formal approach. Finally, Section 2.3 describes the interoperability formal definitions (iop criteria) considered for interoperability test generation.

2.1   Preliminary Definitions

Different kinds of tests exist for testing if protocol implementations will work correctly in an operational environment. For example, conformance testing verifies that an implementation behaves as described in its specification. It considers events observed on the interfaces of the Implementation Under Test (IUT), and compares these events with events foreseen in the specification. The IUT is a black-box: testers do not have any knowledge of its internal structure.

In this paper, we consider another kind of test: interoperability testing. This kind of test has two goals. It verifies that different IUTs (two in this study) can communicate correctly, *and* that they provide the services described in their respective specification while communicating. In an interoperability testing architecture, we can differentiate two kinds of interfaces. The Lower Interfaces are the interfaces used for the interaction while the Upper Interfaces are used for the communication of the implementations with upper layer. Testers are connected to these interfaces but they can *control* (send message) only the upper interfaces. The lower interfaces are only *observable*.

Depending on the access to the interfaces, different architectures can be distinguished. For example, the interoperability testing architecture is called *unilateral* if only the interfaces of one IUT are accessible during the interaction, *bilateral* if the interfaces of both IUTs are accessible but separately, or *global* if the interfaces of both IUTs are accessible with a global view.

2.2   IOLTS Model and Related Definitions

We use IOLTS (Input-Output Labeled Transition System) [12] to model specifications. As usual in the black-box testing context, we also need to model IUTs, even though their behaviors are unknown. They are also modeled by an IOLTS.

Definition 1. *An IOLTS is a tuple $M=(Q^M, \Sigma^M, \Delta^M, q_0^M)$. Q^M is the set of states and $q_0^M \in Q^M$ the initial state. Σ^M denotes the set of observable events on* the interfaces: $p?m \in \Sigma^M$ (resp. $p!m \in \Sigma^M$) stands for an input (resp. output) where p is the interface and m the message. Δ^M is the transition relation.

Based on this model, $\Gamma(q)$ is the *set of observable events* (executed on the interfaces of M) from the state q and $\Gamma(M, \sigma)$ the set of observable events for the system M after the succession of events (*trace*) σ. In the same way, $Out(M, \sigma)$ (resp. $In(M, \sigma)$) is the set of possible outputs (resp. inputs) for M after the trace σ. $Traces(q)$ is the set of possible observable traces from q. We can also define $\bar{\mu}$ as $\bar{\mu} = l_i!a$ if $\mu = l_j?a$ and $\bar{\mu} = l_i?a$ if $\mu = l_j!a$.

An implementation can be *quiescent* in three different situations: either the IUT can be waiting on an input, either it can be executing a loop of internal (non-observable) events, or it can be in a state where no event is executable. For an IOLTS M, a quiescent state q is treated as an observable (practically with timers) output event. An IOLTS with quiescence modeled is noted $\Delta(M)$.

Two other operations need to be modeled: asynchronous interaction and projection. The asynchronous interaction is used to calculate the IOLTS modeling the behavior of a system composed by different communicating entities. For two IOLTS M_1 and M_2, the asynchronous interaction is noted $M_1 \|_A M_2$. The way to obtain this model of the interaction is described in [9]. First, M_1 and M_2 are transformed into IOLTS representing their behavior in an asynchronous environment. Then, these two IOLTS are composed to obtain $M_1 \|_A M_2$.

The projection of an IOLTS on a set of events is used to represent the behavior of the system reduced to specific events (such as events observable on certain interfaces). The projection of M on the set of events executable on its lower interface Σ_L^M is noted M / Σ_L^M.

Conformance formal definition. Contrary to interoperability testing, conformance is precisely formalized with formal definitions [1, 2] and tools for generating automatically tests like TGV [3] or TorX [4]. Among the different formal definitions existing for conformance, the **ioco** conformance relation [2] says that an implementation I is **ioco**-conformant with respect to its specification S if I never produces an output which could not be produced by S after the same trace. Moreover, I may be quiescent only if S can do so. Formally : I **ioco** $S =_\Delta \forall \sigma \in Traces(\Delta(S)), Out(\Delta(I), \sigma) \subseteq Out(\Delta(S), \sigma)$.

2.3   Interoperability Formal Definitions: iop Criteria

Even though some formal definitions exist in [9, 13], there is no precise characterization for interoperability (*iop* for short in the following). Here, we present some formal definitions, called *iop criteria*. They consider different possible architectures for testing the interoperability of two IUTs (one-to-one context) and are based on both purposes of interoperability: verifying that each entity actually receives the outputs sent by the peer entity and that the messages sent by the IUTs on their upper interfaces correspond to the service described in the specifications. Thus, outputs must be verified on both interfaces. As only outputs can be observed, verifying that an input μ is actually received by the peer entity implies to determine the set of outputs that can happen only due to the reception of μ. This set of outputs is calculated based on causal dependencies. The set of outputs on M that are causally dependent of the input μ after the trace σ is noted: $CDep(M, \sigma, \mu)$.

The **global iop criterion** considers both kinds of interfaces and both IUTS globally. It says that, after a trace of the interaction of the specifications, all outputs observed during the interaction of the implementations must be foreseen in the specifications, and that outputs sent by one IUT via its lower interfaces must be effectively received by the interacted IUT. This iop criterion corresponds to the most used testing architecture.

Definition 2 (Global iop criterion iop_G). $I_1 iop_G I_2 =_{def}$
$\forall \sigma \in Traces(S_1\|_A S_2)$, $Out(I_1\|_A I_2, \sigma) \subseteq Out(S_1\|_A S_2, \sigma)$
and $\forall \{i, j\} = \{1, 2\}$, $i \neq j$,
$\forall \sigma \in Traces(S_i\|_A S_j)$, $\sigma_i = \sigma/\Sigma^{S_i} \in Traces(S_i)$, $\sigma_j = \sigma/\Sigma^{S_j} \in Traces(S_j)$,
$\forall \mu \in Out(I_i, \sigma/\Sigma^{S_i})$, $\forall \sigma' \in [(\Sigma^{S_i} \cup \Sigma^{S_j} \cup \{\delta(i), \delta(j)\}) \setminus \bar{\mu}]^* \cup \{\epsilon\}$, $\sigma.\mu.\sigma'.\bar{\mu} \in$
$Traces(S_i\|_A S_j)$, $\bar{\mu} \in In(I_j, \sigma_j.(\sigma'/\Sigma^{I_j})) \Rightarrow Out(I_j, \sigma_j.(\sigma'/\Sigma^{I_j}).\bar{\mu}.\sigma_k) \in$
$CDep(\ S_j, \sigma_j.(\sigma'/\Sigma^{I_j}), \bar{\mu}), \sigma_k \in (\Sigma_I^{S_j})^* \cup \{\epsilon\}$

In [9], we prove the equivalence of the global criterion with the so-called bilateral iop criterion iop_B (defined below via the unilateral iop criterion iop_U) in terms of non-interoperability detection. This equivalence is used for developing our new interoperability test generation method.

The **unilateral iop criterion** iop_U (view I_1) considers interfaces of IUT I_1 while interacting with I_2. It says that, after a trace of S_1 observed during the interaction, all outputs observed in I_1 must be foreseen in S_1, and that I_1 must be able to receive outputs sent by I_2 via its lower interfaces.

Definition 3 (Unilateral iop criterion iop_U). $I_1 iop_U I_2 =_{def}$
$\forall \sigma_1 \in Traces(\Delta(S_1))$, $\forall \sigma \in Traces(S_1\|_A S_2)$,
$\sigma/\Sigma^{S_1} = \sigma_1 \Rightarrow Out((I_1\|_A I_2)/\Sigma^{S_1}, \sigma_1) \subseteq Out(\Delta(S_1), \sigma_1)$
and $\forall \sigma_1 = \sigma/\Sigma^{S_1} \in Traces(\Delta(S_1))$ such that $\sigma \in Traces(\ S_1\|_A S_2)$, $\forall \mu \in$
$Out(I_2, \sigma/\Sigma^{I_2})$, $\forall \sigma' \in [(\Sigma^{S_1} \cup \Sigma^{S_2}) \setminus \bar{\mu}]^* \cup \{\epsilon\}$, $\sigma.\mu.\sigma'.\bar{\mu} \in Traces(S_1\|_A S_2)$, $\bar{\mu} \in$
$In(I_1, \sigma_1.(\sigma'/\Sigma^{I_1})) \Rightarrow Out(I_1, \sigma_1.(\sigma'/\Sigma^{I_1}).\bar{\mu}.\sigma_i) \in CDep(S_1, \sigma_1.(\sigma'/\Sigma^{I_1}), \bar{\mu})$,
$\sigma_i \in (\Sigma_I^{S_1})^* \cup \{\epsilon\}$

The **bilateral total iop criterion** iop_B is verified iff both (on I_1 side and I_2 side) unilateral criteria are verified: $I_1 iop_B I_2 \ (= I_2 iop_B I_1) =_{def} I_1 iop_U I_2 \wedge I_2 iop_U I_1$.

3 Interoperability Test Generation

3.1 Preliminary Definitions

Iop test purpose. In practice, a test purpose is an informal description of behaviors to be tested. Generally it is an incomplete sequence of actions. Formally, a test purpose TP can be represented by a deterministic and complete IOLTS equipped with trap states used to select targeted behaviors. Complete means that each state allows all actions. In this study, we consider simplified iop test purposes with only one possible action after each state ($\forall\ \sigma, |\Gamma(TP,\sigma)| \leq 1$) and one $Accept^{TP}$ trap state used to select the targeted behavior.

Iop test cases. During interoperability tests, three kinds of events are possible for the tester: sending of stimuli to the upper interfaces of the IUTs, receiving inputs from these interfaces, and observing events (input and output) on the lower interfaces. Thus, a test case TC can be represented by $TC = (Q^{TC}, \Sigma^{TC}, \Delta^{TC}, q_0^{TC})$, an extended version of IOLTS. $\{PASS, FAIL, INC\} \subseteq Q^{TC}$ are trap states representing interoperability verdicts. $\Sigma^{TC} \subseteq \{\mu | \bar{\mu} \in \Sigma_U^{S_1} \cup \Sigma_U^{S_2}\} \cup \{?(\mu) | \mu \in \Sigma_L^{S_1} \cup \Sigma_L^{S_2}\}$. $?(\mu)$ denotes the observation of the message μ (that can be an input or an output) on a lower interface.

Iop verdicts. The execution of the test case TC on the system composed of the two IUTs gives an interoperability verdicts. PASS means that no interoperability error was detected during the tests. FAIL stands for the iop criterion is not verified. INC (for Inconclusive) is for the case where the behavior of the SUT seems valid but it is not the purpose of the test case.

3.2   Classical Methods

In practice, most of interoperability test suites are written "by hand". This is done by searching "manually" paths corresponding to the test purpose in the specifications. Considering the number of possible behaviors contained in the specification interaction, this "manual" test derivation is an error-prone task.

Methods for automatic interoperability test generation (as in [7, 8, 14, 15, 16]) also consider algorithms that search paths corresponding to the test purpose in the composition of the specifications (sometimes called reachability graph). The study described in [6, 13] considers an interoperability formal definition that compares events executed by the system composed of the two implementations with events foreseen in the specifications. Thus, traditional methods for deriving interoperability test cases are based on a global approach and on a general interoperability definition corresponding to the formal iop global criterion iop_G. The classical method can be summarized, as in Figure 1(a), by two main steps. The first one is the calculation of the specification interaction (completely or based on the defined test purpose depending on the method). The second step corresponds to the interoperability test case derivation based on the model of the specification interaction and on the test purpose.

The problem with this (these) classical method(s) is that we can have state space explosion when calculating the asynchronous interaction of the specifications [6]. Indeed, the number of states in the specification asynchronous interaction is in the order of $O((n.m^f)^2)$ where n is the number of states in the specifications, f the size of the input FIFO queue on lower interfaces and m the number of messages in the alphabet of possible inputs on lower interfaces. This calculation can be infinite if the size of the input FIFO queues is not bounded.

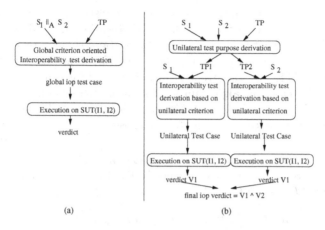

Fig. 1. Interoperability test generation: classical and new methods

3.3 New Approach: Bilateral Criterion Based Method

The equivalence -in terms of non-interoperability detection- between global and bilateral iop criteria (cf. theorem 1 in [9]) suggests that iop tests derived based on the bilateral iop criterion will detect the same non-interoperability situations as tests generated using classical method. Moreover, the bilateral method (see Figure 1(b)) avoids the calculation of the specification interaction.

General principles and verdict management. Let us consider an iop test purpose TP, as described in Section 3.1. The bilateral method can be decomposed in two main steps: cf. Figure 1(b). The first step of the bilateral method is the derivation of two unilateral iop test purposes TP_{S_i} from the global interoperability test purpose TP. Each TP_{S_i} contains only events of S_i and represents the iop test purpose TP in the point of view of S_i. The second step is the unilateral test case derivation. For this step, we can use a conformance test generation tool $\mathcal{F}$ such that $\mathcal{F} : (S_i, TP_{S_i}) \rightarrow TC'_i$, $i \in \{1, 2\}$. The unilateral test cases TC_i are obtained from TC'_i after some modifications due to differences in interface controllability between conformance and interoperability contexts.

As bilateral and global iop criteria are equivalent in terms of non-interoperability detection, we have: $verdict(TC, I_1 \|_A I_2) = verdict(TC_1, I_1 \|_A I_2) \wedge verdict(TC_2, I_1 \|_A I_2)$. The verdicts $verdict(TC, I_1 \|_A I_2)$, $verdict(TC_1, I_1 \|_A I_2)$ and $verdict(TC_2, I_1 \|_A I_2)$ are interoperability verdicts; $verdict(TC, I_1 \|_A I_2)$ is a global interoperability verdict and the two others are unilateral verdicts. The rules for the combination of these two unilateral verdicts to obtain the final bilateral iop_B verdict are obvious: $PASS \wedge PASS = PASS$, $PASS \wedge INC = INC$, $PASS \wedge FAIL = FAIL$, $INC \wedge FAIL = FAIL$, $INC \wedge INC = INC$ and $FAIL \wedge FAIL = FAIL$.

Unilateral Interoperability Test Purposes Derivation. The algorithm of figure 2 shows how to derive two unilateral interoperability test purposes from one global interoperability test purpose. Let us consider an event μ of the test purpose. If μ is an event of the considered specification, it is added to the test purpose. If μ is an event from the other specification, there is two possibilities. Either μ is an event to be executed on lower interfaces: in this case, the mirror event $\bar{\mu}$ is added. Either the event is an event to be executed on the upper interfaces: in this case, the algorithm searches a predecessor of μ, such that this predecessor is an event to be executable on lower interfaces. The algorithm adds the mirror of this predecessor to the test purpose.

Input: S_1, S_2: specification, TP: iop test purpose; **Output:** $\{TP_{S_i}\}_{i=1,2}$;
Invariant: $S_k = S_{3-i}$ (* S_k is the other specification *); $TP = \mu_1...\mu_n$
Initialization: $TP_{S_i} = \epsilon\ \forall i \in \{1,2\}$;
for $(j = 0; j \leq n; j{+}{+})$ **do**
  if $(\mu_j \in \Sigma_L^{S_i})$ **then** $TP_{S_i} = TP_{S_i}.\mu_j$; $TP_{S_k} = TP_{S_k}.\bar{\mu}_j$
  if $(\mu_j \in \Sigma_L^{S_k})$ **then** $TP_{S_i} = TP_{S_i}.\bar{\mu}_j$; $TP_{S_k} = TP_{S_k}.\mu_j$
  if $(\mu_j \in \Sigma_U^{S_i})$ **then** $TP_{S_i} = TP_{S_i}.\mu_j$;
                $TP_{S_k} = $add_precursor$(\mu_j, S_i, TP_{S_k})$
  if $(\mu_j \in \Sigma_U^{S_k})$ **then** $TP_{S_k} = TP_{S_k}.\mu_j$;
                $TP_{S_i} = $add_precursor$(\mu_j, S_k, TP_{S_i})$
  if $(\mu_j \notin \Sigma^{S_k} \cup \Sigma^{S_i})$**then** error(TP not valid : $\mu_j \notin \Sigma^{S_1} \cup \Sigma^{S_2}$)

function add_precursor(μ, S, TP): **return** TP
  $\sigma_1 := TP$; $a_j = $last_event$(\sigma_1)$
  while $a_j \in \Sigma_U^S$ **do** $\sigma_1 = $remove_last$(\sigma_1)$;
                $a_j = $last_event$(\sigma_1)$ **end**
  $M = \{q \in Q^S; \exists\ q'|(q, \bar{a}_j, q') \wedge \sigma = \bar{a}_j.\omega.\mu \in Traces(q)\}$
  if $(\forall q \in M, \sigma \notin Traces(q))$ **then** error(no path to μ)
  while $(e = $last_event$(\omega) \notin \Sigma_L^S \cup \{\epsilon\})$ **do** $\omega = $remove_last$(\omega)$
  if $(e \in \Sigma_L^S)$ **then** $TP_S = TP_{S_i}.\bar{e}$ **end**

Fig. 2. Algorithm to derive TP_{S_i} from TP

Some additional functions are used in the algorithm of figure 2. Let us consider a trace σ and an event a. The function *remove_last* is defined by : remove_last$(\sigma.a) = \sigma$. The function *last_event* is defined by : last_event$(\sigma) = \epsilon$ if $\sigma = \epsilon$ and last_event$(\sigma) = a$ if $\sigma = \sigma_1.a$. The *error* function returns the cause of the error and exits the algorithm.

Deriving unilateral interoperability test cases. The second step of this method is the derivation of two unilateral interoperability test cases based on the obtained test purposes. For this, a conformance test generation tool is used. It takes as inputs a specification S_1 (resp. S_2) and the corresponding test purpose TP_{S_1} (resp. TP_{S_2}) and generates two conformance test cases TC_1' and TC_2' that are modified in order to obtain the unilateral iop test cases TC_1 and TC_2. These unilateral interoperability test cases will be executed unilaterally on the corresponding IUT in the SUT.

The modifications on TC'_1 and TC'_2 to obtain TC_1 and TC_2 are realized to take into account the differences between upper and lower interfaces in interoperability testing. For example, an event $l!m$ (resp. $l?m$) in the obtained test case will be replaced by $?(l?m)$ (resp. $?(l!m)$) in the interoperability test case. This means that the unilateral interoperability tester observes that a message m is received from (resp. sent to) the other IUT on the lower interface l. No changes are made on the test cases for events on the upper interfaces as these interfaces are observable and controllable: a message can be sent (and received) by the tester to the IUT on these interfaces.

Some words about complexity. The first step of this method (algorithm of Figure 2) is linear in the maximum size of specifications. Indeed, the first part of this algorithm is linear as it is a simple traversal of the test purpose graph which is a small automaton compared to a specification graph. The other part of the algorithm (search of predecessor - only if the test purpose event is an event to be executed on a upper interface) is only linear as it is also a simple path search and is based on a stack structure.

The second step corresponds to the test generation. It uses a conformance test generation tool. In our case, we use TGV tool. As TGV [3] is linear in complexity, this step of the method is also linear in complexity.

Thus, the bilateral method costs less than the calculation of $S_1 \|_A S_2$ needed for classical method. Moreover, if an iop test case can be obtained using classical approach, the bilateral method can generate an equivalent bilateral iop test case.

Causal dependency algorithm. One objective of interoperability is to verify the correctness of the communication between the IUTs. Thus, iop test purposes may end with an input. This latter situation occurs in the unilateral test purposes derived by bilateral method. For example, if the iop test purpose ends with an output on lower interface, its mirror event (an input) is added -as last event- to one of the derived test purpose. The unilateral test case derivation generate a test case for which the PASS verdict is affected to an input (a non-observable event). An algorithm based on causal dependencies is used to complete bilateral method. The purpose of this completion is to produce outputs that help in verifying that the input is actually received by the corresponding IUT. The algorithm computes the set of causal dependency events (associated with the paths to these events), based on breadth-first search algorithms of the graph theory. It can also be used for refining interoperability test cases generated by classical methods based on test purpose ending with an input.

4   Applying the New Method to a Connection Protocol

4.1   A Simplified Version of the ISDN Connection Protocol

ISDN (Integrated Services Digital Network) is a set of CCITT/ITU standards for digital transmission over ordinary telephone wire as well as over other media.

This protocol requires a connection. The IUT-T recommendation Q.920 [11] contains the description of the state diagram for sequences of primitives at a point-to-point data link connection endpoint. The specifications of Figure 3 consider a simplified version of this connection protocol. This version is simplified so that it could represent any other communication protocol with request-acknowledge connection negotiation. Two modes are possible: a client/server mode (S_1 and S_2 of Figure 3) or a complete client and server mode (specification S of Figure 3).

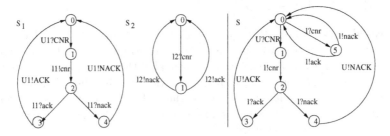

Fig. 3. Examples of specifications: S_1, S_2 and S

Specifications. Let us describe S_1 and S_2 of Figure 3. $U1?CNR$ is a connection request from the upper layer, $l1!cnr$ (resp. $l2?cnr$) the request sent (resp. received) to the peer entity, $l2!ack/l2!nack$ the positive or negative response, and $U1!ACK/U1!NACK$ the response forwarded to the upper layer.

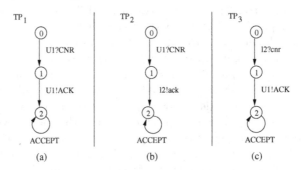

Fig. 4. Test Purpose examples: TP_1, TP_2 and TP_3

Test purposes. A test purpose is an informal description of behaviors to be tested, in general an incomplete sequence of actions. Each state of the test purpose considers a particular event to be executed, but each state allows all actions. Let us consider the three iop test purposes of figure 4. These iop test purposes are applicable to the System Under Test (SUT) composed of two IUTs implementing respectively S_1 and S_2. For example, TP_1 of Figure 4(a) (resp. TP_2 of Figure 4(b)) means that, after the reception by I_1 (implementing S_1) of a connection request on its upper interface $U1$, this IUT I_1 (resp. I_2) must send a connection acknowledgment on its upper interface $U1$ (resp. a connection acknowledgment on its lower interface $l2$).

The three test purposes of figure 4 are also applicable for deriving interoperability test cases executable on a SUT composed of two IUTs implementing the specification S (the complete client and server mode).

4.2 CADP Toolbox Used for Implementing the Method

Both classical and new (bilateral) methods were implemented into the CADP toolbox [10]. CADP is a toolbox for the design of communication protocols and distributed systems. It includes tools for explicit state space manipulation called BCG. BCG (Binary-Coded Graphs) is both a format for the representation of explicit LTSs and a collection of libraries and programs dealing with this format. The BCG format is used for representing specifications and test purposes and for manipulating these LTSs.

One step of the bilateral method is the generation of unilateral interoperability test cases using a conformance test generation tool. In order to automatize conformance test generation, different test tools were developed: TorX [4], TVeda [17], SAMSTAG [18], TGV [3], etc. The conformance tool used in this study is TGV (Test Generation using Verification techniques). TGV is integrated in the CADP toolbox and can take as entries a specification and a test purpose in the BCG format. This conformance test tool is used for the unilateral interoperability test case generation of the bilateral method, but also for the global test case generation of the classical approach with the specification interaction and a test purpose as entries.

4.3 Applying the Classical Approach on the Client/Server Mode

Specification interaction. The first step of classical methods is the calculation of a model of the system behavior, that is to say the calculation of the specification interaction (called also reachability graph). For the interaction of S_1 and S_2 of Figure 3, we obtain the IOLTS of Figure 5.

Global interoperability test case derivation. Based on the model of the behavior of the system composed of the two specifications and on the test purposes of Figure 4, we can derive interoperability global test cases. The TGV tool was used, taking as entries the specification interaction (Figure 5) and a test purpose (Figure 4). Results of this test derivation are shown in Figure 6. These test cases are those obtained after modifications due to controllability differences between interoperability and conformance contexts. Interface $UT1$ is the interface of the tester connected to upper interface $U1$. Thus, $UT1!CNR$ means that a tester sends message CNR to the upper interface $U1$ of I_1. For events on lower interfaces, $?(\mu)$ corresponds to the observation of the event. However, inputs can only be deduced from the corresponding output (sent by the other IUT) and causal-dependency events.

These results are compared with test cases derived for the same test purposes by the bilateral method in next Section.

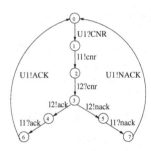

Fig. 5. Interaction of S_1 and S_2

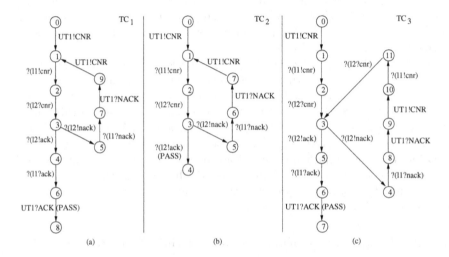

Fig. 6. Test Cases for the interaction of S_1 and S_2

4.4 Applying Our New Method on the Client/Server Version

Unilateral test purpose derivation The first step of the bilateral interoperability test generation method is the derivation of the iop global test purposes into two unilateral iop test purposes. The implemented algorithm corresponds to the algorithm presented in Figure 2.

Applying this step to the iop test purposes of figure 4 and specifications S_1 and S_2 gives as result the unilateral iop test purposes of Figure 7. TP_1^1 and TP_1^2 of Figure 7(a) are the test purposes derived for TP_1 (Figure 4(a)) and respectively specifications S_1 and S_2. In the same way, TP_2^1 and TP_2^2 of Figure 7(b) (resp. TP_3^1 and TP_3^2 of Figure 7(c)) are derived from TP_2 of Figure 4(b) (resp. TP_3 of Figure 4(b)). The same notation will be used for test cases in the following.

When deriving the unilateral iop test purposes, for events on lower interfaces, the returned event is either the event itself, either its mirror. For event $U1!ACK$, as its predecessor is $\mu = l1?ack$, the returned event is $\bar{\mu} = l2!ack$ (TP_1^2 and

TP_3^2) or $U1!ACK$ (TP_1^1 and TP_3^1). The difficulty is for deriving an event from $U1?CNR$ for TP_1^2 (Figure 7(a)) and TP_2^2 (Figure 7(b)). In S_1, this event is the first possible event after the initial state. Its predecessor must be found in the paths bringing back the entity in its initial state. The first predecessor found is $U1!NACK$. As this event is not an event of the interaction, the algorithm continues one more step to find $l1?nack$ as predecessor, and then returns $l2!nack$ (mirror of $l1?nack$).

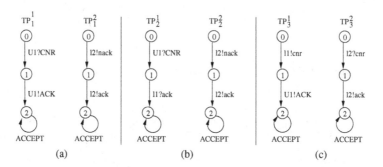

Fig. 7. Unilateral Test Purpose derived for specifications S_1 and S_2

Unilateral test case derivation. The second step of the bilateral interoperability test generation method corresponds to the use of a conformance test tool (here TGV) on a unilateral test purpose and the corresponding specification. TGV will return conformance test cases that we want to reuse in interoperability context after some modifications.

A test case is controllable if the tester does not need to choose arbitrarily between different events. In conformance, inputs on lower interfaces correspond to outputs of the tester: a controllable conformance test case only considers one of the possible inputs on lower interfaces. In interoperability testing, inputs on lower interfaces are sent by the other implementation. An interoperability test case must take into account *all* possible inputs on lower interfaces. The complete test graph is an IOLTS which contains all sequences corresponding to a test purpose: all the inputs of the implementation that correspond to the test purposes are considered in this IOLTS. Thus, to have test cases usable in interoperability context, the conformance tool used in this step (like TGV) for interoperability test generation must compute the complete test graph.

The main modifications to be applied to the obtained conformance test cases concern the types of messages. The messages on lower interfaces are observations in interoperability testing whereas they corresponds to communication with the tester in conformance testing. The results on Figure 8 gives the test cases modified for interoperability.

Interoperability "scenario" and comparison with classical methods. If we calculate the interaction of two unilateral test cases, we obtain the interoperability test case execution scenarios of Figure 9. The scenario obtained for TP_3

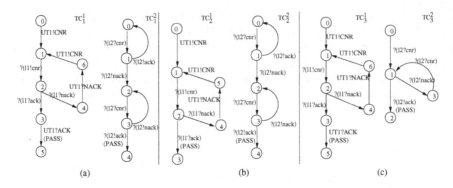

Fig. 8. Unilateral Iop Test Cases for specifications S_1 and S_2

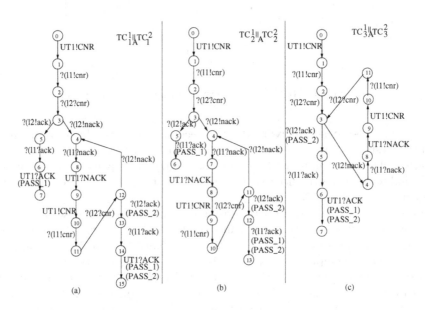

Fig. 9. Test Case interactions

(Figure 9(c)) contains only one deadlock (state 7). This state corresponds to a $PASS$ verdict (PASS_1∧PASS_2=$PASS$).

The scenario obtained for TP_1 (Figure 9(a)) contains two deadlocks: states 15 and 7. State 15 corresponds to a PASS state. But state 7 is only to a verdict state (PASS_1) of TC_1^1, not to a verdict state of TC_1^2 (see test cases on Figure 8(a)). This means that, in this state, TC_1^1 is executed until a verdict state, but TC_1^2 has not reached a such state. The part of TC_1^2 not executed was generated because of the test purpose derivation (calculation of the predecessor of $U1?CNR$ when deriving TP_1^2). We can also remark that the trace executed until state 7 of $TC_1^1\|_A TC_1^2$ verifies the global test purpose TP_1. Thus, even though a deadlock state without both unilateral verdicts exists, the obtained scenario is complete regarding the iop test purpose TP_1.

For TP_2 (scenario on Figure 9(c)), the obtained interoperability test cases end with an input because the unilateral test purpose generated for S_1 ends with an input. To complete these iop test cases (TC_2^1 and $TC_2^1 \|_A TC_2^2$), we can either add a postamble returning to the initial state, either use causal dependency algorithm (using breadth-first search algorithms). It will add paths until outputs that are executed only if the input $l1?ack$ is actually executed. In this simple example (specification S_1), only the event $U1!ACK$ (or $UT1?ACK$ for the tester) will be added with causal dependency event method.

We can observe that the global iop test case generated for TP_3 (Figure 6(c)) corresponds to the scenario obtained by the interaction of the unilateral test cases generated for this iop test purpose (Figure 9(c)). For TP_1 and TP_2 (global test cases on Figures 6 (a) and (b) and scenarios on Figure 9 (a) and (b)), there are more branches. But a look at glance on the traces contained in both the global test case and the scenario from bilateral method shows that they are equivalent in terms of verdicts. Indeed, the same execution paths lead to the same verdicts. These examples confirm the equivalence of both classical (global) and new (bilateral) methods in terms of non-interoperability detection.

4.5   Application to the Complete Client and Server Mode

Both methods were also applied on the specification S describing both client and server parts and using the same test purposes. The interaction $S\|_A S$, calculated for classical approach, is composed of 454 states and 1026 transitions with input queues of each system bounded to one message. Results in the following table are given for a queue size of 3. The table gives the number of states s and transitions t (noted s/t in the table) for different test cases. The first two lines correspond to iop test cases derived with the bilateral method (S as specification 1 and 2) and the third line to the interaction of these test cases. The last line gives results for test cases derived with classical method. For this method, the generated specification interaction has 47546 states and 114158 transitions.

	TP_1	TP_2	TP_3
S as spec. 1 (bilateral method)	9/17	8/16	9/17
S as spec. 2 (bilateral method)	13/24	13/24	12/22
$TC^1\|_A TC^2$	19546/57746	19468/57614	19405/57386
$S\|_A S$ (global method)	54435/120400	18014/40793	54456/120443

We observe that we can derive unilateral test cases via the bilateral method. These test cases can be used for executing interoperability test cases. For classical (global) methods, we faced the state space explosion problem. Indeed, we were not able to compute $S\|_A S$ for a queue size limited to 4 messages (on a system with a 2GHz processor and 1 Gb of memory). This shows that the bilateral method can be used to generate iop test cases even for specifications that produce state space explosion problem. Moreover, these test cases are not dependent of the queue size.

4.6   Summary of the Experimentation Results

The result on the examples can be summarized as follows.

1. In terms of non-interoperability detection, the obtained iop test cases confirm the equivalence of the bilateral and global iop criteria that was proved theoretically in [9].
2. The causal-dependency based algorithm can be used to complete iop test cases generated with both global and bilateral method, particularly when we have test purposes ending with inputs.
3. The bilateral method can be used to generate interoperability test cases even for specifications that produce state space explosion problem with classical methods.

5   Conclusion

In this paper, we present a new method for generating interoperability test cases. The interoperability criterion on which the presented method is based was proved equivalent in terms of non-interoperability detection to another interoperability criterion on which classical methods are generally based. This equivalence was confirmed by experimental results. Moreover, we show that the so-called bilateral interoperability test derivation method allows up to generate interoperability test cases in situations where it would have been impossible with the traditional methods because of state space explosion problem.

As future work, we will study the generalization of the formal interoperability definitions and test generation methods to a context with more than two implementations. We will also study how to apply the described method to a distributed testing architecture.

References

[1] ISO. Information Technology - Open Systems Interconnection Conformance Testing Methodology and Framework - Parts 1-7 (1992) International Standard ISO/IEC 9646/1-7
[2] Tretmans, J.: Testing concurrent systems: A formal approach. In: Baeten, J.C.M., Mauw, S. (eds.) CONCUR 1999. LNCS, vol. 1664, Springer, Heidelberg (1999)
[3] Jard, C., Jéron, T.: Tgv: theory, principles and algorithms, a tool for the automatic synthesis of conformance test cases for non-deterministic reactive systems. Software Tools for Technology Transfer (STTT) (October 2004)
[4] Tretmans, J., Brinksma, E.: Torx: Automated model based testing. In: Hartman, A., Dussa-Zieger, K. (eds.) Proceedings of the First European Conference on Model-Driven Software Engineering, Nurnberg, Germany (December 2003)

[5] Desmoulin, A., Viho, C.: Quiescence Management Improves Interoperability Testing. In: 17th IFIP International Conference on Testing of Communicating Systems (Testcom) Montreal, Canada (May-June 2005)

[6] Castanet, R., Koné, O.: Deriving coordinated testers for interoperability. In: Rafiq, O. (ed.) Protocol Test Systems, Pau-France, IFIP, vol. VI C-19, pp. 331–345. Elsevier, North-Holland (1994)

[7] Seol, S., Kim, M., Kang, S., Ryu, J.: Fully automated interoperability test suite derivation for communication protocols. Comput. Networks 43(6), 735–759 (2003)

[8] El-Fakih, K., Trenkaev, V., Spitsyna, N., Yevtushenko, N.: Fsm based interoperability testing methods for multi stimuli model. In: Groz, R., Hierons, R. (eds.) TestCom. LNCS, Springer, Heidelberg (2004)

[9] Desmoulin, A., Viho, C.: Formalizing interoperability for test case generation purpose. In: IEEE ISoLA Workshop on Leveraging Applications of Formal Methods, Verification, and Validation, Columbia, MD, USA (September 2005)

[10] Garavel, H., Lang, F., Mateescu, R.: An overview of cadp 2001. Technical Report 0254, INRIA (2001)

[11] ITU-T. Digital Subscriber Signalling System No.1 - ISDN User-Network Interface Data Link Layer - General Aspects. ITU-T Recommandation Q.920 (1993)

[12] Verhaard, L., Tretmans, J., Kars, P., Brinksma, E.: On asynchronous testing. In: Bochman, G.V., Dssouli, R., Das, A. (eds.) Fifth inteernational workshop on protocol test systems, North-Holland, IFIP Transactions, pp. 55–66 (1993)

[13] Castanet, R., Kone, O.: Test generation for interworking systems. Computer Communications 23, 642–652 (2000)

[14] Griffeth, N.D., Hao, R., Lee, D., Sinha, R.K.: Integrated system interoperability testing with applications to voip. In: FORTE/PSTV 2000, pp. 69–84. Kluwer, B.V (2000)

[15] Bochmann, G., Dssouli, R., Zhao, J.: Trace analysis for conformance and arbitration testing. IEEE transaction on software engeneering 15(11), 1347–1356 (1989)

[16] Gadre, J., Rohrer, C., Summers, C., Symington, S.: A COS study of OSI interoperability. Computer standards and interfaces 9(3), 217–237 (1990)

[17] Groz, R., Risser, N.: Eight years of experience in test generation from fdts using tveda. In: FORTE, Osaka, Japan (November 1997)

[18] Grabowski, J., Hogrefe, D., Nahm, R.: Test case generation with test purpose specification by mscs. In: SDL Forum, Amsterdam, North-Holland (October 1993)

Component Testing Is Not Enough -
A Study of Software Faults in Telecom Middleware

Sigrid Eldh[1,2], Sasikumar Punnekkat[2], Hans Hansson[2], and Peter Jönsson[3]

[1] Ericsson AB
[2] Mälardalens University
[3] Combitech
Ericsson AB, Kistagången 26, Stockholm, Sweden
sigrid.eldh@ericsson.com

Abstract. The interrelationship between software faults and failures is quite intricate and obtaining a meaningful characterization of it would definitely help the testing community in deciding on efficient and effective test strategies. Towards this objective, we have investigated and classified failures observed in a large complex telecommunication industry middleware system during 2003-2006. In this paper, we describe the process used in our study for tracking faults from failures along with the details of failure data. We present the distribution and frequency of the failures along with some interesting findings unravelled while analyzing the origins of these failures. Firstly, though "simple" faults happen, together they account for only less than 10%. The majority of faults come from either missing code or path, or superfluous code, which are all faults that manifest themselves for the first time at integration/system level; not at component level. These faults are more frequent in the early versions of the software, and could very well be attributed to the difficulties in comprehending and specifying the context (and adjacent code) and its dependencies well enough, in a large complex system with time to market pressures. This exposes the limitations of component testing in such complex systems and underlines the need for allocating more resources for higher level integration and system testing.

Keywords: Software, Fault Classification, Fault distribution, Testing.

1 Introduction

We have investigated a number of failures in a part of a subsystem of a large complex middleware system at Ericsson. Based on this investigation, this paper identifies classes of failures, presents their frequencies, and isolates the failure classes that are caused by software faults. Our overall motivation and planned future work is to create a controlled experiment by re-injecting known faults of different types in the code, to be able to learn more about efficiency and effectiveness of various test techniques [1]. We initially planned to use published information on commonly prevalent software faults, fault classes and their frequencies, but were unable to find sufficient information from existing literature. Since failures are related to software faults in

A. Petrenko et al. (Eds.): TestCom/FATES 2007, LNCS 4581, pp. 74–89, 2007.

complex and intricate manners, we believe that a realistic characterization of their correlation will be helpful in determining effective as well as cost-efficient testing strategies. Our objective is to understand how faults and failures manifest themselves, especially in the context of 'real' faults that occurred in commercial or industrial systems.

Most software industries do not pay enough attention to understand the typical faults that exists in their software. These industries collect every anomaly and complaint, from both verification teams and customers, but seldom faults and failures found by designers. Hence, failures are collected from the later stages in the software process, saved in databases, classified based on priority, status of management (e.g. analyzed, fixed, tested, approved) and classified based on organization or software sub-system where the actual fault is believed to exist. Occasionally deeper analysis and classifications are done and root cause analysis (RCA) are performed, especially when major incidents involving customer complaints occur. Most classifications [8, 9, 30, 35] end prematurely by defining the failure on too high level to understand what software fault caused the failure. We suspect that this is the case in most industry and commercially available software, with the exception of safety critical software..

In this study, we consider the software in a typical telecom system. We have not looked at all the reported failures for the entire node, but focused on one particular part of the software, which we consider to be a typical representative of telecom middleware. From 4769 reported failures, we have selected a sample of 362 failures, which can be considered to be important since they were all corrected. This data has been collected during a period of three years based on 65 different designers and software integrators across the world. We chose these failures since their labels in the configuration management system made it easier to locate the corresponding software fault, compared to repeating the tedious troubleshooting and debugging of the system, which would otherwise have been required. The questions we primarily try to address with this case study are:

- What are the real and important faults in software that have propagated to failures, and subsequently fixed?
- What is the distribution of the failures and faults into various classes? (This classification will allow us to re-inject faults of the same type in the software, thereby providing a basis for our planned evaluation of test techniques.)
- Is there any other specific pattern of faults and failures that would guide us into understanding the software process better?

Our focus is on software faults, but our study has shown that just less than half of the reported failures are not a direct consequence of a software fault that can possibly be re-injected in the code. Instead, failures relate to a variety of problems, e.g. hardware, third party products, process issues, organization, and management issues. We decided to keep all information to give a better perspective for researchers trying to understand problems in the software industry, and better explain them as a part of our case study.

Outline: We discuss and define the terminology used in the next section and then related work in section 2. Section 3 describes the set-up of our case study and data selection. In section 4, we discuss our findings of the failure distribution and the different classes. Validation is discussed in section 5 and future work in section 6. We will end section 7 with the conclusions that we have derived from this case study.

1.1 Terminology

The related terminology in this area (fault, error, cause or reason, failure, bug, defect, and anomaly) is often confusing because these terms are used interchangeably and inconsistently by many in industry and academia; see further discussion in Mohaghegi et al [27]. Therefore we define the following terms with inspiration from earlier work from Avižienis & Laprie [10] and Thane [9], where a *fault* is the static origin in the code, that during dynamic execution propagates (in Figure 1 described as by a solid arrow) to an *error* (which is an intermediate infection of the code). If the error propagates into output and becomes visible during execution, it has caused a failure. An error or failure can both cause another fault to occur. At Ericsson, failures are reported as Trouble Reports (TRs). Occasionally, these TRs gives in their analysis section a more direct explanation of the cause of the failure, but mostly they only describe the symptoms. TRs are not uniquely identifying failures (i.e., several TRs may identify the same failure) and there is not a one to one relationship between a fault and a failure (i.e. different faults may lead to the same failure and some faults may cause multiple failures).

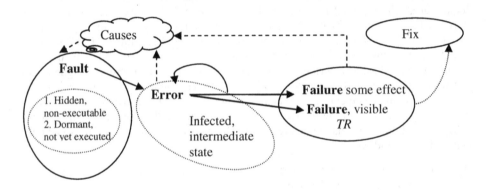

Fig. 1. Terminology mapping

A failure can in turn propagate to another part of the software and be the cause of another fault. One difference compared with Avižienis & Laprie is that we separate the actual cause of the fault from what manifests itself in the software. As an example, a faulty specification can lead to a fault in the software, but to find what code to re-inject in order to represent such a fault is not clear. It might be that the fault specification misses to define a case not implemented in the code, which leads to faulty assumptions and not adding extra paths when needed. We occasionally refer to the term "real fault", meaning, a fault found in a commercial or industrial system.

2   Related Work

Our purpose is to investigate software test techniques is described in our position paper [1]. We noticed that most test technique investigations used small code samples, often with very few faults injected [2, 4]. The faults used as the basis in test technique investigations are often invented and "simple" or made to prove a specific point [20]. This did not match our experience with faults in software for complex systems. Even if there exist attempts to create better faulty programs to use for test techniques research [3, 32, 33, 34], they still do not contain enough data from industry, and are relatively small compared to our complex middleware. Therefore, we argue that the early stages fault investigations for test technique research [5, 6, 7] needs to be updated. Andrews et al. [32], share a similar goal, but uses a different approach. They compared mutant generated faults with hand-seeded real faults, and concluded that the faults were different in nature, but shows statistical promise. Analyzing our result in contrast with theirs, it becomes evident that the type, nature of the fault and fault class, and its distribution *is not sufficiently explored* to draw strong conclusions about test techniques. We tried to find examples of classes and types of code faults to re-inject, but did not find any good list to use, instead we found several papers investigating real faults (and failures) classified for different purposes [8, 9, 11, 12, 13, 14, 15, 16, 18, 20, 23, 24, 35] and not distinguishing faults from failure or cause. In particular, for the more commonly used Orthogonal Defect Classification (ODC) [8], we concluded that the classes are classifying failures, and not faults, e.g. an interface failure (which is an ODC class) could be caused by several software code faults. Thus, these classifications are insufficient to support us in our aim. DeMillo and Marthur [13] have made an attempt to classify real software defects by automatic means. We have used these fault and failure classifications as inspiration to our classification and we will discuss them in the section on future work. Rather than adopting, we strongly propose that different software domains have different sets of fault and fault distributions, depending on organizations, languages, development and testing methods, as well as the ways of measurement. Conclusions on test techniques should be based on first creating a thorough fault analysis particular to the domain, but with known methods. This will provide better understanding of which typical failures and related faults that are relevant for this particular software. We realize that the gap between research and industry is wide [31], but we hope to close this gap by doing controlled research on commercial software in an isolated environment.

Huffman and Rothermel discuss in [29] the semantics of a fault. This is interesting research, since it implies that faults have a variety of impact, depending on the fault. Our research shows that some types of faults that affect the software are a combination of faults, and are definitely involving more than one file, and more than one entry in a file. There is a danger in inserting only single semantic faults even if they are dominating. Single semantic fault injection is the predominant way of injecting faults (and mutations). One fault can propagate into many different symptoms (which is one of the explanations to the high number of duplications). One fault might propagate and behave differently, depending on how "complicated" it is. Hyonsook and Elbaum [3] have with the work on SIR framework, used files from industry (SPACE and Siemens program [33, 34]), but also let experienced designers deliberately insert faults (also used by [32]). There is no way of telling if these faults are representative of common faults in any system, or if they are too simplistic in nature. Our initial reaction when analyzing failures have been that faults that dominate are much more complicated in nature than plain logical or computational faults, which do occur, but not as frequent. Ishoda [21], argues that basing research by capture- recapture (inserting faults, and then estimating how many of them are found) is not a sufficient technique for reliability (*and test evaluations*) analysis, and that correct frequency and type of faults must be known for the software in question. We support this argument, which also lead us to investigate our own frequency and type, to understand the characteristics of our particular type of software. Ostrand, Weuyker, and Bell [22] work in the same domain as us, large middleware systems with similar problems. They have not focused on the fault type in itself, but on occurrences and location, which is similar to our approach. Since they have classified based on their MR (similar to Ericsson's TR), we also assume (but have not verified) that their data suffer from the same problem as our, what is reported (failure) and the connection to the actual fault in some code file needs a much further analysis. Yet, they report interesting results that seem to match our experiences: Most of the faults reside in (or are reported on) 20% of the software, i.e. 80% of the software is more or less fault-free. Furthermore, their distribution seems to match ours, with over 50% of the failures related to missing or spurious code. A comparison with our figures shows that their distribution and frequency is very similar to ours, even if it is done more than 20 years ago. We will discuss this further in our last section: Dicussions and conclusions.

The key problem we would like to address is that there is no recent industry data available for research purposes. In addition, people who measure are often using too high level classifications [8] to be useful for our purpose. We have also studied bug taxonomies [12, 32], but conclude that they mix cause, fault and failure, and are seldom providing obvious support for re-injecting faults, even if they give valuable information about failures.

Our main conclusion is that it is important to regularly collect and report findings of this nature from real industrial and commercially used systems to keep information in tune with development approaches, software and faults. We also assume that there is a large diversity of the frequency of faults, depending on what type of system and domain they reside in. We also suggest that within a domain – or type of system (e.g. a large complex operating system middleware with partially proprietary hardware) it is possible to find similar structures across the entire domain, which could indicate that the results are not limited to only e.g. telecommunication middleware systems.

3   Case Study Process and Data Selection

The process followed in our case study is described in Fig. 2. We started by selecting the Trouble Reports (TRs) for which a link to the corresponding code exist; using a script that automatically linked the fault id into the corrected code as a comment. Then we compared the corrected code with the original version of the code, to identify areas in the code where the fault could reside. This is a non-trivial task, since enhancements and improvements to the code are mixed with corrections. We then classified each Trouble Report into one of the chosen failure classes.

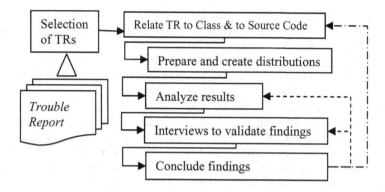

Fig. 2. Overview of the classification process

The distribution of failures was then analyzed, followed by some interviews of the designers, with the aim to validate our findings. The component size was approximately 180-200 000 lines of non-commented code (measured over the three year period). For the entire middleware system, there were 4769 TRs reported and of these we have considered 1191, indicating that it is a central part of the software. From these reported TRs (and also, from the complete set of 4769 TRs), some of the TRs have been analyzed to originate from faults elsewhere, or require corrections in two places. We must understand that not all of the 1191 TRs lead to a correction. There are 181 TRs reported directly on this component's code during this period. Our number is higher (362) which indicates that TRs that reside elsewhere (are reported on other places in the code) affects this code. The TRs within this target are all using a particular labeling function that makes it possible to trace the TR to the actual file (see Fig. 3).

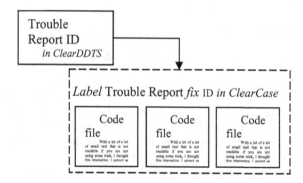

Fig. 3. Failure (Trouble Report) – fault/fix relation, where the label points out involved files

How this subset (of labeling) is related to the 1191 TRs we have not been able to pinpoint, since it would have required serious data mining, taking all change requests in to account and other factors. From the entire set of 362 TRs, we now withdraw all faults that are not software related, e.g. duplications and wrong usage, to come up with our set of 295 unique failures, and of them only 170 (204 adjusted[1]) are software failures. This means that approximate 14% (17% adjusted) of the reported failures (1191 TRs) at least are real software code faults!

Little or no general tool-support exists to trace, categorize and get support in tracking down the fault causing a failure. Currently this has for non-trivial cases to be handled by manual work through code inspection. Manual comparisons of two versions of the source code is often the used method to extract the actual difference, but this is not enough, since enhances to the software and other modifications must be excluded and the fault origin or origins must be isolated. The problem is non-trivial in systems with a multitude of code branches, since it amounts to know where to look without having to scan through a multitude of files and to separate fault corrections from any other code modifications, improvements and change requests.

This is the true complication of relating a "unique" failure to a "unique" fault since they do not have a one to one relationship. What even more complicated our classification was that the original designers (and troubleshooters) were not available to ask of the real origin of the fault. Doing a complete trace could take us between 3 days to two weeks, which explains why this route to identify and gather faults from failures is seldom taken for this type of research. With the labeling function, the connection between the TR-system id and the actual place in the code where the fault could reside could be made, and capturing the correct code became a task of analysis between one hour and 2 days, when done by a person not familiar with the software. We consider this a great improvement.

4 Identified Failure Distributions

This section presents the fault classification, the distribution of failures observed in our case study together with the distribution of the software faults over number of affected code files.

4.1 Fault Classification

In our classification we use the following classes of faults

- *Language Pitfall* are faults that are specific to the programming languages used, e.g. pointer being null, pointer pointing to an invalid address, valid address but pointing at garbage. E.g. arrays is a common type of data structure accessed with index, and if the value falls outside the boundary of the array then the accessed element is unknown, or data might be modified that shouldn't. In addition, overflow and underflow are categorized in this class.
- *Computational/Logical faults* are faults inside a computational expression. A logical fault is similar to a computational fault, except that it is related to a logical expression.

---

[1] Our investigation showed that for many reported failures, the actual fault (represented by the corrections in the code) is not unique and failures and faults do not have a one to one relationship. Therefore, we have adjusted our figure, by adding all software faults that are contributing to the cause of a failure.

- *Fault of omission* is when a fault happens due to missing functionality, i.e. the code that is necessary is missing. E.g. a part, an entire statement or a block of statements missing can be classified as faults of omission, which means missing either of the following: function call, control flow path, computational or logical expression.
- *Spurious faults* are similar to Faults of omission, but in this case there is "too much code", and the correction is to remove one or several statements.
- *Function faults* are faults, such as calling the function with the wrong parameter or calling the wrong function.
- *Data faults* including faults of several types: Primitive data faults, which include defining a variable to the wrong primitive data type, e.g. integer to be unsigned but should not have been; Composite data fault e.g. structure (in C or C++); initialization fault and assignment fault.
- *Resource faults* are faults that deal with some kind of resource, such as memory or a time, therefore this class handles faults from allocation, de-allocation, race-conditions, time issues (dead-lock) and space (memory, stack etc).
- *Static code fault class is* code that does not change after compilation of software or after the first execution, when using an interpreted language. This code is e.g. source code or configuration files which the software when executing uses.
- *Third party faults* are faults in software for which we do not have access to the source code and hence cannot correct ourselves.
- *Hardware faults* and *Documentation faults* are self-explanatory classes, and do not relate to software.

We have grouped the above into four groups, viz., code, process, configuration management (CM) and other.

4.2 Fault Distribution

The selected of 362 trouble reports, were distributed as explained in Table 1 below, where faults of omissions topped the list. The second largest class of Trouble Reports is *Duplicates*. One failure out of three non-software related failures reported is a duplicate. What is interesting is that these duplicates were not identified as duplicates until they were corrected in the code. This means that the TRs were individually decided to be fixed, and assumed to be of different fault origin, since they showed different behavior and were viewed as different failures. Otherwise, these failures would have been omitted before ordering them to be fixed. The high number of duplication of failures is also related to the way testing is done on this complex system, and also to the type of software (which most other parts of the software were dependent upon). This means that many failures in this particular software will be found and reported by several persons. Another aspect is that failure reports are symptoms, and they might not appear in the same way (and be explained similarly) by two different persons. This problem is recently remedied, by enhancing the pre-test on the entire system before release (much as a result of the insight gained by this case study).

Table 1. Failure distribution into classes and their frequency. Second column is the adjusted value (when translating TR to fault classes).

Group	Fault/Failure Class	Failures	Code Fault adjusted
Code	Faults of omission	73	78
Process	Duplicate of TR	67	-
Code	Data fault	22	26
CM	No files associated	21	-
Process	Change requests	21	-
Code	Static code fault	20	21
Code	Spurious faults	17	17
CM	SCM	15	-
Code	Resource fault	13	13
Code	Computational/ Logical fault	11	28
Code	Function fault	11	13
Process	Fault not fixed	9	-
CM	Compile time fault	8	-
Other	Third party fault	5	-
Other	Documentation fault	4	-
Code	Language Pitfall	3	8
Other	Hardware fault	2	-
Process	Not a fault	1	-
Code/Other	Too difficult to classify	39	-
	Sum	**362**	**204**

Disregarding duplicates of failures leading to the same fault, we would focus on the 295 unique failures. However, the classification needs adjustment, in relation to how some failures are reported. For example, if a fault needs two corrections to be fixed, and there are duplicate TRs, there is no way of telling that one TR correction is assigned to one or the other fault correction, and the "duplicate" vice versa. We see this as a disturbance of the data, but have reported the adjustment for the fault classes in software. The adjusted values will if re-inserted cause failures. We have started further investigations to understand the nuances of how faults actually infect the software, and when and how visible they are.

The next interesting class of failures was "too difficult to classify" containing 39 of the failures. The reason is that it is impossible to pinpoint the exact difference, since the entire unit or file was re-designed, and large portions of the code rewritten. This also makes it difficult to pinpoint if the change was only due to the failure, or due to other factors such as updates, expansions, new features etc. We decided that for our purposes is not worth the effort to sort this out, instead we just conclude that 13% of the unique failures lead to a major change in the system. Only 8 (of 19) classes can be considered in our investigation, since these are the ones directly related to software faults. This is only 170 of 362 reported failures (47%), or 170 out of 295 unique failures (58%). This means that as much as 53% can be dismissed due to problems that are either indirect, process or system related, or disturbance in data and that these failures do not uniquely originate from the software. Third party faults are software faults, but they cannot be traced into code, since the source code is not always available to us, and must be corrected by another party. Another category is how the system is built and integrated, including software configuration management (SCM) failures, compile time failures (during the build) and the category of "no files associated". These failures are strongly related to the fact that a major change of build system and product structure happened during the period of data collection, which explains why it constitutes 1 out of 5 non-software related faults.

In Table 2, we present distribution of faults together with the number of files which had to be updated to correct the faults.

Table 2. Distribution of adjusted faults into number of files

Code fault class	Faults	%	Number of files										
			1	2	3	4	5	6	7	8	9	11	25
Faults of omission	78	38.2	50	12	8	4	2	1		1			
Computational/ Logical fault	28	13.7	24	2		2							
Data fault	26	12.7	20	3	1		2						
Static Code Fault	21	10.3	4	7	2	1	1	3	1	1			1
Spurious faults	17	8.3	11	2	1	1	1				1		
Resource fault	13	6.4	8	5									
Function fault	13	6.4	11					1		1			
Language Pitfall	8	3.9	6	1	1								
Summation	**204**	**99.9**	**134**	**32**	**13**	**8**	**6**	**5**	**1**	**2**	**1**	**1**	**1**

The failures found in the case study, can involve between 1 to 64 files. The software faults can be distributed between 1 to 25 files. Analyzing this data shows the majority (66%) are from 1 file, but e.g. that faults of omission involve more than one file for 36%. We have not looked at the details, such as the location of the faults within the file, which is demands a more in depth investigation, or if the files are all "owned" by the same designer or not.

Our result was not what we expected, since we have had the assumption that e.g. more than 4% would be language pitfalls. This is why we felt reporting these findings would aid others in understanding more about the nature of faults. Our most important findings can be summarized as follows:

- Most of the faults were faults of omission or "missing path", meaning, not until execution on higher integration and system levels the lack of code was noted and the failure visible. A designer could clearly not find this, and the cause is most likely insufficient specifications available on lower level or lack of knowledge of the context of the code. A related fault class is spurious faults, where too much code is written, which might be overlapping or creating the problem.
- More than 34% of the corrections involve more than one file and the maximum of involved files for a correction is 25 files – the conclusion is that these faults are not possible to find on component level, and even 100% code-coverage on component level would not reveal the failure.
- Faults are much more complex than simple mistakes; often complicated logic confuses the developer. The semantics of faults are complex.
- Resource faults are not as complicated as e.g. static code faults when it comes to distribution of location.

5 Validation and Threats

This is a case study, which has low control over the environment, we have had some control over the measurements, but they were selected on the basis of possibility to gather (based on the labeling feature), which was random and outside our control. The replication is probably low of the experiment itself, but it should be possible to take any known failure/fault classes and do the same classification with a different outcome, depending on the software, process, organization and situation.

The main argument favoring our study is that, all failures and thus faults are from a live real system, and it is representative of the faults found and fixed, even if it is a not so huge sample. One possible validity threat is the selection of data is created based on the labeling function. The obvious way to perform a study on distribution of failures into fault classes would be to look at all faults, by comparing the difference between code versions. This is not a viable approach, since in systems, fault corrections, changes by adding new and modified code are mixed. Secondly, we have only focused on one small part of the system, since the main purpose was to identify the faults that we could make a copy of and re-inject the faults to use in controlled experiments. We investigated that for every reported failure, finding the actual fault, which represents the correction in the code, is still not completely accurate, since there is not a one to one relationship and this creates a problem in how to classify a fault. This is why we have shown two values, one based on the failure data reported, and one based on the adjusted software fault correction possible to re-inject in the code. The selection was made out of convenience, to find and create a sample to reason about. The failures corrections (labels) selection came from a wide community, of 65 designers from many countries and collected over a long time period and over many versions and changes in the system (3 years). We draw the conclusion that this fact lessens the internal threat, since the bias of interpretation has less impact, but cannot be disregarded.

Conclusion validity relates to subject selection, data collection, measurement reliability, and the validity of the statistical tests. These issues have been addressed in the design of the experiment, and we believe there is a disturbance in the data collection and measure reliability. We have used a nominal scale to classify our Trouble Reports. The classification is rather simple, which could be indicating problems with the internal validity. Classification into another system will yield a different result. Since our distribution is done with little insight of the software and no insight of history, process and organization and by and external party (thesis worker) the bias is minimized. There is no researcher bias put into the investigation, since it was done by a third party, and no guidance was given to what set of faults should be investigated, how the distribution and classification should be used or chosen. Thus, the researcher, who is familiar with some aspects of history, organization, process and software, which could indicate a bias, and a threat to the validity, have made the conclusions and verified the result.

Discussing the generalization of the result, we must look at external threats. The faults selected are on one particular product, but the nature of the product is such that it could probably be representative for lots of industrial software. We cannot conclude that that the result is generic since the distribution is dependent on organization, process, quality awareness, and a lot of other factors. However, earlier studies [12] with similar results supports that for these types of systems the results could be generic and not an isolated result. We do think this is one example of a typical industry software fault distribution. Using our conclusions for any system might be pre-mature, but we suggest that this information can serve as an indication for complex middleware systems, operating systems, and similar large complex systems. All data is still available to pursue further studies.

6 Future Work

For natural reasons, a lot of the designer faults are found by designers themselves and are not registered. Depending on industry, designers are usually responsible for a large amount of the lower-level testing (unit, component, and lower level integration, and even some functional testing). Therefore, data on failures originates from independent test at different integration points, (levels), and from customers reports. We intend to investigate what faults and failures the designers themselves find.

Since we have noticed fault classifications often become failure or cause classifications, we aim to do a further study and make a fault classification (bug taxonomy) that are more useful for fault injection purposes. We believe we have already hinted on a structure (defining our classes and sub-classes), but feel we need to investigate and possibly expand these further. We also need to provide more clear rules for how to classify faults, so they cannot be classified in different classes depending on how the fault is interpreted. The large number of existing fault classifications will also be juxtaposed in such a new classification.

We would like to investigate automatic ways to classify legacy software, but that is not our primary concern, and we invite other researchers into a discussion about the feasibility of automatic classifications, as we have seen a published example of [13].

Our primary concern is to prepare code in many different versions, with real faults injected. These fault-injected code samples are intended to be used for evaluating test techniques in a controlled manner. We aim to inject faults into a system to minimize bias for one test technique over another. This require us to have a clear classification (with relation to the test technique), and there must be a variety of faults that behave (propagate) and is viewed differently, if we are to determine where a test technique would be more efficient. Our interest to find a set of faults that through execution behaved in different ways, and cause different failures in the software. The faults we have found, isolated, classified, and understood are to be re-injected in the code for better comprehension of how each one of them behaves when propagating to failures, as opposed to debugging. Of course, the variety of faults is more interesting, and we are to explore mutation techniques as a complement, and study how they behave and how their fault semantics could look like.

7 Discussions and Conclusions

We now return to the questions posed in the introduction and present our findings based on the collected data.

1. *What are real and important faults in software that have propagated to failures, and are fixed?*

The answer is clearly that faults of omission (38,3% of software faults), together with spurious faults (8,3 % of all software faults), shows that faults related to unclear specifications dominates among the real software faults found in the considered sub-system in a large complex middleware system. The main conclusion and contribution is the fact that the individual designer at component test level do not find these type of faults, and that they must be found at later stages in testing. It underlines how difficult it is for a designer to understand, define, specify and implement code in an environment of complexity, since not enough knowledge of context available. This is supported by looking at distribution of the faults over files, where 36% of these two classes of faults spans over more than one file.

2. *What is the distribution of failures and faults into classes?*

Our observations were as follows:

-   53% of all failures decided to be corrected are not relating to the software and are not possible to re-inject into the software
-   Faults of omission (lack of code) is the dominating class among the software faults (38.3%)
-   Computational/Logical (13.7%) and Data faults (12.7%) followed by Static Code faults (10.3%) are the next largest groups.
-   Language pitfalls are only contributing with 4% of the software fault distribution

3. *Is there any other specific pattern of the faults and failures that would guide us into understanding the software process better?*

-   We have found that as much as 19% of all faults are duplicates that still remain to be corrected (even after management has taken out duplicates). This shows that software failures are often expressed differently and not identified as duplicates until code is corrected.
-   Software configuration management and build related faults together are contributing with as much as 15% of the failures.
-   Conventional faults in software (computational/logical and functional faults, language pitfalls and static code faults) are only 39.6% of the software faults in this complex middleware system.

We think that these findings suggest where effort and cost should be placed. We know for a fact that this software had a targeted component test improvement (unit level) during the year 2005, which greatly improved the quality. We know that from the period 2003-2004 most of the trouble reports came from outside this software organization (applications, customers), but after the improvement most of the trouble reports originated from within the organization. We also know that the testing of these products have improved between 2003 and 2006, with only one test level in 2003, and now with more than 4 test levels. We can also see this reflecting in the number of duplications going down. We have noticed that viewing Trouble Reports (failures) over time, provides us information that many of the changes induced by the trouble reports that we considered were both re-designs for change and expansions purposes and for fixing bad design, and these problems were more common earlier in the development than in the later stages of development. We suggest that changing software configuration and build system will have a great impact on the code (since our study indicated as much as 15% of faults are due to SCM). We think there is much information to be utilized from this study. We do believe it is humanly impossible to understand the entire system, and even if knowing context and teaching about it, this will only remedy a part of the problem, and that unit and component testing alone have no chance of finding a majority of these faults, which is of course already shown and evident. This study strengthens that evidence. Even if we strongly believe component testing is essential for complex systems that need to be robust, we must do testing on many other levels in the system to understand where the important faults hide. Testing is a support to the designer.

We were surprised that the fault categories that are the target of many static analysis tools had so low frequency, which leads us to suggesting a more cost-efficient way might be to work on specifications, understanding the software context, and test-set up. We believe that since the system is build in a "fail-safe" way with the aim to minimize impact of anything going wrong (by duplication of hardware, protocol resending, restarting etc) the impact of simple singular faults are often hidden or dormant. We believe that the improvement on component test are not reported in the same way in the trouble report system, since designers correct their own mistakes when they encounter them, rather than report them, which puts a hidden figure on these types of faults. Most of the component test faults (that is found by e.g. static analysis tools and component test) are not as visible in this study, but still do exist.

We have concluded why there is such a lack of information on code faults – and how good it is to do this analysis and really understand the information, since it gives Ericsson guidance on where efforts of improvements should be targeted. We understand the difficulty to gather this information if traceability of the code is not directly possible from the reported failure. The strength in our study is that it is unbiased to the code, and based solely on applying clear classification rules. We have of course encountered problems with classifications, how distinct the classes are, and how classifications should be applied. We intend to explore this in detail in a future work.

Acknowledgments. We would like to thank Ericsson for their support. We would also like to thank our sponsor The Knowledge Foundation, which via their SAVE-IT program made this research possible.

References

1. Eldh, S., Hansson, H., Punnekkat, S., Pettersson, A., Sundmark, D.: A Framework for Comparing Efficiency, Effectiveness and Applicability of Software Testing Techniques. In: Proc. TAIC, IEEE, New York (2006)
2. Juristo, N., Moreno, A.M., Vegas, S.: Reviewing 25 Years of Testing Technique Experiments. Journal of Empirical Softw. Eng., vol. 9(1-2), pp. 7–44. Springer, Heidelberg (2004)

3. Hyunsook, D., Elbaum, S., Rothermel, G.: Infrastructure support for controlled experimentation with software testing and regression testing techniques. In: Proc. Int. Symp. On Empirical Software Engineering, ISESE '04, pp. 60–70. ACM, New York (2004)

4. Apiwattanapong, T., Santelices, R., Chittimalli, P.V., Orso, A., Harrold, M.J.: Tata: MaTRIX: Maintenance-Oriented Test Requirements Identifier and Examiner. In: Proc. From TAIC, IEEE, New York (2006)

5. Basili, V.R., Selby, R.W.: Comparing the Effectiveness of Software Testing Strategies original 1985, revised dec. 87. In: Boehm, B., Rombach, H.D., Zelkowitz, M.V. (eds.) Foundations of Empirical Software Engineering, The Legacy of Victor R. Basili, Springer, Heidelberg (2005)

6. Myers, G.J.: A controlled experiment in program testing and code walkthroughs inspections, Comm. ACM, pp. 760–768 (September 1978)

7. Hetzel, W.C.: An experimental analysis of program verification methods, PhD dissertation, Univ. North Carolina, Chapel Hill (1976)

8. Chillarege, R., Inderpal, S., Bhandari, J.K., Chaar, M.J., Halliday, Moebus, D.S, Ray, B.K, Wong, M.-Y.: Orthogonal defect classification – a concept for in-process measurements. IEEE Trans. on Soft. Eng 18(11), 943–956 (1992)

9. Thane, H., Wall, A.: Testing Reusable Software Components in Safety-Critical Real-Time Systems, vol. 1(1-2) Artech House Publishers (2002)

10. Avižienis, A., Laprie, J.: Dependable computing: From concepts to design diversity. In: Proceedings of the IEEE, vol. 74, pp. 629–638 (May 1986)

11. Basili, V.R., Perricone, B.T.: Software errors and complexity: An empirical investigation. Communications of the ACM 27(1), 42–52 (1984)

12. Beizer, B.: Software Testing and Quality Assurance. Van Nostrand Reinhold electrical/computer science and engineering series. Van Nostrand Reinhold, NY (1984)

13. DeMillo, R.A., Maihur, A.P.: A grammar based fault classification scheme and its application to the classification of the errors of TEX. Technical Report SERC-TR-165-P, Purdue University, West Lafayette, IN 47907 (1995)

14. Endres, A.: An analysis of errors and their causes in system programs. Technical report, IBM Laboratory, Boebligen, Germany (1975)

15. Johnson, C., et al.: Guide to IEEE standard for classification for software anomalies. In: Technical report, IEEE Computer Society, Washington (1995)

16. Goodenough, J.B., Gerhart, S.L.: Toward a theory of test data selection. In: Proceedings of the international conference on Reliable software, pp. 493–510. ACM Press, New York, USA (1975)

17. Gray, J.: Why do computers stop and what can be done about it? Technical Report, vol. 85(7) Tandem Computers (1985)

18. Harrold, M.J., Offutt, A.J., Tewary, K.: An approach to fault modeling and fault seeding using the program dependence graph. Journal of Systems and Software 36(3), 273–296 (1997)

19. Howden, W.E.: Reliability of the path analysis testing strategy. IEEE Trans. on Software Engineering 2(3), 208–215 (1976)

20. Knuth, D.E.: The errors of TEX. Software Practice and Experience 7, 607–685 (1989)

21. Ishoda, S.: A criticism on the capture-and-recapture method for software reliability assurance. In: Proc. Soft. Eng, IEEE, New York (1995)

22. Ostrand, T.J., Weyuker, E.J, Bell, R.M.: Predicting the Location and Number of Faults in Large Software Systems. IEEE Trans. of Soft. Eng., vol. 31(4) (April 2005)

23. Perry, D.E., Steig, C.S.: Software faults in evolving a large, real-time system: a case study. In: Sommerville, I., Paul, M. (eds.) ESEC 1993. LNCS, vol. 717, pp. 48–67. Springer, Heidelberg (1993)
24. Kaner, C. Falk, J., Nguyen, H.Q: Testing Computer Software. 2nd edn. International Thomson Computer Press (1993)
25. Vaidyanathan, K., Kishor, S., Trivedi, A.: A comprehensive model for software rejuvenation. IEEE Trans. on Dependable and Secure Computing 2(2), 124–137 (2005)
26. Zeil, S.J.: Perturbation techniques for detecting domain errors. IEEE Transactions on Software Engineering 15(6), 737–746 (1989)
27. Mohaghegi, P., Conradi, R., Borretzen, J.A.: Revisiting the problem of Using Problem Reports for Quality Assessments, WQSA, ICSE (2006)
28. Henningsson, K., Wohlin, C.: Assuring fault classification agreement – An Empirical Evaluation. In: Proc. of ISESE'04, IEEE, New York (2004)
29. Offut, A.J., Huffman- Hayes, J.: A Semantic Model of Program Faults. In: Proceedings of ISSTA, pp. 195–200 (1996)
30. Damm, L.O, Lundberg, L., Wohlin, C.: Fault-Slip Through - a concept for measuring the efficiency of the test process. Journal of Software Process: Improvements and Practice, vol. 11(1), pp. 47–59. John Wiley and Sons, New York (2006)
31. Murphy, B., Garzia, M., Suri, N.: Closing the Gap in Failure Analysis. Workshop on Applied SW Reliability-DSN (2006)
32. Andrews, J.H., Briand, L.C., Labiche, Y.: Is Mutation an Appropriate Tool for Testing Experiments? In: ICSE 2005, ACM, New York (2005)
33. Frankl, P.G., Iakounenko, O.: Further Empirical Studies of test Effectiveness. In: Proc. 6th ACM SIGSOFT International Symposium on Foundations of Software Engineering, Orlando, FL, USA, pp. 153–162 (1998)
34. Vokolos, F.I., Frankl, P.G.: Empirical evaluation of the textual differencing regression testing technique. In: Proc. IEEE Int. Conference on Soft. Maint. USA, pp. 44–53 (1998)
35. IEEE Std. 1044 -1993, Standard for classification for software anomalies. IEEE (1993)

Symbolic Model Based Testing
for Component Oriented Systems

Alain Faivre[1], Christophe Gaston[1], and Pascale Le Gall[2,*]

[1] CEA LIST Saclay F-91191 Gif sur Yvette
{alain.faivre,christophe.gaston}@cea.fr
[2] Université d'Évry, IBISC - FRE CNRS 2873,
523 pl. des Terrasses F-91000 Évry
pascale.legall@ibisc.univ-evry.fr

Abstract. In a component oriented approach, components are designed, developed and validated in order to be widely used. However one cannot always foresee which specific uses will be made of components depending on the system they will constitute. In this paper we propose an approach to test each component of a system by extracting accurate behaviours using information given by the system specification. System specifications are defined as input/output symbolic transition systems structured by a communication operator (synchronized product) and an encapsulation operator (hiding communication channels). By projecting symbolic execution of a system on its components, we derive unitary symbolic behaviours to be used as test purposes at the component level. In practice, those behaviours can be seen as typical behaviours of the component in the context of the system. We will illustrate on an example that those behaviours could not have been extracted by reasoning uniquely at the component level.

Keywords: component based system, ioco-based conformance testing, input/output symbolic transition system, symbolic execution.

1 Introduction

In the framework of reactive systems, a component oriented system is constituted of components continuously interacting together and with their environment by means of communication mechanisms. In a first step, basic components are usually specified, implemented and tested: this is called *unitary testing*. Then, the complete system is specified, implemented and tested taking into account the component based structure: this is called *integration testing*. Concerning integration testing, two main approaches can be followed depending on the targeted fault model. In the first approach, the global system is tested according to behaviors involving communication mechanisms, focusing on cases for which those mechanisms are not observable (*i.e* internal communications). Obviously, this approach is used when the targeted fault model

---

* This work was partially supported by the RNRT French project STACS and the RNTL French project EDEN2 .

A. Petrenko et al. (Eds.): TestCom/FATES 2007, LNCS 4581, pp. 90–106, 2007.

mainly deals with communication mechanisms as in [9,5,1]. In the second approach, the global system is tested by selecting behaviors of basic components that are typically activated in the system. It amounts to re-enforce unitary testing with respect to those behaviors. In terms of fault model, the counterpart of this approach is that communication mechanisms are supposed to be correctly implemented and correctly used by programmers. Thus, in this case, a non conformance of the system should only result of uncorrect implementations of components. [13] has proposed a theoretical framework based on these assumptions and has stated results concerning preservation of conformance through component composition. In this contribution, our objective is to re-enforce testing of components and intermediate sub-systems. Now, the question is: how to choose behaviors to re-enforce component and sub-system testing in order to make them more reliable in the context of the system? In fact, when a sub-system is involved in a more complex one, it is very probable that all the sub-system behaviors are not activated. In this paper, the models that we use to denote specifications of communicating systems are made of simple input/output symbolic transition systems (IOSTS) ([4,6,3]) for denoting basic components, and of two structuring operators, namely *composition* and *hiding* (as in [13]). Those models based on input/output symbolic transition systems are equipped with naming mechanisms that allow us to easily retrieve all relevant information concerning sub-systems. Those naming mechanisms together with symbolic execution technics [7] are used to define relevant behaviors of sub-systems. Moreover, we show how to use those behaviors as test purposes in an *ioco*-based [11,12,3,4] conformance testing framework. From a technical point of view, this contribution is an extension of the one presented in [4] for component oriented system testing. As we do not make any assumption concerning the communication mechanisms, a system (implementation) is considered as conformant with respect to a structured specification if it has the same structure, if for each intermediate subspecification, there exists a subsystem corresponding to it, and if each subsystem is conformant according to the ioco conformance relation with respect to the corresponding subspecification.

The paper is organized as follows. In Section 2, we present the IOSTS formalism, the notion of basic component based system and the notion of (sub-)system. In Section 3, we show how to define test purposes from symbolic execution of such systems and how to project them on any sub-system. In Section 4, we define our symbolic test purposes. Section 5 is a conclusion.

2 Structured Input/Output Symbolic Transition Systems

IOSTS are used to represent behaviors of reactive systems. Those behaviors are composed of internal actions and communication actions which are emissions or receptions of values through *channels*. Internal states are modeled by assignments of particular variables called *attributes*.

2.1   Basic Definitions of IOSTS

We use the following set theory notations. The set of functions of domain A and codomain B is denoted B^A. $\coprod$ stands for the disjoint union.

For any set X, $Ident_X$ denotes the identity function on X. For any two functions $f : A \to B$ and $g : C \to D$ such that $A \cap C = \emptyset$, $f|g : A \cup C \to B \cup D$ is the function such that $f|g(x) = f(x)$ if $x \in A$ and $f|g(x) = g(x)$ otherwise. Moreover, for any $E \subseteq A$, $f|_E$ is the restriction of f to E. A data type signature is a couple $\Omega = (S, Op)$ where S is a set of type names, Op is a set of operation names, each of them being provided with a profile $s_1 \cdots s_{n-1} \to s_n$ (for $i \leq n$, $s_i \in S$). Let $V = \bigcup_{s \in S} V_s$ be a set of typed variable names. The set of Ω-*terms* with variables in V is denoted $T_\Omega(V) = \bigcup_{s \in S} T_\Omega(V)_s$ and is inductively defined as usual over Op and V. $Type : T_\Omega(V) \to S$ is the function such that for each $t \in T_\Omega(V)_s$, $Type(t) = s$. In the following, we overload the notation $Type$ by defining $Type(X) = s$ for any set $X \subseteq V_s$. $T_\Omega(\emptyset)$ is simply denoted T_Ω. An Ω-*substitution* is a function of $T_\Omega(V)^V$ preserving types. Any substitution may be canonically extended to terms. The set $Sen_\Omega(V)$ of all typed equational Ω-*formulae* contains the truth values $true$, $false$ and all formulae built using the equality predicates $t = t'$ for $t, t' \in T_\Omega(V)_s$, and the usual connectives $\neg, \vee, \wedge$. A Ω-*model* is a family $M = \{M_s\}_{s \in S}$ with, for each $f : s_1 \cdots s_n \to s \in Op$, a function $f_M : M_{s_1} \times \cdots \times M_{s_n} \to M_s$. We define Ω-*interpretations over V* as applications of M^V preserving types, that are also extended to terms of $T_\Omega(V)$. A model M satisfies a formula φ, denoted by $M \models \varphi$, if and only if, for all interpretations ν, $M \models_\nu \varphi$, where $M \models_\nu t = t'$ iff $\nu(t) = \nu(t')$, and where the truth values and the connectives are handled as usual. Given a model M and a formula φ, φ is said *satisfiable* in M, if there exists an interpretation ν such that $M \models_\nu \varphi$. In the sequel, we suppose that data types of our IOSTS correspond to the generic signature $\Omega = (S, Op)$ and are interpreted in a fixed model M.

$IOSTS$-signatures are composed of a set of particular variables called *Attributes* and of a set of *Channel names*.

Definition 1. (*IOSTS*-**signature**) *An IOSTS-signature is a couple (Att, Chan) such that $Att = \bigcup_{s \in S} Att_s$. For any two IOSTS-signatures $\Sigma_i = (Att_i, Chan_i)$ with $i \in \{1, 2\}$, the union of Σ_1 and Σ_2, denoted $\Sigma_1 \cup \Sigma_2$ is the IOSTS-signature $(Att_1 \coprod Att_2, Chan_1 \cup Chan_2)$.*

Union of signatures does not collapse attributes. Even though Att_1 and Att_2 contain a common variable name x, the union $\Sigma_1 \cup \Sigma_2$ distinguishes the two occurrences of x. On the contrary, channel names are used to synchronize communication actions and thus, are shared by a simple identification in the union.

Definition 2. (*Actions*) *The set of communication actions over $\Sigma = (Att, Chan)$, denoted $Act(\Sigma)$, is the set $Input(\Sigma) \cup Output(\Sigma) \cup \{\tau\}$, where:*
   $Input(\Sigma) = \{c?Y \mid c \in Chan, \exists s \in S, Y \subset Att_s\}$
   $Output(\Sigma) = \{c!t \mid c \in Chan, t \in T_\Omega(Att)\}$

$c?Y$ denotes the awaiting of a value to be received through the channel c and to be stored on all variables of Y. In the sequel, when Y is a singleton $\{y\}$, we can note $c?y$ instead of $c?\{y\}$. $c!t$ denotes the emission of the value t through the channel c and τ is an internal action without any communication action.

We enrich basic-IOSTS of [4] with a naming mechanism associating to each transition a name chosen in a set TN of *transition names*.

Definition 3. (*IOSTS*) *An IOSTS over a signature* $\Sigma = (Att, Chan)$ *is a triple* $G = (State, init, Trans)$ *defined by a set* $State$ *of state names, an* initial *state* $init \in State$, *and a set of* transitions $Trans \subseteq TN \times (State \times Act(\Sigma) \times Sen_\Omega(Att) \times T_\Omega(Att)^{Att} \times State)$. STS *denotes the set of all IOSTS.*

In the sequel, for any transition tr of the form $(n, (q, act, \varphi, \rho, q'))$, $name(tr)$ stands for n and is called the *name of* tr, $source(tr)$ (resp. $target(tr)$) stands for q (resp. q') and is called the *source state of* tr (resp. *target state of* tr), $act(tr)$ stands for act and is called the *communication action of* tr, $guard(tr)$ stands for φ and is called the *guard of* tr, $subst(tr)$ stands for ρ and defines how the attributes are modified when the transition is fired. Finally, $body(tr)$ stands for $(q, act, \varphi, \rho, q')$. For an IOSTS G, $Sig(G)$, $Att(G)$, $Chan(G)$, $State(G)$, $init(G)$ and $Trans(G)$ resp. stand for Σ, Att, $Chan$, $State$, $init$ and $Trans$.

Definition 4. (Runs of a transition) *With notations of Def. 3, let* $tr \in Trans$. *Let us note* $Act(M) = (Chan \times \{?,!\} \times M) \cup \{\tau\}$. *The set* $Run(tr) \subseteq M^{Att} \times Act(M) \times M^{Att}$ *of execution runs of* tr *is s. t.* $(\nu^i, act_M, \nu^f) \in Run(tr)$ *iff:*

- *if* $act(tr)$ *is of the form* $c!t$ *(resp.* τ *) then* $M \models_{\nu^i} guard(tr)$, $\nu^f = \nu^i \circ subst(tr)$ *and* $act_M = c!\nu^i(t)$ *(resp.* $act_M = \tau$ *),*
- *if* $act(tr)$ *is of the form* $c?Y$ *then* $M \models_{\nu^i} guard(tr)$, *there exists* ν^a *such that* $\nu^a(z) = \nu^i(z)$ *for all* $z \notin Y$ *and for any* $x, y \in Y$ $\nu^a(x) = \nu^a(y)$, $\nu^f = \nu^a \circ subst(tr)$ *and* $act_M = c?\nu^a(y)$ *for an arbitrary* $y \in Y$.

For $r = (\nu^i, act_M, \nu^f)$, *we note* $source(r)$, $act(r)$, $target(r)$ *resp.* ν^i, act_M, ν^f.

As in [3], we will use $\delta!$ to denote under which semantic conditions an IOSTS is *quiescent*: quiescence refers to situations for which it is not possible to fire an output transition but only possibly input transitions or τ transitions.

Definition 5. (Suspension traces and *IOSTS* semantics) *The set of finite paths in* G, *denoted* $FP(G)$ *contains all finite sequence* $p = tr_1 \ldots tr_n$ *of transitions in* $Trans(G)$ *such that* $source(tr_1) = init(G)$ *and for all* $i < n$, $target(tr_i) = source(tr_{i+1})$. *The set of* runs *of* p *denoted* $Run(p)$ *is the set of sequences* $r = r_1 \ldots r_n$ *such that for all* $i \leq n$, r_i *is a run of* tr_i *and for all* $i < n$, $target(r_i) = source(r_{i+1})$. *We note* $Tr(r) = act(r_1) \ldots act(r_n)$. *The set of suspension traces of a run* r *of a finite path* p, *with* $r \in Run(p)$, *denoted* $STr(p, r)$ *is the least set s. t.:*

- *If* p *can be decomposed as* $p'.tr$ *with* $tr \in Trans(G)$ *and with* r *of the form* $r'.r_{tr}$ *with* $r_{tr} \in Run(tr)$, *then* $\{m.act(r_{tr}) | m \in STr(p', r')\} \subseteq STr(p, r)$.
- *If there exists no finite path* $p.p'$ *for which there exists* $r.r_1 \cdots r_k \in Run(p.p')$ *with for all* $i \leq k - 1$, $act(r_i) = \tau$ *and* $act(r_k) = c!m$ *for some* c *and* m, *then for any*[1] $\delta_m \in \{\delta!\}^*$, $Tr(r).\delta_m \in STr(p, r)$.

---

[1] A^* denotes the set of finite sequences of elements of A.

The set of suspension traces of a path p is $STr(p) = \bigcup_{r \in Run(p)} STr(p, r)$ and semantics of G are $STr(G) = \bigcup_{p \in FP(G)} STr(p)$.

2.2   Systems

We introduce the concept of *library* which intuitively allows us to characterize a set of IOSTS denoting basic components from which systems can be built. Formally a library is a set of couples, each of them being constituted of an IOSTS name and an IOSTS definition. IOSTS names are chosen in a given set BN whose elements are called *basic-IOSTS names*.

Definition 6. (Library) *A library is a set $\mathcal{B}$ whose elements are of the form (n, G) where $n \in BN$ and $G \in \mathcal{STS}$, s. t. for any two $(n_1, G_1), (n_2, G_2)$ in $\mathcal{B}$, $n_1 = n_2$ iff $G_1 = G_2$. If $G_1 \neq G_2$, for any $t_1 \in Trans(G_1)$ and $t_2 \in Trans(G_2)$, $name(t_1) \neq name(t_2)$. Elements of a library are called* basic-IOSTS.

In the sequel we consider a library $\mathcal{B}$ and we note $BN(\mathcal{B}) = \{n \mid (n, G) \in \mathcal{B}\}$ and $Chan(\mathcal{B}) = \{c \mid \exists(n, G) \in \mathcal{B}, c \in Chan(G)\}$. A *system* over a library $\mathcal{B}$ is built from IOSTS of $\mathcal{B}$ using two structuring mechanisms: *composition* which is used to aggregate two systems by connecting common channels and *hiding* is used to internalize some channels inside the system (they are no more visible from the environment). As for basic-IOSTS, we denote any system by a name and an IOSTS. The name associated to a system reflects the structure of the system. The set $SN(\mathcal{B})$ of *system names over* $\mathcal{B}$ is defined as follows:

- for any $n \in BN(\mathcal{B})$, $n \in SN(\mathcal{B})$, (a basic-IOSTS is also a system),
- for any $n_1, n_2 \in SN(\mathcal{B})$, $(n_1 \otimes n_2) \in SN(\mathcal{B})$ (corresponding to the system obtained by composing two systems named resp. n_1 and n_2),
- for any $n \in SN(\mathcal{B})$ and $C \subseteq Chan(\mathcal{B})$, $Hide(C, n) \in SN(\mathcal{B})$ (corresponding to the system obtained by hiding channels of C in the system named n).

Intuitively, for any system, transitions introduced in its associated IOSTS are defined over transitions of basic-IOSTSs composing the system, mainly by synchronization mechanisms. In order to be able to identify basic transitions involved in system transitions, the name associated to system transitions will explicit the underlying synchronization mechanism. Therefore, those names are of the form $(o, \{i_1, \cdots, i_n\})$ where o is a name of basic output-transition or a τ-transition and $i_1, \cdots, i_n$ are names of basic input-transitions (with possibly $n = 0$). Roughly speaking, the name $(o, \{i_1, \cdots, i_n\})$ generally refers to the synchronization of a basic output-transition named o with basic input-transitions named $i_1, \cdots, i_n$. Let us point out some particular cases. Any transition obtained by synchronizing input-transitions named $i_1, \cdots, i_n$ with an emission of the environment is denoted $(\varepsilon, \{i_1, \cdots, i_n\})$ where ε denotes the absence of output-transition. Any τ-transition in a system has a name of the form $(n, \emptyset)$ where n is the name of some underlying basic τ-transition. The set of *system transition names*, denoted STN, is then the set $(TN \cup \{\varepsilon\}) \times 2^{TN}$ where $\varepsilon \notin TN$.

We now define systems over a library by means of three constructions: renaming to convert a basic-IOSTS into a system, composition and hiding.

Definition 7. (Systems over $\mathcal{B}$) *The set $Sys(\mathcal{B})$ of systems over $\mathcal{B}$ is the subset of $SN(\mathcal{B}) \times \mathcal{IOSTS}$ defined as follows:*

Renaming: *For any $(n, G) \in \mathcal{B}$ and $t \in Trans(G)$, let us define $sn(t) = (name(t), \emptyset)$ if $act(t) = \tau$ or $act(t) \in Output(\Sigma)$ and $sn(t) = (\varepsilon, \{name(t)\})$ otherwise. Let us define $R(Trans(G)) = \bigcup_{t \in Trans(G)} \{(sn(t), body(t))\}$.*
$(n, (State(G), Init(G), R(Trans(G))))$ is in $Sys(\mathcal{B})$.

Composition: *For any two systems (n_1, G_1) and (n_2, G_2) of $Sys(\mathcal{B})$, let us note $G = (State(G_1) \times State(G_2), (init(G_1), init(G_2)), Trans)$ the IOSTS over $Sig(G_1) \cup Sig(G_2)$ where $Trans$ is defined as follows:*

- *If $((o_1, i_1), (q_1, c!t, \varphi_1, \rho_1, q_1')) \in Trans(G_1)$ and $((\varepsilon, i_2), (q_2, c?Y, \varphi_2, \rho_2, q_2')) \in Trans(G_2)$ and such that $type(t) = type(Y)$, then*
  $t = ((o_1, i_1 \cup i_2), ((q_1, q_2), c!t, \varphi_1 \wedge \varphi_2, \rho_1 \mid \rho_2[Y \leftarrow t], (q_1', q_2'))) \in Trans$.
- *If $((o_1, i_1), (q_1, c!t, \varphi, \rho, q_1')) \in Trans(G_1)$, for all $q_2 \in State(G_2)$ let us note $tr_1, \cdots, tr_n$ all transitions of the form $tr_i = (n_i, (q_2, c?Y_i, \varphi_i, \rho_i, q_i''))$ $\in Trans(G_2)$ for which $type(Y_i) = type(t)$. Let us note $guard = \wedge_{i \leq n} \neg \varphi_i$ if $n > 0$ and $guard = true$ otherwise. Then*
  $t = ((o_1, i_1), ((q_1, q_2), c!t, \varphi \wedge guard, \rho \mid Ident_{Att(G_2)}, (q_1', q_2))) \in Trans$.
- *For any two transitions of the form $((\varepsilon, i_1), (q_1, c?Y_1, \varphi_1, \rho_1, q_1')) \in Trans(G_1)$, and $((\varepsilon, i_2), ((q_2, c?Y_2, \varphi_2, \rho_2, q_2')) \in Trans(G_2)$ such that $type(Y_1) = type(Y_2)$, then*
  $t = ((\varepsilon, i_1 \cup i_2), ((q_1, q_2), c?(Y_1 \cup Y_2), \varphi_1 \wedge \varphi_2, \rho_1 \mid \rho_2, (q_1', q_2'))) \in Trans$.
- *If $((\varepsilon, i_1), (q_1, c?Y, \varphi, \rho, q_1')) \in Trans(G_1)$, for all $q_2 \in State(G_2)$, let us note $tr_1, \cdots, tr_n$ all transitions of the form $tr_i = (n_i, (q_2, c?Y_i, \varphi_i, \rho_i, q_i''))$ $\in Trans(G_2)$ for which $type(Y_i) = type(Y)$. Let us note $guard = \wedge_{i \leq n} \neg \varphi_i$ if $n > 0$ and $guard = true$ otherwise. Then*
  $t = ((\varepsilon, i_1), ((q_1, q_2), c?Y, \varphi \wedge guard, \rho \mid Ident_{Att(G_2)}, (q_1', q_2))) \in Trans$.
- *If $((o_1, \emptyset), (q_1, \tau, \varphi_1, \rho, q_1')) \in Trans_1$, for all $q_2 \in State(G_2)$, then*
  $((o_1, \emptyset), ((q_1, q_2), \tau, \varphi_1, \rho \mid Ident_{Att(G_2)}, (q_1', q_2))) \in Trans$.
- *The role of G_1 and G_2 can be permuted in all rules described above.*

$((n_1 \otimes n_2), G)$ is in $Sys(\mathcal{B})$.

Hiding: *For any $(n, G) \in Sys(\mathcal{B})$, for any $C \subseteq Chan(G)$, let us note $G' = (State(G), init(G), Trans')$ where $Trans'$ is defined as follows:*

- *For any $tr \in Trans(G)$ where $act(tr)$ is either of the form τ, $c!t$ or $c?X$ for some $c \notin C$, then $tr \in Trans'$.*
- *For any $tr \in Trans(G)$ where $act(tr)$ is of the form $c!t$ with $c \in C$, then $(name(tr), (source(tr), \tau, guard(tr), subst(tr), target(tr))) \in Trans'$.*

$(Hide(C, n), G')$ is in $Sys(\mathcal{B})$.

Systems inherit all notations from the underlying IOSTS framework: for any system $sys = (n, G)$, $Sig(sys)$ stands for $Sig(G)$, $Att(Sys)$ stands for $Att(G)$... In the same way, semantics of sys are the set of suspension traces of G: $STr(sys) = STr(G)$. Note that for composition, emissions and receptions are not blocking: if no transition can be synchronized with an input (resp. output)-transition tr, then tr is synchronized with the environment. A synchronization involves at most one output-transition: when several output transitions sharing the same source state could be considered at the same time to define a synchronization, this leads to non-determinism. The hiding operation make unobservable actions $c!t$ when c is in C but this operation is non blocking (the output-transition introducing $c!t$ is kept by replacing the communication action by τ). The hiding operation is blocking for inputs $c?X$ for c in C: corresponding transitions are simply removed in $Hiding(C, n)$. We now define sub-systems involved in a given system.

Definition 8. (Sub-systems) *Let* $(n, G) \in Sys(\mathcal{B})$. *The set of sub-systems of* (n, G) *denoted* $SubS((n, G)) \subseteq Sys(\mathcal{B})$ *is inductively defined as follows:*

- *If* $n \in BN$ *then* $SubS((n, G)) = \{(n, G)\}$,
- *If* n *is of the form* $n_1 \otimes n_2$ *then* $SubS((n, G)) = \{(n, G)\} \cup SubS((n_1, G_1)) \cup SubS((n_2, G_2))$ *where* (n_1, G_1) *and* (n_2, G_2) *belongs to* $Sys(\mathcal{B})$,
- *If* n *is of the form* $Hide(C, n')$, *then* $SubS((n, G)) = \{(n, G)\} \cup SubS((n', G'))$ *where* (n', G') *belongs to* $Sys(\mathcal{B})$.

For any sub-system sys' of a system sys, we can identify for any transition tr of sys the underlying transition of sys' involved in the definition of tr. This transition when it exists is called the projection of tr on sys'.

Definition 9. (Projection of a transition) *Let* $sys \in Sys(\mathcal{B})$, $sys' \in SubS$ (sys) *and* $tr = ((o, i), b) \in Trans(sys)$. *The projection of* tr *on* sys' *is the transition, when it is defined,* $tr_{sys'} = ((o', i'), b') \in Trans(G')$ *s. t.* $o' = o$ *or* $o' = \varepsilon$ *and* $i' \subseteq i$.

The naming mechanism for system transitions in Definition 7 makes $((o', i'), b')$ unique when it exists. Intuitively, the name (o, i) captures all the subparts of the system whose state is modified by firing the transition tr. In particular, if (o, i) does not include names of transitions issued from the sub-system sys', it simply means that there is no modification of the state concerning the sub-system sys', and thus that there does not exist a corresponding transition $tr_{sys'}$.

2.3   An Example of a Slot Machine

We consider a simple slot machine, named S and presented in Figure 1. The player can enter a bet into the slot machine and if he/she wins, he/she gets back the amount of his/her bet multiplied by 10. The system S is built from two basic-IOSTS, named resp. Int and SM for Interface and SlotMachine. Those two basic-IOSTS are composed and some channels, used for internal communications, are hidden. Thus the name of S is of the form $Hiding(C, Int \otimes SM)$ where:

- Int corresponds to the basic interface IOSTS between the environment (player) and the slot machine SM. When the system is reset (reception on int_start), the interface IOSTS waits for a bet from the player. The bet is refused when its amount is greater than 100. Otherwise, the IOSTS transmits to SM the amount of the bet and then, waits for a result, win or not, from the SM. Depending of the result, Int transmits to SM which gain should be given to the player.

– *SM* corresponds to the internal mechanism of the slot machine. It manages
  the different functionalities as appropriately updating the bank amount, deciding whether the player wins or not, and in the relevant cases, delivering
  cash to the player. For simplicity sake, the algorithm used to decide whether
  the player wins or not, is abstracted by a boolean non initialized variable w.
– *C* corresponds to all the channels used by *Int* and *SM* to communicate.
  That is, $C = \{int\_start, int\_bet, int\_wim, int\_amount, int\_cash\}$.

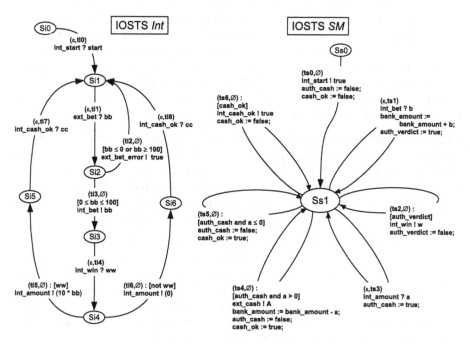

Fig. 1. An example of a slot machine

3 System Based Test Purposes for Sub-systems

We show how we define for any system, some test purposes dedicated to test its
sub-systems. Those test purposes will capture behaviors of sub-systems that typically occur in the whole system. This is done by combining symbolic execution
technics and projection mechanisms.

3.1 Symbolic Execution

We call a *symbolic behavior* of a system *sys* any finite path p of *sys* for which
$STr(p) \neq \emptyset$. In order to characterize the set of suspension traces of a symbolic
behavior we propose to use a *symbolic execution* mechanism. Symbolic execution
has been first defined for programs [7] and mainly consists in replacing concrete
input values and initialization values of variables by symbolic ones in order to

compute constraints induced on these variables by the execution of the program. Symbolic execution applied to IOSTS-based systems follows the same intuition considering guards of transitions as conditions and assignments together with communication actions as instructions. Herein, symbolic execution is presented as an adaptation of [4]. In the sequel, we assume that a set of fresh variables $F = \bigcup_{s \in S} F_s$ disjoint from the set of attribute variables $\coprod_{(n,G) \in \mathcal{B}} Att(G)$ is given. We now give the intermediate definition of *symbolic extended state* which is a structure allowing to store information about a symbolic behavior: the system current state (target state of the last transition of the symbolic behavior), the path condition which characterizes a constraint on symbolic variables to reach this state, and the symbolic values associated to attribute variables. As compared to [4], we also add a fourth stored information: it is given in the form of a constraint on symbolic variables which is not computed during the symbolic execution of the system. It is called an *external constraint* and in practice it will be inherited from a projection mechanism.

Definition 10. (Symbolic extended state) *A symbolic extended state of sys is a quadruple $\eta = (q, \pi, f, \sigma)$ where $q \in State(sys)$, $\pi \in Sen_{\Omega}(F)$ is called a* path condition, *$f \in Sen_{\Omega}(F)$ is called an external constraint and $\sigma \in T_{\Omega}(F)^{Att(sys)}$ is called a symbolic assignment of variables. $\eta = (q, \pi, f, \sigma)$ is said to be satisfiable if $\pi \wedge f$ is satisfiable[2]. One notes $\mathcal{S}(sys)$ (resp. $\mathcal{S}_{sat}(sys)$) the set of all the (resp. satisfiable) symbolic extended states over F.*

For any symbolic extended state η of the form (q, π, f, σ), q is denoted $state(\eta)$, π is denoted $pc(\eta)$, σ is denoted $sav(\eta)$ (for symbolic assignment of variables) and f is denoted $ec(\eta)$. Now, we show how to give symbolic counterparts to transitions of a system. The idea is to consider any symbolic extended state defined over the source state of the transition, and to construct a new target symbolic extended state defined over the target state of the transition. The external constraint of the target symbolic extended state is a conjunction formed with the external constraint of the source symbolic extended state and a new external constraint (denoted ct in the following Definition). In the sequel, for any system sys, $Sig(sys, F)$ stands for the signature $(F, Chan(sys))$.

Definition 11. (Symbolic execution of a transition) *With notations of Definition 10, for any $\eta \in \mathcal{S}(sys)$, for any $tr \in Trans(sys)$ such that $source(tr) = state(\eta)$, a symbolic execution of tr from η is a triple $st = (\eta, sa, \eta') \in \mathcal{S}(sys) \times Act(Sig(sys, F)) \times \mathcal{S}(sys)$ such that there exists $ct \in Sen_{\Omega}(F)$ for which:*

- *if $act(tr) = c!t$ then sa is of the form $c!z$ for some $z \in F$ and $\eta' = (target(tr), pc(\eta) \wedge sav(\eta)(guard(tr)) \wedge z = sav(\eta)(t), ec(\eta) \wedge ct, sav(\eta) \circ subst(tr))$,*
- *if $act(tr) = c?Y$ then sa is of the form $c?z$ for some $z \in F$ and $\eta' = (target(tr), pc(\eta) \wedge sav(\eta)(guard(tr)), ec(\eta) \wedge ct, sav(\eta) \circ (y \mapsto z)_{y \in Y} \circ subst(tr))$,*
- *if $act(tr) = \tau$ then $sa = \tau$ and $\eta' = (target(tr), pc(\eta) \wedge sav(\eta)(guard(tr)), ec(\eta) \wedge ct, sav(\eta) \circ subst(tr))$.*

---

[2] Here $\pi \wedge f$ is *satisfiable* if and only if there exists $\nu \in M^F$ such that $M \models_{\nu} \pi \wedge f$.

The definition of st only depends on tr, η, ct and the chosen variable z. Therefore, it is conveniently denoted $SE(tr, \eta, ct, z)$ (if $act(tr) = \tau$, z is useless). For any $st = (\eta, sa, \eta')$, $source(st)$ stands for η, $target(st)$ stands for η' and $act(st)$ stands for sa.

We now define symbolic execution of systems. Intuitively, a symbolic execution of a system sys is seen as a rooted tree whose paths are composed of sequences of symbolic executions of transitions which are consecutive in sys. The root is a symbolic extended state made of the initial state $init(sys)$, the path condition $true$, an arbitrary initialization σ_0 of variables of $Att(sys)$ in F, and an external constraint reduced to $true$ (no constraint at the beginning of the execution). Moreover, if a transition is symbolically executed with an external constraint ct, then it is also executed with the external constraint $\neg ct$.

Definition 12. (Symbolic execution of a system) *A full symbolic execution of sys over F is a triple $sys_{symb} = (\mathcal{S}(sys), init, R)$ with $init = (init(sys), true, true, \sigma_0)$ where σ_0 is an injective substitution in $F^{Att(sys)}$ and $R \subseteq \mathcal{S}(sys) \times Act(Sig(sys, F)) \times \mathcal{S}(sys)$ satisfies the following properties:*

- *for any $\eta \in \mathcal{S}(sys)$, for all $tr \in Trans(sys)$ such that $source(tr) = state(\eta)$, there exists exactly two constrained symbolic executions of tr in R respectively of the form $SE(tr, \eta, ct, z)$ and $SE(tr, \eta, \neg ct, z)$. Those two transitions are said to be* complementary.
- *for any $(\eta^i, c\natural x, \eta^f) \in R$ with $\natural \in \{!, ?\}$, $\forall a \in Att(sys)$, then $\sigma_0(a) \neq x$,*
- *for any $(\eta^i, c\natural x, \eta^f) \in R$ and $(\eta'^i, d\sharp x, \eta'^f) \in R$ with $\natural, \sharp \in \{!, ?\}$ which are not complementary, then $x \neq y$.*

The symbolic execution of sys over F associated to sys_{symb} is the triple $SE(sys) = (\mathcal{S}_{sat}(sys), init, R_{sat})$ where R_{sat} is the restriction of R to $\mathcal{S}_{sat}(sys) \times Act(Sig(sys, F)) \times \mathcal{S}_{sat}(sys)$.

We use the notation $FP(SE(sys))$ to denote the set of finite paths of $SE(sys)$. To define a run of a finite path p, we proceed as follows. We choose an interpretation $\nu : F \to M$ such that $M \models_\nu pc(\eta_f) \wedge ec(\eta_f)$ where η_f is the last symbolic extended state of p. Then for each (η, act, η') of p we associate a run $(\nu(sav(\eta)), act_M, \nu(sav(\eta')))$ where $act_M = \tau$ if $act = \tau$ and $act_M = c\natural\nu(z)$ if act is of the form $c\natural z$ with $\natural \in \{!, ?\}$. The sequence of such formed triples constitute a run of p. Note that the set of all runs of all finite paths of $FP(SE(sys))$ is exactly the set of all runs of all finite paths of sys in the sense of Definition 5 and this set is independent of the external constraints chosen to execute transitions. Those external constraints are simply used to partition symbolic behaviors. A trivial partitioning can be characterized by choosing $true$ as external constraints for executing any transition from any symbolic state. In this case the obtained symbolic execution is isomorphic to the one described in [4] which does not contain any external constraint. Besides note that any finite path p of a symbolic execution of sys characterizes a set of suspension traces obviously determined by its set of runs and the finite path corresponding to p in sys (See Definition 5). Therefore any symbolic execution of sys characterizes a set of suspension traces which can be easily proven to be this associated to sys in the sense of

Definition 5. Now, since internal actions are not observable in black box testing, we propose to eliminate them as follows.

Definition 13. (τ-reduction of a constrained symbolic execution) *The τ-reduction of $SE(sys)$ is the triple $SE(sys)_\tau = (\mathcal{S}_{sat}(sys), init, R^\tau_{sat})$, where $R^\tau_{sat} \subseteq \mathcal{S}_{sat}(sys) \times Act(Sig(sys, F)) \times \mathcal{S}_{sat}(sys)$ is such that for any sequence $st_1 \cdots st_n$ where for all $i \leq n$ $st_i \in R_{sat}$:*
- *for all $i \leq n-1$ $act(st_i) = \tau$, $source(st_{i+1}) = target(st_i)$ and $act(st_n) \neq \tau$,*
- *either $source(st_1) = init$ or there exists $st \in R_{sat}$ such that $target(st) = source(st_1)$ and $act(st) \neq \tau$,*

then $(source(st_1), act(st_n), target(st_n)) \in R^\tau_{sat}$.

Note that $SE(sys)$ and $SE(sys)_\tau$ characterize the same suspension traces. However, we need in the sequel to be able to symbolically identify situations in which quiescence is allowed. This is done by adding symbolic transitions labeled by $\delta!$ in the $SE(sys)_\tau$.

Definition 14. (Quiescence enrichment) *Quiescence enrichment of $SE(sys)$ is the triple $SE(sys)_\delta = (\mathcal{S}_{sat}(sys), init, R_\delta)$ where $R_\delta = R^\tau_{sat} \cup \Delta R_\delta$ with $\Delta R_\delta \subseteq \mathcal{S}_{sat}(sys) \times \{\delta!\} \times \mathcal{S}_{sat}(sys)$ is such that for any $\eta \in \mathcal{S}_{sat}(sys)$, if we note $out_\eta = \{tr_1, \cdots, tr_n\}$ the set of all transitions $tr_i \in R^\tau_{sat}$ such that $act(tr_i) \in output(Sig(sys, F))$, if we note $f \in Sen_\Omega(F)$ the formula of the form true if out_η is empty and of the form $\bigwedge_{i \leq n} \neg(pc(target(tr_i)) \wedge ec(target(tr_i)))$ otherwise, if we note $\eta' = (state(\eta), pc(\eta) \wedge f, ec(\eta), sav(\eta))$ then $(\eta, \delta!, \eta') \in \Delta R_\delta$.*

An example of a slot machine: symbolic execution Figure 2 shows a sub-tree of the symbolic execution of the slot machine system presented in Figure 1, as carried out by the AGATHA tool ([8,2]).

External constraints for any two complementary transitions are resp. *true* and *false* in the corresponding full symbolic execution. They never appear in the figure. We use the so-called *inclusion criteria* to end this execution. This criteria allows to stop symbolic execution when it detects that an encountered symbolic extended state is *included* in another already computed one. Intuitively, (q, π, f, σ) is included

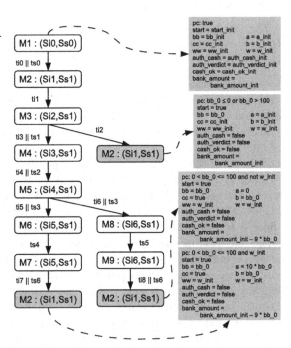

Fig. 2. Symbolic execution of the slot machine

in (q', π', f', σ') if $q' = q$ and the constraints induced on $Att(sys)$ by σ and $\pi \wedge f$ are stronger than those induced by σ' and $\pi' \wedge f'$. The interested readers can refer to [10,4] for more formal definitions. Let us point out that the symbolic sub-tree of S computes three characteristic symbolic behaviors. The left path corresponds to a winning bet, the middle path corresponds to a lost bet, and finally the right path corresponds to a forbidden bet. The initial and ending states are annotated with symbolic values of all attribute variables.

3.2 Symbolic Behavior Projections

For any finite path p of a symbolic execution of sys and a sub-system sys' of sys, we characterize the symbolic behavior $p_{sys'}$ of sys' involved in p. For this purpose, we begin by defining the *projection of a symbolic transition*.

Definition 15. (Projection of a symbolic transition) *Let sys be a system of $Sys(\mathcal{B})$. Let $sys' \in SubS(sys)$. Let $tr \in Trans(sys)$ such that $tr_{sys'}$ is defined. Let us note $st = SE(tr, \eta, ct, z)$ a symbolic execution of tr and $\eta_{sys'} \in \mathcal{S}_{sat}(sys')$ such that $state(\eta_{sys'}) = source(tr_{sys'})$. The projection of st on sys' of source $\eta_{sys'}$ is $SE(tr_{sys'}, \eta_{sys'}, pc(target(st)) \wedge ec(target(st)), z)$.*

The external constraint of the target state of the projection represents the constraints induced by the nature of the interactions of the sub-system with the other parts of the whole system. Now we generalize to symbolic behaviors.

Definition 16. (Projection of a path) *Let $p \in FP(SE(Sys))$. The projection of p on sys' denoted $p_{sys'} \in (\mathcal{S}_{sat}(Sys') \times Act(Sig(sys', F)) \times \mathcal{S}_{sat}(Sys'))^*$ together with its associated target state denoted $target(p_{sys'})$ are inductively mutually defined as follows:*

- *if p is of the form $st = SE(tr, init, ct, z) \in R_{sat}$ then let us note $\eta_{sys'} = (init(sys'), true, true, sav(init)|_{Att(sys')})$ then $p_{sys'}$ is the projection $st_{sys'}$ of st on sys' of source $\eta_{sys'}$ when it is defined, and in this case $target(p_{sys'}) = target(st_{sys'})$. Otherwise $p_{sys'}$ is the empty path and $target(p_{sys'}) = \eta_{sys'}$.*
- *if p is of the form $p'.st$ with $st = SE(tr, \eta, ct, z)$ then either the projection $st_{sys'}$ of st on sys' of source $target(p'_{sys'})$ is defined and: $p_{sys'} = p'_{sys'}.st_{sys'}$ and $target(p_{sys'}) = target(st_{sys'})$. Otherwise, $p_{sys'} = p'_{sys'}$ and $target(p_{sys'}) = target(p'_{sys'})$.*

Thus from any symbolic behavior of a system we can identify by projection symbolic behaviors of any sub-system whose external constraints reflect a usage of the sub-system in the whole system. Those projected behaviors are then good candidates to become behaviors to be tested on sub-systems: thus they will be chosen to construct test purposes.

4 Symbolic Execution Based Conformance Testing

4.1 Conformance Testing and System-Based Test Purposes

Model-based testing supposes that a conformance relation formally defines how are linked the specification G and the system under test SUT. Our work is based on the widely used *ioco* relation, initially designed for labeled transition systems [11] and afterwards adapted for symbolic transition systems [6,3,4]. All the ioco-based testing settings consider that the SUT is a black-box system which can be observed only by its behavior given as input/output sequences. These sequences of observations may include the special output $\delta!$ indicating that the SUT is in a quescient state. The set of all traces, possibly including suspension transitions, which can be observed from SUT is denoted $STr(SUT)$. When dealing with IOSTS, data handled in these sequences are concrete values denoted by ground terms of T_Ω. By test hypothesis, the SUT is modeled as a labeled transition system S for which transitions are emissions (outputs), receptions (inputs) carrying concrete values and such that the set of suspension traces of S coincide with $STr(SUT)$. Moreover, as usual, the SUT is supposed to accept all inputs in all states (hypothesis of input-enabled system). Intuitively a SUT conforms to its specification G with respect to *ioco* if any SUT output (including $\delta!$) is specified in G provided that the sequence of input/output preceding the considered observation is also specified in G.

Definition 17. (*ioco*) *An input-enabled system SUT conforms to G iff for any* $tra \in STr(G) \cap STr(SUT)$, *if there exists* $act \in Act(M) \cup \{\delta!\}$ *of the form* $c!t$ *or* $\delta!$ *such that* $tra.act \in STr(SUT)$, *then* $tra.act \in STr(G)$.

A test execution consists in executing a transition system, called a test case, on the SUT in order to produce test verdicts. The test case and the SUT are synchronized by coupling emissions and receptions. Test purposes are used to select some behaviors to be tested. In a previous work [4], we have proposed to model test purposes as finite trees extracted from symbolic executions of G. Such a symbolic execution describes all the possible behaviors of G. Therefore it is equivalent to test the SUT by selecting paths in G or in a symbolic execution of G. Indeed, we have demonstrated the following completeness result : if an SUT does not conform to a specification G, then there exists a test purpose such that our corresponding testing algorithm can emit a verdict $FAIL$. The main advantage of characterizing test purposes from a symbolic execution of G is that the testing process can be expressed as a simultaneous traversal of both the symbolic execution and the test purpose. Verdicts are emitted according to the fact that the observed behavior, in the form of a sequence of inputs (stimulations) and outputs (observations), does or does not belong to the test purpose and to the symbolic execution. We have defined 4 verdicts: $WeakPASS$ when the behavior belongs to the test purpose and to at least one path of the symbolic execution which is not in the test purpose, $PASS$ when the behavior belongs to the test purpose and not to any path of the symbolic execution which does not belong to the test purpose, $INCONC$ (for inconclusive) when the behavior belongs to the symbolic execution and not to the test purpose, and finally $FAIL$ when the behavior belongs neither to the test purpose nor to the symbolic execution. In the sequel, we slightly adapt the framework described in [4] to our purpose. Behaviors of any sub-system sys' to be tested are obtained by projecting behaviors of a symbolic execution of the whole system. It remains to define test purposes dedicated to test such projected behaviors. As basic-

IOSTS and hiding mechanism introduce τ-transitions, then such a projected behavior $p_{sys'}$ may contain τ-transitions. Since such internal transitions cannot be observed during testing, we construct test purposes from a τ-reduced symbolic execution enriched by quiescence. We identify all its finite paths whose last transitions are output-transitions (including δ-transitions) and which result of the τ-reduction of a path whose $p_{sys'}$ is a prefix. Those τ-reduced finite paths become behaviors to be tested.

Definition 18. (Test purpose) *Let $SE(sys')$ be a symbolic execution of sys' such that $p_{sys'} \in FP(SE(sys'))$. Let us note $ext_o(p_{sys'})$ the set $\{p_{sys'}\}$ if $p_{sys'}$ is of the form $p.(\eta, act, \eta')$ with $act \in Output(Sig(sys', F))$ and whose elements are all paths of the form $p_{sys'}.(\eta_1, act_1, \eta_1') \cdots (\eta_n, act_n, \eta_n')$ with $act_i = \tau$ for $i < n$ and $act_n \in Output(Sig(sys', F))$ otherwise. Let us note $T \subseteq S_{sat}(sys')$ the set of all the target states of all the finite paths of $ext_o(p_{sys'})$. A symbolic test purpose for $p_{sys'}$ and $SE(sys')$ is an application $TP : S_{sat} \to \{skip, accept, \odot\}$ such that:*

- *for all $\eta \in T$, $TP(\eta) = accept$,*
- *for all finite path $st_1 \cdots st_n$ such that for all $i \leq n$, $st_i \in R_\tau$ and $TP(target(st_n)) = accept$, then $TP(source(st_i)) = skip$,*
- *If $ext_o(p_{sys'}) = \{p_{sys'}\}$ then all other states η verify $TP(\eta) = \odot$,*
- *if $ext_o(p_{sys'}) \neq \{p_{sys'}\}$ and the last transition of $p_{sys'}$ is an input-transition st then if there exists a transition $st_\delta \in \Delta R_\delta$ s. t. $source(st_\delta) = target(st)$ then $TP(target(st_\delta)) = accept$ and all other states $\eta \in R_\delta$ verify $TP(\eta) = \odot$,*
- *if $ext_o(p_{sys'}) \neq \{p_{sys'}\}$ and the last transition of $p_{sys'}$ is a τ-transition then all other states $\eta \in R_\delta$ verify $TP(\eta) = \odot$.*

Definition 18 introduces the notion of symbolic test purpose, which extends the notion of test purposes as defined in [4] by considering a symbolic execution of a system which incorporates constraints issued from a surrounding system. Let us remark that constraint symbolic executions allow us to characterize test purposes in the same way: a test purpose is a finite sub-tree of a δ-enriched symbolic execution whose leaves are target states of output transitions (identified by means of the labeling function which associates *accept* to those states). The algorithm of test case generation given in [4] can directly be applied.

An example of a slot machine: projection Let us consider p the left path of Figure 2, corresponding to the winning case. In Figure 3, the left path represents p. The right path is the projection p_{SM} of p on SM. Nearby each symbolic extended state name Ss_i we indicate in the grey box the content of the symbolic state, up to simplifications in path conditions and external constraints for sake of readability. The behavior denoted by p_{SM} corresponds intuitively to the following scenario: after the initialization, a bet is received for amount greater to 0 and less or equal to 100 (this is a constraint induced by the interface). Then SM sends a verdict stating that the player has won, the value to be removed of the bank account is received and correspond to 10 times the bet. The amount is sent to the interface and effectively removed from the bank account. Finally,

SM sends an ending operation message to the *Int*. A test purpose *L* for this behavior would label N_6 by *accept* and $N0$ to $N5$ by *skip*. On the right part of the figure, $N'2$ and $N'4$ are target states of the complementary transitions of respectively $(N\_1, int\_bet?bb\_0, N\_2)$ and $(N\_3, int\_amount?a\_0, N4)$. $N'2$ characterizes cases for which the received bet is out of the range allowed by the interface. N'_4 characterizes situation for which the gain does not correspond to 10 times the bet contrarily to the information sent by the interface.

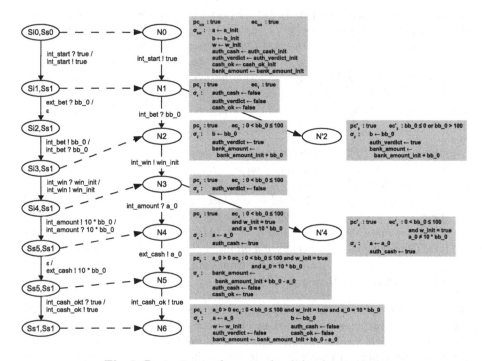

Fig. 3. Projection in the example of the slot machine

Those two situations are possible for *SM* but note relevant in the frame of the whole system. Therefore *L* would label $N'2$ and $N'4$ with $\odot$. To conclude, let us point out that such a test purpose cannot be deduced only from the knowledge of *SM*: it clearly depends on the way *SM* is used in the whole system *S*. This exemplifies our initial goal of eliciting from a system dedicated test purposes for each subsystem.

5   Conclusion and Future Works

We have extended the framework of IOSTS introduced in [4], in order to deal with component-based system specifications and we have used symbolic execution mechanisms in order to compute behaviors of sub-systems constrained by systems in which they are involved. Then, we have defined test purposes from those constrained behaviors. The definition of dedicated methodologies for component based systems should clearly rely on the targeted fault models. We plan to study fault models that mainly deal with communication mechanisms as in [5]. For such fault models, a testing methodology would probably preconize to construct test purposes for behaviors involving a lot of internal communication synchronizations. Besides, we also plan to target fault models that mainly deal with basic components. As in [13], we could consider that composition and hiding mechanisms are well implemented such that an appropriate testing methodology would only consider test purposes directly defined at the component level. More generally, our next goal is to provide testing methodologies for component based systems which take advantage of the fact that some components or subsystems have been previously intensively tested such that a large class of tests becomes useless in the context of the whole system.

References

1. Berrada, I., Castanet, R., Félix, P.: Testing Communicating Systems: a Model, a Methodology, and a Tool. In: Khendek, F., Dssouli, R. (eds.) TestCom 2005. LNCS, vol. 3502, pp. 111–128. Springer, Heidelberg (2005)
2. Bigot, C., Faivre, A., Gallois, J.-P., Lapitre, A., Lugato, D., Pierron, J.-Y., Rapin, N.: Automatic test generation with AGATHA. In: Garavel, H., Hatcliff, J. (eds.) ETAPS 2003 and TACAS 2003. LNCS, vol. 2619, pp. 591–596. Springer, Heidelberg (2003)
3. Frantzen, L., Tretmans, J., Willemse, T.A.C.: A Symbolic Framework for Model-Based Testing. In: Havelund, K., Núñez, M., Roşu, G., Wolff, B. (eds.) Formal Approaches to Software Testing and Runtime Verification. LNCS, vol. 4262, pp. 40–54. Springer, Heidelberg (2006)
4. Gaston, C., Le Gall, P., Rapin, N., Touil, A.: Symbolic Execution Techniques for Test Purpose Definition. In: Uyar, M.Ü., Duale, A.Y., Fecko, M.A. (eds.) TestCom 2006. LNCS, vol. 3964, pp. 1–18. Springer, Heidelberg (2006)
5. Gotzhein, R., Khendek, F.: Compositional Testing of Communication Systems. In: Uyar, M.Ü., Duale, A.Y., Fecko, M.A. (eds.) TestCom 2006. LNCS, vol. 3964, pp. 227–244. Springer, Heidelberg (2006)
6. Jeannet, B., Jéron, T., Rusu, V., Zinovieva, E.: Symbolic test selection based on approximate analysis. In: Halbwachs, N., Zuck, L.D. (eds.) TACAS 2005. LNCS, vol. 3440, pp. 349–364. Springer, Heidelberg (2005)
7. King, J.-C.: A new approach to program testing. In: Proc. of the Int. Conference on Reliable software, vol. 21(23), pp. 228–233 (1975)
8. Lugato, D., Rapin, N., Gallois, J.-P.: Verification and tests generation for SDL industrial specifications with the AGATHA toolset. In: Larsen, K.G., Nielsen, M. (eds.) CONCUR 2001. LNCS, vol. 2154, pp. 1404–3203. Springer, Heidelberg (2001) ISSN 1404-3203
9. Pelliccione, P., Muccini, H., Bucchiarone, A., Facchini, F.: TeStor: Deriving Test Sequences from Model-based Specification. In: Heineman, G.T., Crnković, I., Schmidt, H.W., Stafford, J.A., Szyperski, C.A., Wallnau, K. (eds.) CBSE 2005. LNCS, vol. 3489, pp. 267–282. Springer, Heidelberg (2005)
10. N. Rapin, C. Gaston, A. Lapitre, and J.-P. Gallois. Behavioural unfolding of formal specifications based on communicating automata. In: Proc. of the 1th Int. Workshop ATVA 2003 (2003)

11. Tretmans, J.: Conformance Testing with Labelled Transition Systems: Implementation Relations and Test Generation. Computer Networks and ISDN Systems 29, 49–79 (1996)
12. Tretmans, J.: Test generation with inputs, outputs and repetitive quiescence. Software—Concepts and Tools 17(3), 103–120 (1996)
13. van der Bijl, M., Rensink, A., Tretmans, J.: Compositional Testing with IOCO. In: Petrenko, A., Ulrich, A. (eds.) FATES 2003. LNCS, vol. 2931, pp. 86–100. Springer, Heidelberg (2004)

A Compositional Testing Framework Driven by Partial Specifications

Yliès Falcone[1], Jean-Claude Fernandez[1], Laurent Mounier[1], and
Jean-Luc Richier[2]

[1] Vérimag Laboratory, 2 avenue de Vignate 38610 Gières, France
[2] LIG Laboratory, 681, rue de la Passerelle, BP 72, 38402 Saint Martin d'Hères
Cedex, France
{Ylies.Falcone,Jean-Claude.Fernandez,Laurent.Mounier,
Jean-Luc.Richier}@imag.fr

Abstract. We present a testing framework using a compositional approach to generate and execute test cases. Test cases are generated and combined with respect to a partial specification expressed as a set of requirements and elementary test cases. These approach and framework are supported by a prototype tool presented here. The framework is presented here in its LTL-like application, besides other specification formalisms can be added.

1 Introduction

Testing is a popular validation technique which purpose is essentially to find defects on a system implementation, either during its development, or once a final version has been completed. Therefore, and even if lots of work have already been carried out on this topic, improving the effectiveness of a testing phase while reducing its cost and time consumption remains a very important challenge, sustained by a strong industrial demand.

From a practical point of view, a test campaign consists in producing a test suite (test generation), and executing it on the target system (test execution). Automating test generation means deriving the test suite from some initial description of the system under test. The test suite consists in a set of test cases, where each test case is a set of interaction sequences to be executed by an external tester. Any execution of a test case should lead to a test verdict, indicating if the system succeeded or not on this particular test (or if the test was not conclusive).

The initial system description used to produce the test cases may be for instance the source code of the software, some hypothesis on the sets of inputs it may receive (user profiles), or some requirements on its expected properties at run-time (i.e., a characterization of its (in)-correct execution sequences). In this latter case, when the purpose of the test campaign is to check the correctness of some behavioral requirements, an interesting approach for automatic test generation is the so-called model-based testing technique. Model-based testing

A. Petrenko et al. (Eds.): TestCom/FATES 2007, LNCS 4581, pp. 107–122, 2007.

is rather successful in the communication protocol area, especially because it is able to cope with some non-determinism of the system under test. It has been implemented in several tools, see for example [1] for a survey. However, it suffers from some drawbacks that may prevent its use in other application areas. First of all, it strongly relies on the availability of a system specification, which is not always the case in practice. Moreover, when it exists, this specification should be complete enough to ensure some relevance of the test suite produced. Finally, it is likely the case that this specification cannot encompass all the implementation details, and is restricted to a given abstraction level. Therefore, to become executable, the test cases produced have to be *refined* into more concrete interaction sequences. Automating this process in the general case is still a challenging problem [2], and most of the time, when performed by hand, the soundness of the result cannot be fully guaranteed.

We propose here an alternative approach to produce a test suite dedicated to the validation of behavioral requirements of a software (see Fig. 1). In this framework the requirements $\mathcal{R}$ are expressed by logical formulas φ built upon a set of (abstract) predicates P_i describing (possibly non-atomic) operations performed on the system under test. A typical example of such requirements could be for instance a security policy, where the abstract predicates would denote some high-level operations like "user A is authenticated", or "message M has been corrupted". The approach we propose relies on the following consideration: a perfect knowledge of the implementation details is required to produce elementary test cases Tc_i able to decide whether such predicates hold or not at some state of the software execution. Therefore, writing the test cases dedicated to these predicates should be left to the programmer (or tester) expertise when a detailed system specification is not available. However, correctly orchestrating the execution of these "basic test cases" and combining their results to deduce the validity of the overall logical formula is much easier to automate since it depends only of the semantics of the operators used in this formula. This step can therefore be produced by an automatic test generator, and this test generation can even be performed in a compositional way (on the structure of the logical formula). More precisely, from the formula φ, a test generation function automatically produces an (abstract) tester AT_φ. This tester consists of a set of communicating *test controllers*, one for each operator appearing in φ. Thus, AT_φ depends only on the structure of formula φ. AT_φ is then instantiated using the elementary test cases Tc_i to obtain a concrete tester T_φ for the formula φ. Execution of this tester on the implementation I produces the final verdict.

We believe that this approach is general enough to be instantiated with several logic formalisms commonly used to express requirements on execution traces (e.g., extended regular expressions or linear temporal logics).

This works extends some preliminary descriptions on this technique [3,4] in several directions: first we try to demonstrate that it is general enough to support several logical formalisms, then we apply it for the well-known LTL temporal logic, and finally we evaluate it on a small case study using a prototype tool under development.

In addition to the numerous works proposed in the context of model-based test generation for conformance testing, this work also takes credits from the community of run-time verification. In fact, one of the techniques commonly used in this area consists in generating a monitor able to check the correctness of an execution trace with respect to a given logical requirement (see for instance [5,6] or [7] for a short survey). In practice, this technique needs to *instrument* the software under verification with a set of observation points to produce the traces to be verified by the monitor. This instrumentation should of course be correlated with the requirement to verify (i.e., the trace produced should contain enough information). In the approach proposed here, these instrumentation directives are replaced by the elementary test cases associated to each elementary predicates. The main difference is that these test cases are not restricted to pure observation actions, but they may also contain some active testing operations, like calling some methods, or communicating with some remote process to check the correctness of an abstract predicate.

The rest of the paper is organized as follows: Sect. 2 introduces the general approach, while Sect. 3 details its sound-proved application for a particular variant of the linear temporal logic LTL. Section 4 and 5 respectively describe the architecture of a prototype tool based on this framework, and its application on a small case study. The conclusion and perspectives of this work are given is Sect. 6.

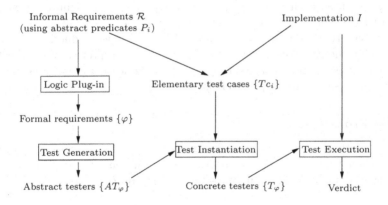

Fig. 1. Test generation overview

2   The General Approach

We describe here more formally the test generation approach sketched in the introduction. As it has been explained, this approach relies on the following steps:

- generation of an *abstract tester* AT_φ from a formal requirement φ;
- instantiation of AT_φ into a concrete tester A_φ using the set of elementary testers associated to each atomic predicate of φ;
- execution of T_φ against the System Under Test (SUT) to obtain a test verdict.

2.1   Notations

A *labelled transition system* (LTS, for short) is a quadruplet $S = (Q, A, T, q_0)$ where Q is a set of states, A a set of labels, $T \subseteq Q \times A \times Q$ the transition relation and $q_0 \in Q$ the initial state. We will denote by $p \xrightarrow{a}_T q$ (or simply $p \xrightarrow{a} q$) when $(p, a, q) \in T$. A *finite execution sequence* of S is a sequence $(p_i, a_i, q_i)_{\{0 \leq i \leq m\}}$ where $p_0 = q_0$ and $p_{i+1} = q_i$. For each finite execution sequence λ, the sequence of actions $(a_0, a_1, \ldots, a_m)$ is called a *finite execution trace* of S. We denote by $Exec(S)$ the set of all finite execution traces of S. For an execution trace $\sigma = (a_0, a_1, \ldots, a_m)$, we denote by $|\sigma|$ the length $m + 1$ of σ, by $\sigma^{k \cdots l}$ the sub-sequence $(a_k, \ldots, a_l)$ when $0 \leq k \leq l \leq m$, and by $\sigma^{k \cdots}$ the sub-sequence $(a_k, \ldots, a_m)$ when $0 \leq k \leq m$. Finally, $\sigma_{\downarrow X}$ denotes the *projection* of σ on action set X. Namely, $\sigma_{\downarrow X} = \{a_0 \cdots \cdots a_m \mid \forall i \cdot a_i \in X \wedge \sigma = w_0 \cdot a_0 \cdots w_m \cdot a_m \cdot w_{m+1} \wedge w_i \in (A \setminus X)^*\}$.

2.2   Formal Requirements

We assume in the following that the formal requirements φ we consider are expressed using a logic $\mathscr{L}$. Formulas of $\mathscr{L}$ are built upon a finite set of *n-ary operators* F^n and a finite set of *abstract predicates* $\{p_1, p_2, \ldots, p_n\}$ as follows:

$$\text{formula} ::= F^n(\text{formula}_1, \text{formula}_2, \ldots, \text{formula}_n) \mid p_i$$

We suppose that each formula of $\mathscr{L}$ is interpreted over a finite execution trace of a LTS S, and we say that S satisfies φ (we note $S \models \varphi$) iff *all* sequences of $Exec(S)$ satisfy φ. Relation $\models$ is supposed to be defined inductively on the syntax of $\mathscr{L}$ in the usual way: abstract predicates are interpreted over $Exec(S)$, and the semantics of each operator $F^n(\varphi_1, \ldots, \varphi_n)$ is defined in terms of sets of execution traces satisfying respectively $\varphi_1, \ldots, \varphi_n$.

2.3   Test Process Algebra

In order to outline the compositionality of our test generation technique, we express a tester using an algebraic notation. We recall here the dedicated "test process algebra" introduced in [4], but other existing process algebras could also be used.

Syntax. Let Act be a set of *actions*, $\mathcal{T}$ be a set of *types* (with $\tau \in \mathcal{T}$), Var a set of *variables* (with $x \in Var$), and Val a set of *values* (union of values of types $\mathcal{T}$). We denote by $expr_\tau$ (resp. x_τ) any expression (resp. variable) of type τ. In particular, we assume the existence of a special type called $Verdict$ which associated values are $\{pass, fail, inconc\}$ and which is used to denote the *verdicts* produced during the test execution. The syntax of a test process t is given by the following grammar:

$$t ::= [b]\,\gamma \circ t \mid t + t \mid nil \mid recX\ t \mid X$$
$$b ::= true \mid false \mid b \vee b \mid b \wedge b \mid \neg b \mid expr_\tau = expr_\tau$$
$$\gamma ::= x_\tau := expr_\tau \mid !c(expr_\tau) \mid ?c(x_\tau)$$

In this grammar t denotes a basic tester (*nil* being the empty tester doing nothing), b a boolean expression, c a channel name, γ an action, $\circ$ is the *prefixing* operator, $+$ the *choice* operator, X a term variable, $recX$ allows *recursive* process definition (with X a term variable)[1]. When the condition b is true, we abbreviate $[true]\gamma$ by γ. Atomic actions performed by a basic tester are either internal assignments ($x_\tau := expr_\tau$), value emissions ($!c(expr_\tau)$) or value receptions ($?c(x_\tau)$) over a channel c[2].

Semantics. We first give a semantics of basic testers (t) using rewriting rule between uninterpreted terms in a CCS-like style (see Fig. 2).

$$
\frac{\gamma \in Act}{[b]\gamma \circ t \overset{[b]\gamma}{\rightharpoonup} t}\,(\circ)
\qquad
\frac{t[recX \circ t/X] \overset{[b]\gamma}{\rightharpoonup} t' \qquad \gamma \in Act}{recX \circ t \overset{[b]\gamma}{\rightharpoonup} t'}\,(rec)
$$

$$
\frac{\gamma \in Act \qquad t_1 \overset{[b]\gamma}{\rightharpoonup} t_1'}{t_1 + t_2 \overset{[b]\gamma}{\rightharpoonup} t_1'}\,(+)_l
\qquad
\frac{\gamma \in Act \qquad t_2 \overset{[b]\gamma}{\rightharpoonup} t_2'}{t_1 + t_2 \overset{[b]\gamma}{\rightharpoonup} t_2'}\,(+)_r
$$

Fig. 2. Rules for term rewriting

The semantics of a basic test process t is then given by means of a LTS $S_t = (Q^t, A^t, T^t, q_0^t)$ in the usual way: states Q^t are "configurations" of the form (t, ρ), where t is a term and $\rho : Var \to Val$ is an *environment*. States and transition of S_t (relation $\longrightarrow$) are the smallest sets defined by the rules given in Fig. 3 (using the auxiliary relation $\rightharpoonup$ defined in Fig. 2). The initial state q_0^t of S is the configuration (t_0, ρ_0), where ρ_0 maps all the variables to an undefined value. Finally, note that actions A^t of S_t are labelled either by internal assignments ($x_\tau := v$) or external emission ($!c(v)$). In the following we denote by $A_{\text{ext}}^t \subseteq A^t$ the external emissions and receptions performed by the LTS associated to a test process t.

Complex testers are obtained by parallel composition of test processes with synchronisation on a channel set cs (operator $\|_{cs}$), or using a so-called "join-exception" operator ($\ltimes^{\mathcal{I}}$), allowing to interrupt a process on reception of a communication using the interruption channel set $\mathcal{I}$. We note $\|$ for $\|_\emptyset$ and Act_chan(s) all possible actions using a channel in the set s. To tackle with communication in our semantics, we give two sets of rules specifying how LTSs are composed relatively to the communication operators ($\|_{cs}, \ltimes^{\mathcal{I}}$). These rules aim to maintain asynchronous execution, communication by *rendez-vous*. Let $S_i^t = (Q_i^t, A_i^t, T_i^t, q_{0i}^t)$ be two LTSs modelling the behaviours of two processes t_1 and t_2, we define the LTS $S = (Q, A, T, q_0)$ modelling the behaviours of $S_1^t \|_{cs} S_2^t$

---

[1] We will only consider *ground* terms: each occurrence of X is bound to $recX$.

[2] To simplify the calculus, we supposed that all channels exchange one value. In the testers, we also use "synchronisation channels", without exchanged argument, as a straightforward extension.

$$\frac{\rho(expr_\tau) = v \qquad t \overset{[b]x_\tau := expr_\tau}{\longrightarrow} t' \qquad \rho(b) = true}{(t, \rho) \overset{x_\tau := v}{\longrightarrow} (t', \rho[v/x_\tau])} \ (:=)$$

$$\frac{\rho(expr_\tau) = v \qquad t \overset{[b]!c(expr_\tau)}{\longrightarrow} t' \qquad \rho(b) = true}{(t, \rho) \overset{!c(v)}{\longrightarrow} (t', \rho)} \ (!)$$

$$\frac{v \in Dom(\tau) \qquad t \overset{[b]?c(x_\tau)}{\longrightarrow} t' \qquad \rho(b) = true}{(t, \rho) \overset{!c(v)}{\longrightarrow} (t, \rho[v/x_\tau])} \ (?)$$

Fig. 3. Rules for environment modification

$$\frac{p_1 \overset{a}{\longrightarrow} p_1' \quad a \notin \text{Act\_chan}(cs)}{(p_1, p_2) \overset{a}{\longrightarrow} (p_1', p_2)} \ (\overline{\|^l_{cs}}) \qquad \frac{p_2 \overset{a}{\longrightarrow} p_2' \quad a \notin \text{Act\_chan}(cs)}{(p_1, p_2) \overset{a}{\longrightarrow} (p_1, p_2')} \ (\overline{\|^r_{cs}})$$

$$\frac{p_1 \overset{a}{\longrightarrow} p_1' \quad p_2 \overset{a}{\longrightarrow} p_2' \quad a \in \text{Act\_chan}(cs)}{(p_1, p_2) \overset{a}{\longrightarrow} (p_1', p_2')} \ (\|_{cs})$$

$$\frac{p_1 \overset{a}{\longrightarrow} p_1' \quad a \notin \text{Act\_chan}(\mathcal{I})}{(p_1, p_2) \overset{a}{\longrightarrow} (p_1', p_2)} \ (\overline{\ltimes^{\mathcal{I}}}) \qquad \frac{p_2 \overset{a}{\longrightarrow} p_2' \quad a \in \text{Act\_chan}(\mathcal{I})}{(p_1, p_2) \overset{a}{\longrightarrow} (\bot, p_2')} \ (\ltimes^{\mathcal{I}})$$

Fig. 4. LTS composition related to $\|_{cs}$ and $\ltimes^{\mathcal{I}}$

and $S_1^t \ltimes^{\mathcal{I}} S_2^t$ as the product of S_1^t and S_2^t where $Q \subseteq (Q_1^t \cup \{\bot\}) \times Q_2^t$ and the transition rules are given in Fig. 4.

2.4 Test Generation

Principle. The test generation technique we propose aims to produce a tester process t_φ associated to a formal requirement φ and it can be formalized by a function called *GenTest* in the rest of the paper ($GenTest(\varphi) = t_\varphi$). This generation step depends of course of the logical formalism under consideration, but it is compositionally defined in the following way:

- a basic tester t_{p_i} is associated with each abstract predicate p_i of φ;
- for each sub-formula $\phi = F^n(\phi_1, \cdots, \phi_n)$ of φ, a test process t_ϕ is produced, where t_ϕ is a parallel composition between test processes $t_{\phi_1}, \ldots, t_{\phi_n}$ and a test process C_{F^n} called a *test controller* for operator F^n.

The purpose of test controllers C_{F^n} is both to schedule the test execution of the t_{ϕ_k} (starting, stopping or restarting their execution), and to combine their verdicts to produce the overall verdict associated to ϕ. As a result, the architecture of a tester t_φ matches the abstract syntax tree corresponding to formula φ: leaves are basic tester processes corresponding to abstract predicates p_i of φ, intermediate nodes are controllers associated with operators of φ.

Hypothesis. To allow interactions between the internal sub-processes of a tester t_φ, we assume the following hypotheses:

Each tester sub-process t_{ϕ_k} (basic tester or controller) owns a special variable used to store its *local verdict*. This variable is supposed to be set to one of these values when the test execution terminates – its intuitive meaning is similar to the conformance testing case:

- *pass* means that the test execution of t_{ϕ_k} did not reveal any violation of the sub-formula associated to t_{ϕ_k};
- *fail* means that the test execution of t_{ϕ_k} did reveal a violation of the sub-formula associated to t_{ϕ_k};
- *inconc* (inconclusive) means that the test execution of t_{ϕ_k} did not allow to conclude about the validity of the sub-formula associated to t_{ϕ_k}.

Each tester process t_{ϕ_k} (basic tester or controller) owns a set of four dedicated communication channels $cs_k = \{c\_start_k, c\_stop_k, c\_loop_k, c\_ver_k\}$ used respectively to start its execution, to stop it, to resume it from its initial state and to deliver a verdict. In the following, we denote by $\mathsf{C}(cs, cs_1, \cdots, cs_n)$ each controller C where cs is the channel set dedicated to the communication with the embracing controller whereas the (cs_i) are the channel sets dedicated to the communication with the sub-test processes. Finally, a "starter" process is also required to start the topmost controller associated to t and to read the verdict it delivered.

Each basic tester process t_{p_i} associated to an LTS $S_{t_{p_i}}$ is supposed to have a subset of actions $A_{ext}^{t_{p_i}} \subseteq A^{t_{p_i}}$ used to communicate with the SUT. Considering $t = GenTest(\varphi)$, the set A_{ext}^t is defined as the union of the $A_{ext}^{t_{p_i}}$ where p_i is a basic predicate of φ.

Test generation function definition (GenTest). *GenTest* can then be defined as follows using GT as an intermediate function:

$GenTest(\varphi) \overset{\text{def}}{=} GT(\varphi, cs) \parallel_{\{c\_start, c\_ver\}} (!c\_start() \circ ?c\_ver(x) \circ nil)$
where cs is the set $\{c\_start, c\_stop, c\_loop, c\_ver\}$ of channel names associated to t_φ.

$GT(p_i, cs) \overset{\text{def}}{=} Test(t_{p_i}, cs)$

$GT(F^n(\phi_1, \ldots, \phi_n), cs) \overset{\text{def}}{=} (GT(\phi_1, cs_1) \parallel \cdots \parallel GT(\phi_n, cs_n)) \parallel_{cs'} \mathsf{C}_{F^n}(cs, cs_1, \ldots, cs_n)$
where $cs_1, \ldots, cs_n$ are sets of fresh channel names and $cs' = cs_1 \cup \cdots \cup cs_n$.

$Test(t_p, \{c\_start, c\_stop, c\_loop, c\_ver\}) \overset{\text{def}}{=}$

$recX \, (?c\_start() \circ t_p \circ !c\_ver(ver) \circ ?c\_loop() \circ X) \ltimes^{\{c\_stop\}} (?c\_stop() \circ nil)$

2.5   Test Execution and Test Verdicts

As seen in the previous subsections, the semantics of a tester represented by a test process t is expressed by a LTS $S_t = (Q^t, A^t, T^t, q_0^t)$ where $A_{ext}^t \subseteq A^t$ denotes the external actions it may perform. Although the system under test I is not described by a formal model, its behaviour can also be expressed by a LTS $S_I = (Q^I, A^I, T^I, q_0^I)$. A *test execution* is a sequence of interactions (on

A_{ext}^t) between t and I in order to deliver a *verdict* indicating whether the test succeeded or not. We define here more precisely these notions of test execution and test verdict.

Formally speaking, a test execution of a test process t on a SUT I can be viewed as an execution trace of the parallel product $\otimes_{A_{\text{ext}}^t}$ between LTSs S_t and S_I with synchronizations on actions of A_{ext}^t. This product is defined as follows:
$S_t \otimes_{A_{\text{ext}}^t} S_I$ is the LTS (Q, A, T, q_0) where $Q \subseteq Q^t \times Q^I$, $A \subseteq A^t \cup A^I$, $q_0 = (q_0^t, q_0^I)$, and
$T = \{(p^t, p^I) \xrightarrow{a} (q^t, q^I) \mid (p^t, a, q^t) \in T^t \wedge (p^I, a, q^I) \in T^I \wedge a \in A_{\text{ext}}^t\} \cup \{(p^t, p^I) \xrightarrow{a} (q^t, p^I) \mid (p^t, a, q^t) \in T^t \wedge a \in A^t \setminus A_{\text{ext}}^t\} \cup \{(p^t, p^I) \xrightarrow{a} (p^t, q^I) \mid (p^I, a, q^I) \in T^I \wedge a \in A^I \setminus A_{\text{ext}}^t\}$.

For any test execution $\sigma \in \text{Exec}(S_t \otimes_{A_{\text{ext}}^t} S_I)$, we define the verdict function: $\text{VExec}(\sigma) = pass$ (resp. *fail*, *inconc*) iff $\sigma = c\_start() \cdot \sigma' \cdot c\_ver(pass)$ (resp. $\sigma = c\_start() \cdot \sigma' \cdot c\_ver(fail)$, $\sigma = c\_start() \cdot \sigma' \cdot c\_ver(inconc)$) and c_start (resp. c_ver) is the starting (resp. the verdict) channel associated to the topmost controller of t.

3   Application to Variant of LTL

This section presents an instantiation of the previous framework for a (non atomic) action-based version of LTL-X, the next-free variant of LTL [8].

3.1   The Logic

Syntax. The syntax of a formula φ is given by the following grammar, where the atoms $\{p_1, \ldots, p_n\}$ are action predicates.

$$\varphi ::= \neg \varphi \mid \varphi \, \mathcal{U} \, \varphi \mid \varphi \wedge \varphi \mid p_i$$

Semantics. Formulas φ are interpreted over the finite execution traces $\sigma \in A^*$ of a LTS. We introduce the following notations.

To each atomic predicate p_i of φ we associate a subset of actions A_{p_i} and two subsets L_{p_i} and $L_{\overline{p_i}}$ of $A_{p_i}^*$. Intuitively, A_{p_i} denotes the actions that influence the truth value of p_i, and L_{p_i} (resp. $L_{\overline{p_i}}$) the set of finite execution traces satisfying (resp. non satisfying) p_i. We suppose that the action sets A_{p_i} are such that $\{(A_{p_i})_i\}$ forms a partition of A, that for all i, j, $L_{p_i} \cap L_{\overline{p_i}} = \emptyset$ and $(L_{p_i} \cup L_{\overline{p_i}}) \cap (L_{p_j} \cup L_{\overline{p_j}}) = \emptyset$. The sets of actions for a predicate are easily extended to sets of actions for a formula: $A_{\neg \varphi} = A_\varphi$, $A_{\varphi_1 \wedge \varphi_2} = A_{\varphi_1 \mathcal{U} \varphi_2} = A_{\varphi_1} \cup A_{\varphi_2}$.

The truth value of a formula is given in a three-valued logic matching our notion of test verdicts: a formula φ can be evaluated to *true* on a trace σ ($\sigma \models_T \varphi$), or it can be evaluated to *false* ($\sigma \models_F \varphi$), or its evaluation may remain inconclusive ($\sigma \models_I \varphi$).

The semantics for a formula φ is defined by three sets. The set of sequences that satisfy (resp. violate) the formula φ is noted $[\![\varphi]\!]^T$ (resp. $[\![\varphi]\!]^F$). We also note $[\![\varphi]\!]^I$ the set of sequences for which the satisfaction remains inconclusive.

$$- [\![p_i]\!]^T = \{\omega \mid \exists \omega', \omega'' \cdot \omega = \omega' \cdot \omega'' \wedge \omega'_{\downarrow A_{p_i}} \in L_{p_i}\}$$
$$ [\![p_i]\!]^F = \{\omega \mid \exists \omega', \omega'' \cdot \omega = \omega' \cdot \omega'' \wedge \omega'_{\downarrow A_{p_i}} \in L_{\overline{p_i}}\}$$
$$- [\![\neg\varphi]\!]^T = [\![\varphi]\!]^F$$
$$ [\![\neg\varphi]\!]^F = [\![\varphi]\!]^T$$
$$- [\![\varphi_1 \wedge \varphi_2]\!]^T = \{\omega \mid \exists \omega', \omega'' \cdot \omega = \omega' \cdot \omega'' \wedge \omega'_{\downarrow A_{\varphi_1}} \in [\![\varphi_1]\!]^T \wedge \omega'_{\downarrow A_{\varphi_2}} \in [\![\varphi_2]\!]^T\}$$
$$ [\![\varphi_1 \wedge \varphi_2]\!]^F = \{\omega \mid \exists \omega', \omega'' \cdot \omega = \omega' \cdot \omega'' \wedge \omega'_{\downarrow A_{\varphi_1}} \in [\![\varphi_1]\!]^F \vee \omega'_{\downarrow A_{\varphi_2}} \in [\![\varphi_2]\!]^F\}$$
$$- [\![\varphi_1 \mathcal{U} \varphi_2]\!]^T = \{\omega \mid \exists \omega_1, \ldots, \omega_n, \omega' \cdot \omega = \omega_1 \cdots \omega_n \cdot \omega'$$
$$\wedge \forall i < n \cdot \omega_{i \downarrow A_{\varphi_1}} \in [\![\varphi_1]\!]^T \wedge \omega_{n \downarrow A_{\varphi_2}} \in [\![\varphi_2]\!]^T\}$$
$$ [\![\varphi_1 \mathcal{U} \varphi_2]\!]^F = \{\omega \mid \exists \omega_1, \ldots, \omega_n, \omega' \cdot \omega = \omega_1 \cdots \omega_n \cdot \omega'$$
$$\wedge \Big(\forall i \leq n \cdot \omega_{i \downarrow A_{\varphi_2}} \in [\![\varphi_2]\!]^F \vee (\exists l \leq n \cdot \omega_{l \downarrow A_{\varphi_2}} \in [\![\varphi_2]\!]^T \wedge \exists k < l \cdot \omega_{k \downarrow A_{\varphi_1}} \in [\![\varphi_1]\!]^F) \Big)\}$$
$$- [\![\varphi]\!]^I = A^* \setminus ([\![\varphi]\!]^P \cup [\![\varphi]\!]^F)$$

Finally we note $\sigma \models_T \varphi$ (resp. $\sigma \models_F \varphi$, $\sigma \models_I \varphi$) for $\sigma \in [\![\varphi]\!]^T$ (resp. $\sigma \in [\![\varphi]\!]^F, \sigma \in [\![\varphi]\!]^I$).

3.2   Test Generation

Following the structural test generation principle given in Sect. 2.4, it is possible to obtain a *GenTest* function for our LTL-like logic. The *GenTest* definition can be made explicit simply by giving controller definitions. So, we give a graphical description of each controller used by *GenTest*. To simplify the presentation, the *stop* transitions are not represented: the receptions all lead from each state of the controller to some "sink" state corresponding to the nil process, and emissions are sent by controllers to stop sub-tests when their execution is not needed anymore for the verdict computation.

The $C_\neg (\{c\_start, c\_loop, c\_ver\}, \{c\_start', c\_loop', c\_ver'\})$ controller is shown on Fig. 5. It inverts the verdict received by transforming *pass* verdict into *fail* verdict (and conversely) and keeping *inconc* verdict unchanged.

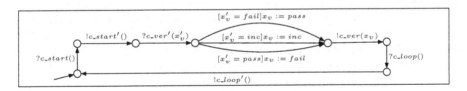

Fig. 5. The $C_\neg$ controller

The $C_\wedge (\{c\_start, c\_loop, c\_ver\}, \{c\_start_l, c\_loop_l, c\_ver_l\}, \{c\_start_r, c\_loop_r, c\_ver_r\})$ controller is shown on Fig. 6. It starts both controlled sub-tests and waits for their verdict returns, and sets the global verdict depending on received values.

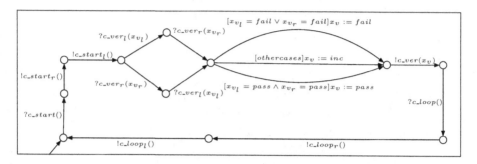

Fig. 6. The $C_\wedge$ controller

The $C_\mathcal{U}(\{c\_start, c\_loop, c\_ver\}, \{c\_start_l, c\_loop_l, c\_ver_l\}, \{c\_start_r, c\_loop_r, c\_ver_r\})$ controller is shown on Fig. 7 and Fig. 8. It is composed of three sub-processes executing in parallel and starting on the same action $?c\_start()$. The first sub-process C_m is represented on Fig. 7. The second and third ones corresponds to two instantiations

$$C_l(\{c\_start, c\_loop, c\_ver\}, \{c\_start_l, c\_loop_l, c\_ver_l\}),$$
$$C_r(\{c\_start, c\_loop, c\_ver\}, \{c\_start_r, c\_loop_r, c\_ver_r\})$$

of $C_x(\{c\_start, c\_loop, c\_ver\}, \{c\_start_x, c\_loop_x, c\_ver_x\})$ for the two controlled sub-test for the two sub-formulas. An algebraic expression of this controller could be

$$C_\mathcal{U}(\cdots) = (C_l(\cdots) \parallel C_r(\cdots)) \parallel_{\{r\_fail, l\_fail, r\_pass, l\_pass\}} C_m(\cdots)$$

One could understand C_l and C_r as two sub-controllers in charge of communicating with the controlled tests that send relevant information to the "main" sub-controller C_m deciding the verdict. The reception of an inconclusive verdict from a sub-test process interrupts the controller which emits an inconclusive verdict (not represented on the figure). If no answer is received from the sub-processes after some finite amount of time, then the tester delivers its verdict (*timeout* transitions). For the sake of clarity we simplified the controller representation. First, we represent the emission of the controller verdict and the return to the initial state under a reception of a loop signal ($?c\_loop()$) by a state which name represents the value of the emitted verdict. Second, we do not represent *inconc* verdict, the controller propagates it.

3.3   Soundness Proposition

We express that an abstract test case produced by the *GenTest* function is always *sound*, *i.e.* it delivers a *pass* (resp. *fail*) verdict when it is executed on a SUT behavior I only if the formula used to generate it is satisfied (resp. violated) on I. This proposition relies on one hypothesis, and two intermediate lemmas.

Hypothesis 1. *Each test case t_{p_i} associated to a predicate p_i is strongly sound in the following sense:*

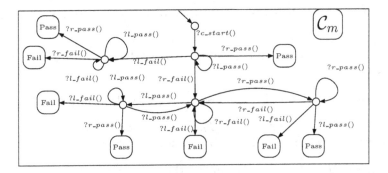

Fig. 7. The $\mathcal{C}_\mathcal{U}$ controller, the $\mathcal{C}_m$ part

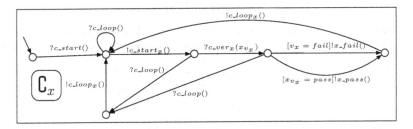

Fig. 8. The $\mathcal{C}_\mathcal{U}$ controller, the $\mathcal{C}_x$ part

$$\forall \sigma \in \mathrm{Exec}(t_{p_i} \otimes_{A_{p_i}} I), \mathrm{VExec}(\sigma) = pass \Rightarrow \sigma \models_T p_i$$
$$\forall \sigma \in \mathrm{Exec}(t_{p_i} \otimes_{A_{p_i}} I), \mathrm{VExec}(\sigma) = fail \Rightarrow \sigma \models_F p_i$$

The lemmas state that the verdict computed by t_φ on a sequence σ only depends on actions of σ belonging to A_φ.

Lemma 1. *All execution sequences with the same projection on a formula φ actions have the same satisfaction relation towards φ. That is:*

$$\forall \sigma, \sigma' \cdot \sigma_{\downarrow A_\varphi} = \sigma'_{\downarrow A_\varphi} \Rightarrow (\sigma \models_T \varphi \Leftrightarrow \sigma' \models_T \varphi) \wedge (\sigma \models_F \varphi \Leftrightarrow \sigma' \models_F \varphi)$$

Lemma 2. *For each formula φ, each sequence σ, the verdicts pass and fail of a sequence do not change if we project it on φ's actions. That is:*

$$\forall \varphi, \forall \sigma \cdot \sigma \models_T \varphi \Rightarrow \sigma_{\downarrow A_\varphi} \models_T \varphi$$
$$\forall \varphi, \forall \sigma \cdot \sigma \models_F \varphi \Rightarrow \sigma_{\downarrow A_\varphi} \models_F \varphi$$

These lemmas come directly from the definition of our logic and the controllers used in *GenTest*. Now we can formulate the proposition.

Theorem 1. *Let φ be a formula, and $t=GenTest(\varphi)$, S a LTS, $\sigma \in Exec(t \otimes_{A_\varphi} S)$ a test execution sequence, the proposition is:*

$$VExec(\sigma) = pass \Longrightarrow \sigma \models_T \varphi$$
$$VExec(\sigma) = fail \Longrightarrow \sigma \models_F \varphi$$

Sketch of the soundness proof. The proof is done by structural induction on φ. We give the proof for two cases.

For the predicates. The proof relies directly on predicate strong soundness (Hypothesis 1).

For the negation operator. Let suppose $\varphi = \neg\varphi'$. We have to prove that:

$$\forall \sigma \in \text{Exec}(GT(\neg\varphi', \mathcal{L}) \otimes_{A_\varphi} I), \text{VExec}(\sigma) = pass \Rightarrow \sigma \models_T \neg\varphi'$$
$$\forall \sigma \in \text{Exec}(GT(\neg\varphi', \mathcal{L}) \otimes_{A_\varphi} I), \text{VExec}(\sigma) = fail \Rightarrow \sigma \models_F \neg\varphi'$$

Let $\sigma \in \text{Exec}(GT(\neg\varphi', \mathcal{L}) \otimes_{A_\varphi} I)$ suppose that $\text{VExec}(\sigma) = pass$. By definition of GT,

$$GT(\neg\varphi', \mathcal{L}) = GT(\varphi', \mathcal{L}') \parallel_{\mathcal{L}'} C_\neg(\mathcal{L}, \mathcal{L}')$$

Since controller $C_\neg$ does not trigger the *c_loop* transition of its subtest when it is used as a main tester process, execution sequence σ is necessarily in the form:

$$c\_start() \cdot \sigma_I \cdot \sigma' \cdot \sigma_I \cdot$$
$$([x_v = pass]x_{v_g} := fail \mid [x_v = fail]x_{v_g} := pass \mid [x_v = inconc]x_{v_g} := inconc) \cdot \sigma_I \cdot c\_ver(x_{v_g})$$

with $\sigma' \in \text{Exec}(GT(\varphi', \mathcal{L}') \otimes_{A_{\varphi'}} I)$, σ_I denoting SUT's actions, and $\omega \cdot (a \mid b) \cdot \omega'$ denoting the sequences $\omega \cdot a \cdot \omega'$ and $\omega \cdot b \cdot \omega'$.

As the controller emits a *pass* verdict ($!c\_ver(x_{v_g})$ with x_{v_g} evaluated to *pass* in the $C_\neg$'s environment) it means that it necessarily received a *fail* verdict ($[x_v = fail]x_{v_g} := pass$) on c_ver' from the sub-test corresponding to $GT(\varphi', \mathcal{L}')$. So we have $\sigma' \in \text{Exec}(GT(\varphi', \mathcal{L}') \otimes_{A'_\varphi} I)$ and $\text{VExec}(\sigma') = fail$.

The induction hypothesis implies that $\sigma' \models_F \varphi'$. The Lemma 2 gives that $\sigma'_{\downarrow A_{\varphi'}} \models_F \varphi'$. And we have:

$$\sigma'_{\downarrow A_{\varphi'}} = \sigma'_{\downarrow A_\varphi} \quad (\forall \varphi, A_\varphi = A_{\neg\varphi})$$
$$= \sigma_{\downarrow A_\varphi} \quad (c\_start, \sigma_I \notin A_\varphi{}^*)$$

So $\sigma_{\downarrow A_\varphi} \models_F \varphi'$. We conclude using the Lemma 1 that $\sigma \models_T \varphi'$ that is $\sigma \models_T \neg\varphi'$. The proof for $\forall \sigma \in \text{Exec}(GT(\neg\varphi', \mathcal{L}) \otimes_{A_\varphi} I), \text{VExec}(\sigma) = fail \Rightarrow \sigma \models_F \neg\varphi'$ is similar.

Others operators. Proofs for the other operators follow the same principle and can be found in [9].

4   Java-CTPS

We now present Java-CTPS, a prototype of a testing framework tool for the Java environment which follows our approach. We just describe an abstract view of the tool. Interested readers can refer to [9] which contains a more detailed description.

Java-CTPS contains a test generator using the compositional approach for Java, *i.e.* the tester is generated in Java, for a SUT written in the same language. An interface is provided for the user to write a library of elementary test cases from the SUT interface. Indeed, the interface defines a set of methods that can be called. Elementary test cases are terms of our test calculus which external actions correspond to these methods: execution of an external action on the tester leads to a complete execution of the method on the SUT (from the call to the return). An elementary test case execution on the tester leads to the execution of some methods in the SUT interface.

Afterwards, using our method, the tool transforms a specification in a given formalism in a abstract test case. Then it is combined with the library to provide an executable test case.

Synthesis algorithms of controlled tests for different formalisms have been defined and implemented. Two interfaces are provided to the user: a command-line mode and a graphic interface. A simplified version of the test generation and execution is depicted on Fig. 9.

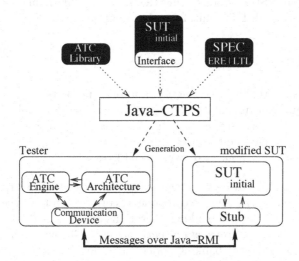

Fig. 9. Simplified working principle

Elementary test cases library establishment. The SUT's architecture description provides the set of controllable and observable actions on the system. The user can compose them, with respect to the test calculus, and write new elementary test cases. Programming is eased by the abstraction furnished by the SUT interface.

Specification as a set of requirements. Java-CTPS offers several formalisms to express requirements on the system.

- *Temporal logics.* Temporal logics [8] are frequently used to express specification for reactive systems. Their use has proved to be useful and appreciated for system verification. Our case studies have shown that many concepts in security policies are in the scope of these logics.

- *Regular Expression.* Regular expressions [10] allows to define behaviour schemes expressed on a system traces. They are commonly used and well-understood by engineers for their practical aspect.

Test of a system. Java-CTPS translates the specification into abstract test cases following the specification formula structure. Depending on the used specification formalism and the expressed requirement, the tool generates a test architecture whose test cases are coming from the controller library in accordance with *Gen-Test*. An execution engine is also generated. So, the generated tester can execute different test cases translated into a unique representation formalism on the test calculus engine. The initial SUT is also modified by adding a stub to communicate with the tester. This component provides means to launch method calls on the reception of specific signals. Thus, abstract test cases executing on the tester guide concrete test cases execution on the modified SUT. Communication between tester and SUT is done using the Java-RMI mechanism as we plan to support distributed SUT in a future evolution of our tool.

5   Case Study

We present a case study illustrating the approach presented above. From some credit card security documents [11], we established a security policy and a credit card model. We applied our method with the security policy as a partial specification and the executable credit card model as a SUT. The credit card model and part of its security policy are overviewed here.

The card. The architecture of the credit card is presented on Fig. 10. The interface is modeled by the *Device* component, corresponding to the possible action set on the card. Several banking operations are proposed, *e.g. provide_pin, change_pin, init_session, transaction, close_session*. Choice was made to use a Java interface to model the banking application one. The *Device* component interacts with a *Memory Abstraction* component providing, as its name indicates, some basic operations on the memory's areas. The *Memory* is just the credit card memory represented as a fixed size integer array.

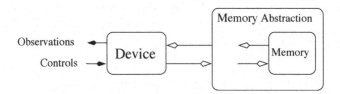

Fig. 10. The credit card architecture

The security policy. Our study allowed us to extract several security requirement specific to the credit card security domain. These requirements concerned several specification formalisms: regular expressions, and temporal logics. Some examples of properties that we were able to test can be expressed at an informal level:

1. After three failed authentications, the card is blocked, i.e. no action is permitted anymore.
2. If the card is suddenly removed the number of remaining authentications tries is set to 0.
3. No action is permitted without an identification.

For example one could see the first property formalised in our logic as several rules, one for each possible action:

$$try\_3\_authentications(all\_failed) \implies action(blocked)$$

The third one could be reasonably understood as:

$$action(blocked)\ \mathcal{U}\ authentication(success)$$

With this formalisation, these properties were tested with test cases that use elementary combinations of card interface actions. For example we wrote an abstract test case leading to three failed authentications using actions *provide_pin* and *init_session*.

6 Conclusion

In this work we have proposed a testing framework allowing to produce and execute test cases from a partial specification of the system under test. The approach we follow consists in generating the test cases from some high-level requirements on the expected system behaviour (expressed in a trace-based temporal logic), assuming that a concrete elementary tester is provided for each abstract predicate used in these requirements. This "partial specification" plays a similar role to the instrumentation directives currently used in run-time verification techniques, and we believe that they are easier to obtain in a realistic context than a complete operational specification. Furthermore, we have illustrated how this approach could be instantiated on a particular logic (an action-based variant of LTL-X), while showing that it is general enough to be applied to other similar trace-based logics. Finally, a prototype tool implementing this framework is available and preliminary experiments have been performed on a small case study.

Our main objective is now to extend this prototype in order to deal with larger examples. A promising direction is to investigate how the so-called MOP technology [6] could be used as an implementation platform. In particular, it already offers useful facilities to translate high-level requirements (expressed in various logics) into (passive) observers, and to monitor the behaviour of a program under test using these monitors. A possible extension would then be to replace these observers by our active basic testers (using the aspect programming techniques supported by MOP).

Acknowledgement. The authors thank the referees for their helpful remarks.

References

1. Hartman, A.: Model based test generation tools survey. Technical report, AGEDIS Consortium (2002)
2. van der Bijl, M., Rensink, A., Tretmans, J.: Action refinement in conformance testing. In: Khendek, F., Dssouli, R. (eds.) Testing of Communicating Systems (TESTCOM) LNCS, vol. 3205, pp. 81–96. Springer, Heidelberg (2005)
3. Darmaillacq, V., Fernandez, J.C., Groz, R., Mounier, L., Richier, J.L.: Test generation for network security rules. In: TestCom, pp. 341–356 (2006)
4. Falcone, Y., Fernandez, J.C., Mounier, L., Richier, J.L.: A test calculus framework applied to network security policies. In: Havelund, K., Núñez, M., Roşu, G., Wolff, B. (eds.) Formal Approaches to Software Testing and Runtime Verification. LNCS, vol. 4262, pp. 55–69. Springer, Heidelberg (2006)
5. Havelund, K., Rosu, G.: Synthesizing monitors for safety properties. In: Katoen, J.-P., Stevens, P. (eds.) ETAPS 2002 and TACAS 2002. LNCS, vol. 2280, pp. 342–356. Springer, Heidelberg (2002)
6. Chen, F., D'Amorim, M., Roşu, G.: Checking and correcting behaviors of java programs at runtime with java-mop. In: Workshop on Runtime Verification (RV'05), ENTCS, vol. 144(4), pp. 3–20 (2005)
7. Artho, C., Barringer, H., Goldberg, A., Havelund, K., Khurshid, S., Lowry, M., Pasareanu, C., Rosu, G., Sen, K., Visser, W., Washington, R.: Combining test case generation and runtime verification. Theor. Comput. Sci. 336(2-3), 209–234 (2005)
8. Manna, Z., Pnueli, A.: Temporal verification of reactive systems: safety. New York, Inc. Springer, Heidelberg (1995)
9. Falcone, Y., Fernandez, J.C., Mounier, L., Richier, J.L.: A partial specification driven compositional testing method and tool. Technical Report TR-2007-04, Vérimag Research Report (2007)
10. Kleene, S.C.: Representation of events in nerve nets and finite automata. In: Shannon, C.E., McCarthy, J. (eds.) Automata Studies, pp. 3–41. Princeton University Press, Princeton, New Jersey (1956)
11. Mantel, H., Stephan, W., Ullmann, M., Vogt, R.: Guideline for the development and evaluation of formal security policy models in the scope of itsec and common criteria. Technical report, BSI,DFKI (2004)

Nodes Self-similarity to Test Wireless Ad Hoc Routing Protocols

Cyril Grepet and Stephane Maag

Institut National des Télécommunications
CNRS SAMOVAR
9 rue Charles Fourier
F-91011 Evry Cedex
{Cyril.Grepet, Stephane.Maag}@int-evry.fr

Abstract. In this paper we present a new approach to test the conformance of a wireless ad hoc routing protocol. This approach is based on a formal specification of the DSR protocol described by using the SDL language. Test scenarios are automatically generated by a tool developed in our laboratory. A method enabling to execute them on an implementation into a real network is illustrated. Indeed, an important issue is to execute some generated test scenarios on a dynamic network in which the links are routinely modified. Therefore, the concept of self-similarity is presented to reduce the number of nodes by collapsing them in a real network. This enables to execute the test scenarios in defining a relationship between the network and specification topologies.

1 Introduction

A wireless mobile ad hoc network (MANET) is a collection of mobile nodes which are able to communicate with each other without relying on predefined infrastructures. In these networks, there is no administrative node and each node participates in the provision of reliable operations in the network. The nodes may move continuously leading to a volatile network topology with interconnections between nodes that are often modified. As a consequence of this infrastructureless environment, each node communicates using their radio range with open transmission medium and some of them behave as routers to establish multi-hop connections. Due to these aspects and the limited resources of the mobile nodes, efficient routing in ad hoc networks is a crucial and challenging problem for the quality of the communication systems.

From these unique characteristics of ad hoc networks, many requirements for routing protocol design are raised. Protocols can be classified mainly into three categories: the *proactive*, *reactive* and *hybrid* protocols. Classes such as *hierarchical*, *geographical* or *multicasting* protocols also emerge.

The techniques used by the ad hoc network experts to design and ensure the quality of their protocols essentially rely on descriptions for simulations and/or emulations. These methods provide an idea of the real behavior of the

A. Petrenko et al. (Eds.): TestCom/FATES 2007, LNCS 4581, pp. 123–137, 2007.

implemented protocol in a node. However, the testing coverage is rather low and is only restricted to the simulation context.

Formal description techniques and their testing tools are rarely applied in such kind of networks. The main reason is the difficulty to take into account the MANET protocol characteristics and the mobility of nodes in the test sequences generation and their execution. Therefore, our work focuses on a new testing technique to check the conformance of these ad hoc routing protocols. We present in this paper a formal specification of the Dynamic Source Routing (DSR) protocol from which we may generate some test scenarios. Nevertheless, the execution of these scenarios is currently an issue. Indeed there is often a gap between the dynamic topology designed in a specification and the one of a real case study. Therefore, we illustrate in this paper the concept of Node self-similarity in order to execute generated test scenarios on a real wireless ad hoc routing protocol taking into account the network topologies.

In the Section 2, we present the related works dealing with formal methods enabling to test ad hoc routing protocols. In Section 3, our DSR formal model is described and the conformance testing approach is depicted. Then, in Section 4, the concept of self-similarity to combine real nodes is illustrated and in Section 5 an example is developed. Finally we conclude the paper.

2   Related Works

Conformance testing for ad hoc routing protocols is crucial to the reliability of those networks. Paradoxically, few works currently exist on the formal specifications to analyse routing protocol testing [1]. The majority of these works rely on non-formal models provided as input to simulators such as NS-2 [2] or OpNet [3]. However, as is often noted, the simulation and emulation techniques do not replace a real case study [4]. Indeed, normal or ideal behaviors obtained by simulation may be proved erroneous in the real case. This is why formal description techniques are required to test this kind of protocols.

In [5] and [6], two routing protocols are validated using a testbed and analysing the network performances. But many constraints are applied and no formal model is used. In [7], a formal validation model named RNS (Relay Node Set) is studied. It is well suited to the conformance testing but denotes two drawbacks. First, only experts of the proposed languages may establish the tests which does not facilitate the test of other protocols. Secondly, only metrics of the protocol may be tested (overhead control, power consumption, etc.) and no basic functionalities (reception of a RteReply, etc.). Moreover, interactions between nodes may not be tested with this method.

A formal model using Distributed Abstract Machines allows to specify the LTLS protocol [8]. Nevertheless, the specifications are not executable and no functional analysis of the protocol is realized. In addition, the authors do not consider the testing process of an implementation from these models. Even if the syntax and the semantic are interesting, this formal description is still unusable for the conformance testing.

The game theory is also used in order to specify and analyse ad hoc routing protocols [9]. But two main inconveniences appear. First, non determinism is not allowed in this model and random behavior of nodes is not specified. Secondly the inherent constraints of this kind of networks are not considered. Indeed, a very strong assumption in this work is that every node needs to have a global knowledge of the network topology which is unusual in real case studies.

In our work we propose a new approach relying on well-known formal methods in order to enable conformance testing of ad hoc routing protocols.

3   Conformance Testing of an Ad Hoc Routing Protocol

Testing techniques can be divided in two categories: active testing which relies on stimulation and observation of an implementation, and passive testing which only observes the system without interactions [10], [11], [12]. Our research focuses on active testing of ad hoc routing protocols.

Conformance testing usually relies on the comparison between the behavior of an implementation and the formal specification of a given protocol i.e a conformed implementation has to behave as its specification.

The conformance testing procedure follows these steps :

- Step 1. Define a *testing architecture* with respect to the characteristics of the system under test and its possible implementations. This step could impact on each following step and has to be defined according to the context.
- Step 2. Make some *assumptions* that are sometimes required to enable the test.
- Step 3. Construct a precise *formal specification* of the system to be tested. This specification takes into account the system functionalities as well as the data specific to the test environment (test architecture, test interface, etc.).
- Step 4. *Select* the appropriate tests. This step corresponds to the definition of the test purposes. A test purpose can be a specific property of the system such as tasks or assignments with regard to values of variables, or the behavior of a specific component of the system taking into account the current values of the variables.
- Step 5. *Generate* the test scenarios. The test purposes are used as a guide by an algorithm based on simulation to produce the test scenarios. As a result, our algorithm computes a test scenario that can be applied to the implementation under test to verify the test purpose. A scenario is a sequence of interactions (between the system and the environment) that includes the interactions that represent a test purpose.
- Step 6. *Format* the test scenarios i.e to produce test scenarios in some accepted formalism as *Message Sequence Charts* (MSC), a formalism widely used in industry to describe message exchanges, or in *Testing and Test Control Notation* (TTCN), the ITU-TS standard language used for test specification.

Problematic. The main goal tackled in this paper is to provide a reliable method to test a routing protocol in a network in which we do not control neither the number of nodes nor the mobility scenario. Therefore, three main and relevant problems may be defined.

Mobility representation: The mobility in ad hoc networks implies that a specification has to represent more than one node communicating with each other. Thus the specification has to allow the creation or the suppression of a link between a pair of nodes in order to represent their mobility.

Test sequences generation: Another objective is to try to maximize the automation of the test sequences generation from classical tools avoiding the well-known state space explosion problem.

Test execution: The mobility of nodes and the hazard of radio communications can lead to many inconclusive verdicts or even to prevent the test. Moreover, the testing architecture to be used is also subject to the same problems. This problematic has to be study in order to provide a reliable verdict to the implementation conformity.

In the remaining of the paper we present our solutions to these problems with respect to the six steps aforementioned.

3.1 Testing Architecture

Some testing architectures are proposed by the ISO standard [13]. The coordinated, remote and distributed test architectures need reliable communications between the implementation under test (IUT) and the other components of the test. Due to the nodes mobility and in order to provide a general approach, we can not ensure that both sites can always communicate with each other. Due to the inherent constraints of ad hoc networks, a local testing architecture is chosen. We describe the different components of our architecture as follows. First, in order to observe packets, we need some Points of Observation (PO), whereas to observe and control these packets (if the IUT allows it (white/grey/black box testing)) we need Points of Control and Observation (PCO). These points are connected with the upper and lower testers (UT and LT) which are controlled by the test coordination procedure (TCP). The PO/PCO connected to UT aim to control the packets between the IUT and the upper layer (IP for example) whereas those connected to LT aim to control communications between the IUT and the lower layer (the link layer for instance). Each time one of the testers observes a packet, the TCP checks if it is the one expected regarding the specification. The testing architecture is depicted in Figure 1.

This architecture enables to observe and to control the message exchanges between the implementation and the strict upper and lower layers. In our case study (see section 5) the IUT is a DSR implementation. The UT controls the packet between DSR implementation and IP layer whereas the LT controls the communication between DSR implementation and the link layer.

3.2 Testing Assumptions

In order to execute test sequences in a real network and to provide reliable verdicts, some testing assumptions are required. First, we make the classical

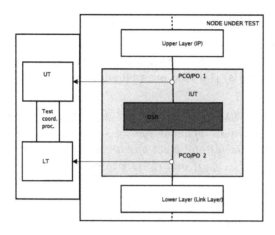

Fig. 1. Local test architecture applied on IUT

assumption that the implementation could be tested, i.e we can install some Points of Control (PCO) or Points of Observation (PO) (see section 3.1) into the IUT. The specific constraints of ad hoc networks imply that we have to make some assumptions to take into account the mobility of the nodes. Six hypothesis are defined:

1. *Existing destination nodes:* Each destination node (D) of the packets used by our test scenarios exists and is or will be connected by the network to the IUT.
2. *Connectivity of node D:* We assume that in a reasonable time, one or more paths will enable to execute the test scenarios between IUT and D.
3. *Stability of routes:* The routes which allow IUT and D to communicate with each other will remain stable during the execution of the test scenarios. This assumption is necessary to realize conformance testing and relies on the fact that an ad hoc network is created in order to allow communication for a community. If the communications are reliable enough for this purpose, we can suppose that the routes will allow to execute the test scenarios.
4. *Replay:* Despite the connectivity and stability assumptions, the test could sometimes fail or be inconclusive due to the radio or topological hazards. Thus, to avoid wrong decisions, we have to replay test scenarios before giving the verdict.
5. *Fail:* If the test is too many times "inconclusive" then we may consider that the test has failed. The testing replay number has to be decided according to the implementation and the test context.
6. *Implementation choices:* We assume that the implementation under test has the same options as the context. Capability testing techniques can be applied to check that aspect [14]. This assumption is necessary in conformance testing as well as in interoperability testing to prevent wrong decisions.

These assumptions allow to ensure a reliable testing environment by reducing wrong final verdicts.

3.3   Formal Specification

The third step is to formally describe the system to be tested. It is necessary to choose a formal model to reach this objective. We select the Extended Finite State Machines (EFSMs) [14] that are well adapted to describe communication protocols.

Definition 1. *An EFSM M is defined as : $M = (I, O, S, \overrightarrow{x}, T)$ with I, O, S, $\overrightarrow{x}$ and T respectively a set of input symbols, a set of output symbols, a set of states, a vector of variables and a set of transitions. Each transition $t \in T$ is a 6-tuple defined as : $t=(s_t, q_t, i_t, o_t, P_t, A_t)$ where*

- s_t *is the current state,*
- q_t *is the next state,*
- i_t *is an input symbol,*
- o_t *is an output symbol,*
- $P_t(\overrightarrow{x})$ *a predicate on the values of the variables,*
- $A_t(\overrightarrow{x})$ *an action on the variables.*

The language selected to provide the specification is the Specification and Description Language (SDL) standardized by ITU-T [15]. This is a widely used language to specify communicating systems and protocols, based on the semantic model of EFSM. Its goal is to specify the behavior of a system from the representation of its functional aspects. It allows to describe the architecture of the system i.e the connection and organization of the elements (blocks, processes, etc.) with the environment and between them. The behaviors of the entities in terms of their interactions with the environment and among themselves may also be designed. These interactions are described by tasks, transitions between states, and are based on the EFSMs.

All along our work, we assume that the conformance testing of a routing protocol could be performed in an unknown network topology. SDL allows to describe network topologies by using node instances, although it is impossible to guarantee that the topology of the real network will match the one of the specification. In that way, this work aims to reduce the number of nodes required to generate the test sequences and also to take into account their eventual mobility, according to the test objectives. The minimization of the specification helps to avoid the state space explosion problem. Nevertheless it is necessary to map it with the implementation in a real network.

This is the subject of our next section.

4   Self-similarity of Nodes

In this section we present the self-similarity of nodes used to reduce the number of nodes in the specification. Furthermore it allows the resulting test sequences to be executed on large class of real topologies. This approach takes into account the strong constraint of mobility in MANETs adapting the known self-similarity to this specific context as it is presented hereafter.

4.1   Definition of Self-similarity

Node self-similarity is inspired by [16] where fixed nodes in a wired network are merged in one single node with the main assumption that the communications are reliable between the combined nodes. In our work, we deal with wireless ad hoc nodes and unreliable links. We take into account these inherent constraints in the remaining of this work. First we have to define the combination of EFSMs.

Definition 2. *Let* $\{N_i\}_{i \in E}$ *where* $E \in [1..n]$ *and* $n \in \mathbb{N}$ *be a collection of model that can be described as EFSMs. We note* $N_1 \circ ... \circ N_n$ *the combination of all* N_i *defined as :*

$O(N) = \bigcup_{i \in E} O(N_i)$
$I(N) = \bigcup_{i \in E} I(N_i) - \bigcup_{i \in E} O(N_i)$
$S(N) = \Pi_{i \in E} S(N_i)$
$\overrightarrow{x}(N) = \Pi_{i \in E} \overrightarrow{x}(N_i)$
$T(N) = (s, s', e, o, P(\overrightarrow{x}), A(\overrightarrow{x}))$ *if* $(s_i, s'_i, e, o, P_i(\overrightarrow{x}), A_i(\overrightarrow{x})) \in T(N_i)$
where $P_i(\overrightarrow{x}) \equiv P_i(\overrightarrow{x_i})$, $A_i(\overrightarrow{x}) \equiv A_i(\overrightarrow{x_i})$, $e \in I(N_i)$ *and* $o \in O(N_i)$
Let $\Phi \subset O(N)$, *we define* $ActHide_\Phi(N)$ *as the obtained EFSM from* N *where each action of* Φ *becomes an internal one. This application transforms the communications between the different components of* N *into non-observable actions.*

Thus we can define the self-similarity of two nodes as :

Definition 3. *Let two possible actions for a node be* $send(Message, n, m)$ *and* $receive(Message, n', m')$ *where* n *(respectively* m'*) the observed node,* m *(respectively* n'*) the destination of the packet (respectively sender), and* $Message$ *the whole possible contents of a packet. Let* N *be a node specification. We note* $Tr(N)$ *the set of observable traces, a trace being an input/output sequence. Beside,* $Tr(N)$ *is a finite set, indeed the variable domains of the EFSM are discrete and finite (as most of the communication protocols).*

Some $N_{i \in I}$ *are self-similar if :*
$Tr(ActHide_\Phi(N_1 \circ N_2)) \subseteq Tr(N)$,
where $\Phi = \{send(Message, N_1, N_2), send(Message, N_2, N_1),$
$receive(Message, N_1, N_2), receive(Message, N_2, N_1)\}$

The self-similarity approach is easily applied in a wired network but due to mobility and the unreliable communications in a MANET, it can not be performed directly. The combination of mobile nodes is impossible, indeed the trace property could not exist if for example two consecutive nodes on a route can not communicate each other anymore.

Therefore, we use the self-similarity with three restrictions:

1. The self-similarity is applied from the point of view of a single node which is the IUT.
2. The self-similarity is applied each time a packet of the test sequences is received or sent in order to simplify the possible topologies known by the IUT.

3. The self-similarity is applied only for a specific communication on a defined route (not considering all communications in the network) between the IUT and another node.

The main idea, by reducing the number of nodes in the specification, is to identify all the different possible node behaviours according to the test purposes to define a minimal topology required and sufficient to generate test sequences for each test purpose. In order to illustrate our approach we choose the *Dynamic Source Routing* (DSR) protocol as a real case study.

5   A Case Study: DSR

5.1   Dynamic Source Routing Protocol

Dynamic Source Routing (DSR) is a reactive protocol that discovers and maintains routes between nodes on demand [17]. It relies on two main mechanisms, *Route Discovery* and *Route Maintenance*. In order to discover a route between two nodes, DSR floods the network with a *Route Request* packet. This packet is forwarded only once by each node after concatenating its own address to the path. When the targeted node receives the *Route Request*, it piggybacks a *Route Reply* to the sender and a route is established. Each time a packet follows an established route, each node has to ensure that the link is reliable between itself and the next node. DSR provides three successive steps to perform this maintenance: link layer acknowledgment, passive acknowledgment, and network layer acknowledgment. If a route is broken, the node which detects the failure sends by piggybacking a *Route Error* packet to the original sender.

5.2   DSR Formal Model

DSR is specified using SDL and the formal model describes the DSR draft 10 with the *Flow State Extension* (our specification is detailed in [18]). We do not specify all the possible features. Only some basic features for *Route Maintenance*, *Route Discovery* as the *Cached Route Reply* and all the main structures required by DSR (as the *Route Cache* or the *Send Buffer*) are described.

In order to represent the different links between nodes, we use a special block, called *Transmission* that contains a matrix where we define the connectivity in the networks. The matrix could be easily updated by sending information to create or remove a link. It means that we may modify dynamically the topology of the network in the purpose of representing the node mobility by changing their neighborhood. Details are given in [18]. Besides, we do not specify how to support multiple interfaces or security concepts.

5.3   Specification Reduction Using Nodes Self-similarity

First, a node may behave as a source S, a router N_i or a destination D. A route is defined as a succession of S, N_i $where$ $i \in [1..n]$ and D (Figure 2). We consider the nodes in the route from the point of view of S which is the IUT. Two possible cases arise during a communication between nodes on a particular route: either the communication between two successive nodes N_i and N_{i+1} succeeds, or it fails. We consider a communication as a success if a packet received by N_i is forwarded to N_{i+1} and forwarded after to N_{i+2} without provoking a *RteError* regardless of the meaning used for the acknowledgment. In the following N_o illustrates the abstract node defined as:

$$N_o = \begin{cases} N_\varnothing & \text{where } N_\varnothing \text{ is the neutral element.} \\ N_x \circ N_{x+1} & \text{where } N_x \in \{N_\varnothing\} \cup \bigcup_{i=1}^n \{N_i\}. \end{cases}$$

$N_\varnothing$ represents a node that only forwards the packets without modifying anything in the packet.

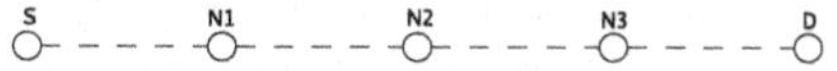

Fig. 2. A simple route between S and D

The process of nodes self-similarity may be illustrated as follows:

– *Transmission success:* If a transmission between N_i and N_{i+1} succeeds, we combine these two nodes in a new node N_o. The communications between N_i and N_{i+1} are considered as N_o internal actions. If the communication between N_o and N_{i+2} succeeds, we iterate the process and so on. Thus, in case that the packet from S reaches D without causing a *RteError*, we may combine all the intermediate nodes as illustrated in Figure 3.

Fig. 3. Combination by self-similarity when all communications succeed

– *Transmission failure:* If a communication fails between N_i and N_{i+1}, it means that all the previous communications have succeeded. So the nodes between N_1 and N_i are combined. Finally, all the nodes after N_{i+1}, including D have the same behavior for an observer placed on the IUT. We therefore combine all the nodes from N_{i+1} to D into a new node D (Figure 4).

We can note two exceptions: if the failure occurs between S and N_1 or S and D when a direct link exists.

However, from the point of view of S and with respect to our test sequences, the length of the main route does not matter. Indeed, a direct connection or a multi-hop route (except for a route sorting in the RouteCache) are observationnaly equivalent to S. Therefore, we handle the exceptions as it follows.

- If a direct link between S and D is broken, we introduce $N_o = N_\varnothing$ as a node on the route matching our specification described in section 5.4

Fig. 4. Combination by self-similarity when a communication fails

- if the broken link takes place between S and N_1, we have combined all the N_i and D in a new node D. Thus this leads to a similar situation than in the first case, we also introduce $N_o = N_\emptyset$ between S and the combined node D

The topology is reduced by using self-similarity in each case for the DSR protocol. The nodes self-similarity allows to represent a large class of topologies with a small number of nodes and to execute test sequences regardless of the number of intermediate nodes. Thereby we can reduce the number of nodes used in our specification in order to generate test scenarios which is an important issue in the testing activities especially for wireless networks.

5.4   Test Scenarios Equivalence

In order to generate adjustable test scenarios to a large class of topology taking into account the nodes mobility, we minimize our specification. Due to nodes self-similarity that allows to reduce a route between two nodes in a two-hop one, we can keep only the smallest number of nodes required to generate a test sequence according to specific test objectives. To test functional properties of DSR, no test objectives requiring more than two routes in the network were found out. Then, our specification can be reduced into four nodes, S, N_0, N_1 and D which compose two routes $[S, N_0, D]$ and $[S, N_1, D]$ as represented in Figure 5.

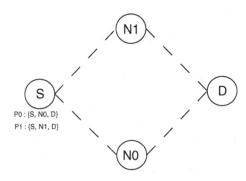

Fig. 5. Specification topology

Despite nodes self-similarity allows to reduce the length of route, the specification describes only two routes and the real network could have more than two between the IUT and D. The main idea is therefore to create a relation between the specification and the implementation by two sets $P0$ and $P1$ defined as:

Definition 4. *Definition of sets*

Let S_{spec} and D_{spec} be respectively the representations of S and D in the specification Spec and S_{imp}, D_{imp} their representations in the implementation Imp. Let $(p_n(x) \mid (x, n) \in \mathbb{N})$ be the nth route chosen by S_{imp} to reach D_{imp} and composed by x nodes.

– *In Spec:*
- $P0_{spec} = \{(S_{spec}, N0, D_{spec})\}$
- $P1_{spec} = \{(S_{spec}, N1, D_{spec})\}$
– *In Imp:*
- $p_n(x) \in P(n \bmod 2)_{imp}$ *i.e all the route with 0 or an even subscript will be in the set $P0_{imp}$ and those with an odd subscript will be in the set $P1_{imp}$. Thus, if there's a RouteError we preseve the route alternance between the set.*

All along the test execution, the *Test Coordination Procedure* (*TCP*, see section 3.1) will preserve a relation between $P0_{spec}$ and $P0_{imp}$, and also between $P1_{spec}$ and $P1_{imp}$. For instance, if the routing metric is "minimal hop count" (assumed in the rest of the paper), *TCP* will affect in $P0_{imp}$ the shortest path as $p_0(x)$, in $P1_{imp}$ the second as $p_1(y)$, in $P0_{imp}$ the third as $p_2(z)$ and so on with $x \leqslant y \leqslant z \leqslant ...etc.$ Both sets save the theoretical *RouteCache* in the *TCP*. Here "theoretical" is used because the *RouteCache* could eventually be filled in an other way during the *RouteDiscovery* mechanism. This problem will not be discussed in this paper because we focus here our study on functional properties.

With respect to *Spec*, $P0_{imp}$ and $P1_{imp}$ match possible routes described in the specification. For instance, if a test sequence implies that $P0_{spec}$ disappears: the *TCP* will detect the RouteError packet as an input, will erase the first element of $P0_{imp}$ i.e $p_0(x)$ and will select $p_1(y) \in P1_{imp}$ as the new route that *IUT* must use.

We have to underline the fact that a node N could belong to several routes (a *GratuitousRouteReply* sent to S by the node involved into different routes).

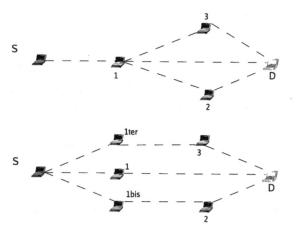

Fig. 6. Self-Similarity of a node involved in different routes

In this case, with respect to our sets, we have to duplicate this node *nbr* times where *nbr* is the number of routes containing N in the network representation of the *TCP*. Thus we create *nbr* routes in order to apply nodes self-similarity to each one and to split up these routes into our two sets $P0_{imp}$ and $P1_{imp}$ (Figure 6).

5.5   Experiment Context

Our approach is applied on a experimentation through the DSR-UU implementation [19]. The test sequences are provided by one of our tools TESTGEN-SDL [20] and some test purposes. Direct emulation is used. It allows to use a real implementation of a protocol stack with a simulator to represent the mobility and to manage the communications.

The direct emulation is performed on a focal machine with the following characteristics:

- Pentium©M 1,6 GHz
- 512 Mo Ram
- Fedora-2.6.15 kernel with skas patch
- TUN/TAP interfaces activated

We use *User Mode Linux* [21] to create virtual machines with existing prepared kernel and file system [22]. DSR-UU was added in the kernel.

NS-2 patched for emulation was performed to manage mobility and wireless communication between the virtual machines. We may note that a maximum of six virtual nodes is possible on a same focal machines. If we want a larger collection of nodes, it is necessary to distribute the virtual machines on more than one focal computer. The proposed emulation and testing architecture are depicted in Figure 7.

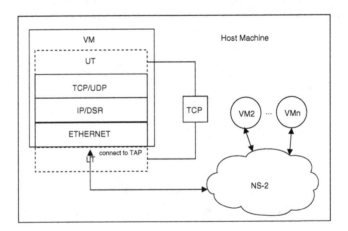

Fig. 7. Direct emulation and testing architecture

5.6   Sets Management for Unexpected Messages

In case of a broken link or detected unexpected *RouteError* from a node belonging to multiple routes, or unexpected *RouteError*, an inconclusive verdict could be obtained. The experimentations have shown that despite our assumptions, the number of inconclusive verdicts is important depending on the mobility. Thus, we automatize an error recovery for this kind of messages. An algorithm is defined to maintain the relation between $P0_{imp}$ and $P1_{imp}$ with the routes of *Spec* taking into account unexpected *RouteError* packets. A received *RouteError* could be characterized by two criterias:

1. expected/unexpected packet
2. the failure is/is not on the route used by the test scenario

---

Global Set Management Algorithm receiving a RouteError

Let $P_{imp} = P0_{imp} \cup P1_{imp}$ (w.r.t *spec*). Let P be a pointer in the TCP selecting the set containing the route of the test scenario and $\bar{P}$ its complement.
Let p be a pointer in the TCP on the first element of P.
Let lr be a broken link built from the address couple (i,j) of identified nodes carried by the RouteError packet.
Let $a = 1$ if the RouteError is expected else 0

1. **If** $lr \in p$ et $a = 1$ **then**
   (a) $P_{imp} := P_{imp} \setminus \{p_n(x) | lr \in p_n(x)\}$.
   (b) **if** $P = P0_{imp}$ **then** index each route by $n \in [1..m]$ **else** $n \in [0..m-1]$, with m the number of known routes.
   (c) To build sets $P0_{imp}$ and $P1_{imp}$.
   (d) $P := \bar{P}$.
   (e) $p := P(1)$ the first element of P.
   (f) To continue the test if it is possible.(*)
2. **If** $lr \in p$ and $a = 0$ **then**
   (a) $P_{imp} := P_{imp} \setminus \{p_n(x) | lr \in p_n(x)\}$.
   (b) **if** $P = P0_{imp}$ **then** index each route by $n \in [0..m-1]$ **else** $n \in [1..m]$, with m the number of known routes.
   (c) To build sets $P0_{imp}$ and $P1_{imp}$.
   (d) $P := P$.
   (e) $p := P(1)$.
   (f) To restart the test one step before(*)
3. **If** $lr \notin p$ and a **then**
   (a) $P_{imp} := P_{imp} \setminus \{p_n(x) | lr \in p_n(x)\}$.
   (b) **if** $P = P0_{imp}$ **then** index each route by $n \in [0..m-1]$ **else** $n \in [1..m]$, with m the number of known routes.
   (c) To build $P0_{imp}$ and $P1_{imp}$.
   (d) $P := P$.
   (e) $p := p$.
   (f) To continue the test.

(*) i.e the previous SourceRoute sent. If the test scenario needs a route to send a message and $P_{imp} = \varnothing$, an inconclusive verdict will be obtained.

---

6 Conclusion

In this paper we present a new approach to test the conformance of a wireless ad hoc routing protocol, namely DSR. This approach is based on a formal specification of the protocol described in SDL. This work has as a main advantage to formally test such kind of protocols and to take into account the nodes mobility and the network volatility aspect as well. Test scenarios are automatically generated by a tool developed in our laboratory and a method is illustrated enabling to execute them on a real implementation into a real network. This technique called the nodes self-similarity allows to bridge the gap between the dynamic topologies of a real network and the ones of the specification. This allows to reduce the number of nodes in a specification in order to generate the sequences and then avoiding the eventual state space explosion. The main advantage of this method is the possibility to execute a test sequence generated from a usable specification on an implementation running in a real mobile ad hoc network. An algorithm is depicted in order to illustrate the relationship during the testing process between the tester, the specification and the IUT. Finally, an application with an emulator is illustrated in which only four nodes are necessary to generate some test scenarios.

References

1. Obradovic, D.: Formal Analysis of Routing Protocols. PhD thesis, University of Pennsylvania (2002)
2. NS: The network simulator (2004) http://www.isi.edu/nsnam/ns
3. OPNet: The opnet modeler (2005) http://www.opnet.com/products/modeler/home.html
4. Bhargavan, K., Gunter, C., Lee, I., Sokolsky, O., Kim, M., Obradovic, D., Viswanathan, M.: Verisim: Formal analysis of network simulations. IEEE Trans. Softw. Eng. 28(2), 129–145 (2002)
5. Yi, Y., Park, J.S., Lee, S., Lee, Y., Gerla, M.: Implementation and validation of multicast-enabled landmark ad-hoc routing (m-lanmar) protocol. In: IEEE MILCON'03 (2003)
6. Bae, S., Lee, S.J., Gerla, M.: Multicast protocol implementation and validation in an ad hoc network testbed. In: Proc. IEEE ICC, pp. 3196–3200 (2001)
7. Lin, T., Midkiff, S., Park, J.: A framework for wireless ad hoc routing protocols. In: Proc. of IEEE Wireless Communications and Networking Conf (WCNC) (2003)
8. Glasser, U., Gu, Q.P.: Formal Description and Analysis of a Distributed Location Service for Mobile Ad Hoc Networks. Frazer Univ. (2003)
9. Zakkiudin, I.: Towards a game theoretic understanding of ad-hoc routing. In: ENTCS. vol. 119, pp. 67–92 (2005)
10. Lee, D., Chen, D., Hao, R., Miller, R., Wu, J., Yin, X.: A formal approach for passive testing of protocol data portions. In: Proceedings of the IEEE International Conference on Network Protocols, ICNP'02 (2002)
11. Alcalde, B., Cavalli, A., Chen, D., Khuu, D., Lee, D.: Network protocol system passive testing for fault management - a backward checking approach. In: de Frutos-Escrig, D., Núñez, M. (eds.) FORTE 2004. LNCS, vol. 3235, pp. 150–166. Springer, Heidelberg (2004)

12. Arnedo, J., Cavalli, A., Nunez, M.: Fast testing of critical properties through passive testing. In: Hogrefe, D., Wiles, A. (eds.) TestCom 2003. LNCS, vol. 2644, pp. 295–310. Springer, Heidelberg (2003)
13. ISO: information technology - Open Systems Interconnections - Conformance testing methodology and framework (1992)
14. Lee, D., Yannakakis, M.: Principles and methods of testing finite state machines - a survey. IEEE Transactions on Computers 84, 1090–1123 (1996)
15. ITU-T: Recommandation Z.100: CCITT Specification and Description Language (SDL) Technical report, ITU-T (1999)
16. Djouvas, C., Griffeth, N., Lynch, N.: Using self-similarity for effecient network testing. Technical report, Lehman College (2005)
17. Johnson, D., Maltz, D., Hu, Y.C.: The Dynamic Source Routing Protocol for Mobile Ad Hoc Networks (DSR) - Experimental RFC. IETF MANET Working Group (2004) http://www.ietf.org/internet-drafts/draft-ietf-manet-dsr-10.txt
18. Maag, S., Grepet, C., Cavalli, A.: Un Modèle de Validation pour le Protocole de Routage DSR. In: Hermes, (ed.) CFIP 2005, pp. 85–100. Bordeaux, France (2005)
19. Nordstrom,   E.:   Dsr-uu   v0.1.   Uppsala   University, http://core.it.uu.se/core/index.php/DSR-UU
20. Cavalli, A., Lee, D., Rinderknecht, C., Zaidi, F.: Hit-or-jump: An algorithm for embedded testing with application to IN services. In: Wu, J., Chanson, S., Gao, Q. (eds.) Formal Method for Protocol Engineering and Distributed Systems, FORTE XII/PSTV XIX'99. IFIP Conference Proceedings, Beijing, China, vol. 156, Kluwer, Dordrecht (1999)
21. Dike, J.: user-mode-linux, http://user-mode-linux.sourceforge.net/
22. Wehbi, B.: Dynamic remote access solution to a hot-zone. Master's thesis, Université Pierre et Marie Curie (2005)

Testing and Model-Checking Techniques for Diagnosis

Maxim Gromov[1] and Tim A.C. Willemse[2]

[1] Institute for Computing and Information Sciences (ICIS)
Radboud University Nijmegen – The Netherlands
m.gromov@cs.ru.nl
[2] Design and Analysis of Systems Group,
Eindhoven University of Technology – The Netherlands
t.a.c.willemse@tue.nl

Abstract. Black-box testing is a popular technique for assessing the quality of a system. However, in case of a test failure, only little information is available to identify the root-cause of the test failure. In such cases, additional diagnostic tests may help. We present techniques and a methodology for efficiently conducting diagnostic tests based on explicit fault models. For this, we rely on *Model-Based Testing* techniques for *Labelled Transition Systems*. Our techniques rely on, and exploit differences in outputs (or inputs) in fault models, respectively. We characterise the underlying concepts for our techniques both in terms of mathematics and in terms of the modal μ-calculus, which is a powerful *temporal logic*. The latter characterisations permit the use of efficient, off-the-shelf model checking techniques, leading to provably correct algorithms and pseudo decision procedures for diagnostic testing.

1 Introduction

Testing has proved to be a much-used technique for validating a systems behaviour, but in itself it is a quite labour-intensive job. Formal approaches to testing, collectively known as *Model-Based Testing*, have been touted as effective means for reducing the required effort of testing by allowing for automation of many of its aspects. However, MBT provides only a partial answer to the validation problem, as in most cases its automation ceases at the point where a test failure has been detected; pinpointing the root-cause of the test failure remains a laborious and time-consuming task. Finding this root-cause is known as the *diagnosis* problem, and it is tightly linked to testing.

Formal approaches to the diagnosis problem rely on the use of models of the system-under-diagnosis, and are often referred to as *Model-Based Diagnosis* techniques. While MBD has been studied extensively in the formal domain of *Finite State Machines* (see e.g. [3,4,6,11]), the topic is little studied in the setting of Labelled Transition Systems. The advantage of many LTS-based theories over FSM-based theories is that the assumptions under which they operate are more liberal, which makes them easier to apply in practice. In this paper, we advocate an LTS-based MBD approach for non-deterministic, reactive systems. The techniques that we put forward in this paper operate under the liberal LTS-based testing hypothesis of **ioco**-based testing [13]; our methods rely on explicit models describing the faulty behaviour, henceforth referred to as *fault models*.

A. Petrenko et al. (Eds.): TestCom/FATES 2007, LNCS 4581, pp. 138–154, 2007.

The problem that we consider consists of identifying "correct" fault models among a given (but large) set of possible fault models. By "correct", we understand that no evidence of a mismatch between the malfunctioning system and the fault model can be found. This can be asserted by e.g. testing. Note that even though this problem is readily solved by testing the malfunctioning system against each fault model separately, this is a daunting task which is quite expensive in terms of resources, even when fully automated. The main contributions of this paper are twofold:

1. inspired by classical FSM-based diagnosis approaches we present diagnostic concepts and techniques to make the fault model selection process more effective in an LTS-based setting. In particular, we adopt and modify the notion of *distinguishability* (see e.g. [11]) from FSMs to fit the framework of LTSs. Secondly, we introduce a novel notion, called *orthogonality* which helps to direct test efforts onto isolated aspects of fault models. Both notions are studied in the setting of **ioco**-based testing.
2. we link our diagnostic concepts and techniques to *model-checking* problems. This gives rise to effective and provably correct automation of our approach, and leads to a better understanding of all involved concepts.

Note that the problem of constructing the set of fault models is left outside the scope of this paper; in general, there are an infinite number of fault models per implementation. While this is indeed a very challenging problem, for the time being, we assume that these have been obtained by manually modifying e.g. a given specification, based on modifications of subcomponents of the specifications. Such modifications can be driven by the observed non-conformance between the specification and the implementation, but also fault injection is a good strategy.

Related work. In [8], Jéron *et al* paraphrase the diagnosis problem for discrete event systems (modelled by LTSs), as the problem of finding whether an observation of a system contains forbidden sequences of actions. Their approach takes a description of the structure of a system as input; the sequences of forbidden actions are specified using patterns. They subsequently propose algorithms for, a.o., synthesising a diagnoser which tells whether or not a pattern occurred in the system. A variation on this approach is given in [10], in which all actions are unobservable except for special "warning" actions. The problem that is solved is finding explanations for the observations of observed warning actions. Both works view the diagnosis problem as a supervisory problem.

Apart from the above mentioned works in the setting of LTSs, there is ample literature on diagnosis based on FSMs. Guo *et al*, in [6] focus on heuristics for fault diagnosis, which helps to reduce the cost of fault isolation and identification. El-Fakih *et al* [4] define a diagnostic algorithm for nets of FSMs, and in [3] these techniques are extended; the effectiveness of (a minor modification of) that algorithm is assessed in [5]. Most FSM-based approaches consist of two steps, the first step being the generation of a number of candidate fault models (often referred to as *candidates*), and the second step being a selection of appropriate candidates. The first step relies on strict assumptions, which in general are not met in an LTS-based setting.

In [12] the emphasis is on diagnosing non-reactive systems, mostly hardware, although their techniques have also been applied to software. Based on the topology of a system, explanations for a system's malfunctioning are computed and ranked according to likeliness. The techniques underlying the diagnosis are based on propositional logic and satisfiability solvers.

Structure of the paper. In Section 2 we repeat the **ioco**-based testing theory and the modal μ-calculus [2], the latter being our carrier for linking diagnosis to the problem of model-checking. The basic concepts for diagnosis, and their link to model-checking problems is established in Section 3. In Section 4, we provide an algorithm and a semi-decision procedure that implement the techniques and concepts of Section 3.

2   Background

In this section, we briefly recall the testing theory **ioco** as defined in [13]. The **ioco** framework and its associated testing hypotheses serve as the basic setting for our diagnosis techniques. Furthermore, we introduce the modal μ-calculus [2], which is a modal logic that we will use as a tool for characterising our diagnostic techniques.

Definition 1. *A Labelled Transition System (LTS) with inputs Act_I and outputs Act_U is a quintuple $\langle S, Act_I, Act_U, \rightarrow, \bar{s} \rangle$, where S is a non-empty set of states with initial state $\bar{s} \in S$; Act_I and Act_U are disjoint finite sets representing the set of input actions and output actions, respectively. We denote their union by Act. As usual, $\tau \notin Act$ denotes an internal non-observable action, and we write Act_τ for $Act \cup \{\tau\}$. The relation $\rightarrow \subseteq S \times Act_\tau \times S$ is the transition relation.*

Let $L = \langle S, Act_I, Act_U, \rightarrow, \bar{s} \rangle$ be a fixed LTS. Let $s, s', \dots$ range over S. Throughout this paper, we use the following conventions: for all actions a, we write $s \xrightarrow{a} s'$ iff $(s, a, s') \in \rightarrow$, and $s \xnrightarrow{a}$ iff for all $s' \in S$, not $s \xrightarrow{a} s'$.

Ioco-*based testing theory.* The notion of *quiescence* is added to an LTS as follows: a state s is quiescent — notation $\delta(s)$ — iff $s \xnrightarrow{\tau}$ and for all $a \in Act_U$, $s \xnrightarrow{a}$. Informally, a quiescent state is a state that is "stable" (it does not allow for internal activity) and it refuses to provide outputs. Let $\delta \notin Act_\tau$ be a fresh label representing the possibility to observe quiescence; Act_δ abbreviates $Act \cup \{\delta\}$. Let $\sigma, \sigma', \dots$ range over Act_δ^*, actions a range over Act_δ, and $S', S'', \dots \subseteq S$. We generalise the transition relation $\rightarrow$ to $\Longrightarrow \subseteq S \times Act_\delta^* \times S$, and write $s \xLongrightarrow{\sigma} s'$ iff $(s, \sigma, s') \in \Longrightarrow$. We define $\Longrightarrow$ as the smallest relation satisfying the following four rules:

$$\frac{}{s \xLongrightarrow{\epsilon} s} \qquad \frac{s \xLongrightarrow{\sigma} s' \quad s' \xrightarrow{\tau} s''}{s \xLongrightarrow{\sigma} s''} \qquad \frac{s \xLongrightarrow{\sigma} s' \quad s' \xrightarrow{a} s''}{s \xLongrightarrow{\sigma \cdot a} s''} \qquad \frac{s \xLongrightarrow{\sigma} s' \quad \delta(s')}{s \xLongrightarrow{\sigma \cdot \delta} s'}$$

Analogously to $\rightarrow$, we write $s \xLongrightarrow{\sigma}$ for $s \xLongrightarrow{\sigma} s'$ for some s'. For ease of use, we introduce the following functions and operators.

1. $[s]_\sigma \stackrel{\text{def}}{=} \{s' \in S \mid s \xLongrightarrow{\sigma} s'\}$; generalised: $[S']_\sigma \stackrel{\text{def}}{=} \bigcup_{s \in S'} [s]_\sigma$;
2. $\mathbf{out}(s) \stackrel{\text{def}}{=} \{a \in Act_U \mid s \xrightarrow{a}\} \cup \{\delta \mid \delta(s)\}$; generalised: $\mathbf{out}(S') \stackrel{\text{def}}{=} \bigcup_{s \in S'} \mathbf{out}(s)$,
3. $s\text{-}traces(s) \stackrel{\text{def}}{=} \{\sigma \in Act_\delta^* \mid s \xLongrightarrow{\sigma}\}$,
4. $traces(s) \stackrel{\text{def}}{=} s\text{-}traces(s) \cap Act^*$,
5. $\mathbf{der}(s) \stackrel{\text{def}}{=} \bigcup_{\sigma \in Act^*} [s]_\sigma$; generalised: $\mathbf{der}(S') \stackrel{\text{def}}{=} \bigcup_{s \in S'} \mathbf{der}(s)$.

Note 1. Our notation $[S']_\sigma$ is a deviation from the standard **ioco**-notation, where $[S']_\sigma$ is written as S' **after** σ. While we are not in favour of changing common notation, our main motivation for using our notation is brevity in definitions, theorems and algorithms, in support of readability.

Definition 2. *We say that:*

- *L is* image finite *if for all $\sigma \in Act^*$, $[\bar{s}]_\sigma$ is finite,*
- *L is* deterministic *if for all $s' \in S$ and all $\sigma \in Act^*$, $|[s']_\sigma| \leq 1$,*
- *L is* strongly converging *if there is no infinite sequence of τ transitions,*
- *A state $s \in S$ is* input-enabled *if for all $s' \in \mathbf{der}(s)$ and all $a \in Act_I$, we have $s' \overset{a}{\Longrightarrow}$. L is* input-enabled *if $\bar{s}$ is input-enabled.*

Throughout this paper, we restrict to image finite, strongly converging LTSs. The *testing hypothesis* for **ioco** states that *implementations* can be modelled using *input-enabled* LTSs. Note that this does not imply that the theory requires that this LTS is known. The *conformance* relation **ioco** is defined as follows:

Definition 3. *Let $L_i = \langle S_i, Act_I, Act_U, \rightarrow_i, \bar{s}_i \rangle$ (for $i = 1, 2$) be two LTSs. Let $s_1 \in S_1$ and $s_2 \in S_2$. Then s_1 is* **ioco***-conforming to s_2 – notation s_1* **ioco** s_2 *– when s_1 is input-enabled and*

$$\forall \sigma \in \textit{s-traces}(s_2) : \mathbf{out}([s_1]_\sigma) \subseteq \mathbf{out}([s_2]_\sigma)$$

We sometimes write L_1 **ioco** L_2 *instead of $\bar{s}_1$* **ioco** $\bar{s}_2$.

Note that *proving* **ioco**-conformance is generally not feasible, as there is no guarantee that we have seen all the behaviours of an implementation (because of non-determinism). In practice, we settle for *confidence* in **ioco**-conformance, which is obtained by testing the implementation with a large set of successfully executed test-cases. A sound and complete algorithm for **ioco** for deriving test-cases from a specification is proved correct in [13]; it is implemented in e.g. TorX [1] and TGV [7].

Modal μ-calculus. The modal μ-calculus is a powerful logic which can be used to express complex temporal properties over dynamic systems. Next to its modal operators $\langle a \rangle \phi$ and $[a]\phi$, it is equipped with least and greatest fixpoint operators. The grammar for the modal μ-calculus, given directly in positive form is as follows:

$$\phi ::= \mathtt{tt} \mid \mathtt{ff} \mid X \mid \phi \wedge \phi \mid [a]\phi \mid \langle a \rangle \phi \mid \phi \vee \phi \mid \mu X.\phi \mid \nu X.\phi$$

where $a \in Act_\tau$ is an action and X is a propositional variable from a set of propositional variables $\mathcal{X}$. A formula ϕ is said to be in *Positive Normal Form* (PNF) if all its propositional binding variables are distinct. We only consider formulae in PNF. A formula ϕ is interpreted relative to an LTS $L = \langle S, Act_I, Act_U, \rightarrow, \bar{s} \rangle$ and a *propositional environment* $\eta : \mathcal{X} \rightarrow 2^S$ that maps propositional variables to sets of states. The semantics of ϕ is given by $[\phi]_\eta^L$, which is defined as follows:

$$
\begin{aligned}
[\text{tt}]_\eta^L &= S \\
[\text{ff}]_\eta^L &= \emptyset \\
[\phi_1 \wedge \phi_2]_\eta^L &= [\phi_1]_\eta^L \cap [\phi_2]_\eta^L \\
[\phi_1 \vee \phi_2]_\eta^L &= [\phi_1]_\eta^L \cup [\phi_2]_\eta^L \\
[X]_\eta^L &= \eta(X) \\
[[a]\phi]_\eta^L &= \{s \in S \mid \forall s' \in S : s \xrightarrow{a} s' \Rightarrow s' \in [\phi]_\eta^L\} \\
[\langle a \rangle \phi]_\eta^L &= \{s \in S \mid \exists s' \in S : s \xrightarrow{a} s' \wedge s' \in [\phi]_\eta^L\} \\
[\mu X.\phi]_\eta^L &= \bigcap \{S' \subseteq S \mid [\phi]_{\eta[S'/X]}^L \subseteq S'\} \\
[\nu X.\phi]_\eta^L &= \bigcup \{S' \subseteq S \mid S' \subseteq [\phi]_{\eta[S'/X]}^L\}
\end{aligned}
$$

where we write $\eta[S'/X]$ for the environment that coincides with η on all variables $Y \neq X$, and maps variable X to value S'. A state $s \in S$ satisfies a formula ϕ, written $s \models_L \phi$ when $s \in [\phi]_\eta^L$. We write $L \models \phi$ when $\bar{s} \models_L \phi$.

The operator $\langle a \rangle \phi$ is used to express that there must exist an a transition from the current state to a state satisfying ϕ. Dually, the operator $[a]\phi$ is used to express that *all* states that can be reached by executing an a action satisfy property ϕ. Remark that when an a transition is impossible in a state s, the property $[a]\phi$ is trivially satisfied in state s. These operators are well-understood and can be found in early logics such as Hennessy-Milner Logic. In this paper, we use the following additional conventions: for sets of actions A we define:

$$
[A]\phi \stackrel{\text{def}}{=} \bigwedge_{a \in A} [a]\phi \qquad\qquad\qquad \langle A \rangle \phi \stackrel{\text{def}}{=} \bigvee_{a \in A} \langle a \rangle \phi
$$

Moreover, for a formula ϕ, we denote its dual by $\overline{\phi}$. Such a dual formula always exists and is readily obtained by simple transformations and renamings, see e.g. [2].

The major source for the expressive power of the modal μ-calculus is given by the fixpoint operators μ and its dual ν. Technically, a least fixpoint $\mu X.\phi$ is used to indicate the smallest solution of X in formula ϕ, whereas the greatest fixpoint $\nu X.\phi$ is used for the greatest solution of X in formula ϕ. These fixpoint expressions are generally understood as allowing one to express *finite looping* and *looping*, respectively.

Example 1. A system that can always perform at least one action is said to be *deadlock-free* (note that we do not require this to be a visible action). This can be expressed in the modal μ-calculus using a greatest fixpoint: $\nu X. [\mathcal{Act}_\tau]X \wedge \langle \mathcal{Act}_\tau \rangle \text{tt}$. Informally, the formula expresses that we are interested in the largest set of states (say this would be $\overline{X}$) that satisfies the property that from each reachable state s ($s \in \overline{X}$), at least one action is enabled, and all enabled actions lead to states s' ($s' \in X$) that also have this property.

For a more detailed account we refer to [2], which provides an excellent treatment of the modal μ-calculus.

3   Techniques and Heuristics for Diagnostic Testing

Testing is a much used technique to validate whether an implementation conforms to its specification. Upon detection of a non-conformance, all that is available is a trace, also known as a *symptom*, that led to this non-conformance. Such a symptom is often insufficient for locating the root-cause (or causes) of the non-conformance; for this, often additional tests are required. We refer to these additional tests as *diagnostic tests*.

In a Model-Based Testing setting, the basis for conducting diagnostic tests is given by a set of *fault models*. Each fault model provides a possible, formal explanation of the behaviour of the implementation; one may consider it a possible specification of the faulty implementation. Remark that we here appeal to the testing hypothesis of **ioco**, stating that there is an input enabled LTS model for every implementation. The different fault models describe different fault situations. The diagnostics problem thus consists of selecting one or more fault model(s) from the given set of fault models that best explain the behaviour of the implementation.

Formally, the diagnostics problem we are dealing with is the following: given a specification S, a non-conforming implementation I and a non-empty set of fault models $F = \{F_1, F_2, \ldots, F_n\}$. A *diagnosis* of I is given by the largest set $D \subseteq F$ satisfying I **ioco** F_i for all $F_i \in D$. The focus of this paper is on two techniques for obtaining D efficiently, viz. distinguishability and orthogonality. Note that given the partiality of the **ioco**-relation, the fault models in D are –generally– all unrelated.

In Sections 3.1 and 3.2, we introduce the notions of (strong and weak) *distinguishability* and (strong and weak) *orthogonality*, respectively. We provide alternative characterisations of all notions in terms of modal logic, which 1) provides a different perspective on the technique and, 2) enables the use of efficient commonplace tool support. The discussion on how exactly the theory and results described in this section can be utilised for diagnostic testing is deferred to Section 4.

3.1 Distinguishability

Given two fault models F_1 and F_2 and an implementation I. Chances are that during naive testing for I **ioco** F_1 and I **ioco** F_2, there is a large overlap between the test-cases for F_1 and F_2, as both try to model to a large extent the same implementation. This means that F_1 and F_2 often agree on the outcome of most test-cases. An effective technique for avoiding this redundancy is to exploit the *differences* between F_1 and F_2. In particular, when, after conducting an experiment σ on I, F_1 and F_2 predict different outputs, this provides the opportunity to remove at least one of the two fault models from further consideration. When one or more such experiments exist, we say that the fault models are *distinguishable*. Two types of distinguishability are studied: weakly and strongly distinguishable fault models.

We next formalise the above concepts. At the root of the distinguishability property is the notion of an *intersection* of fault models. Intuitively, the intersection of two fault models contains exactly those behaviours that are shared among the two fault models.

Definition 4. *Let* $F_i = \langle S_i, Act_I, Act_U, \rightarrow_i, \bar{s}_i \rangle$, *for* $i = 1, 2$ *be two LTSs. Assume* $\Delta \notin Act$ *is a fresh constant, and denote* $Act_U \cup \{\Delta\}$ *by* Act_U^Δ. *Likewise,* Act^Δ. *The intersection of* F_1 *and* F_2, *denoted* $F_1 \| F_2$, *is again an LTS defined by* $\langle (2^{S_1} \setminus \emptyset) \times (2^{S_2} \setminus \emptyset), Act_I, Act_U^\Delta, \rightarrow, ([\bar{s}_1]_\epsilon, [\bar{s}_2]_\epsilon) \rangle$, *where* $\rightarrow$ *is defined by the following rules:*

$$\frac{\emptyset \neq q_1 \subseteq S_1 \quad \emptyset \neq q_2 \subseteq S_2 \quad a \in Act}{(q_1, q_2) \xrightarrow{a} ([q_1]_a, [q_2]_a)} \qquad \frac{\emptyset \neq q_1 \subseteq S_1 \quad \emptyset \neq q_2 \subseteq S_2}{(q_1, q_2) \xrightarrow{\Delta} ([q_1]_\delta, [q_2]_\delta)}$$

Remark that no transitions lead to, or start in an element $(q, \emptyset)$ or $(\emptyset, q)$ since these are no elements of the state-space of the intersection of two LTSs.

The intersection of two LTSs extends the alphabet of output actions of both LTSs with the symbol Δ. This action captures the synchronisation of both LTSs over the observations of quiescence, which in the **ioco**-setting is treated as an output of the system. A "true" quiescent state in the intersection of two LTSs indicates that the output actions offered by both LTSs are strictly disjoint. In order to facilitate the mapping between the sets Act_δ and Act^Δ, we use a *relabelling* function. Let $\mathfrak{R} : Act^\Delta \to Act_\delta$ be the following bijective function:

$$\mathfrak{R}(a) \overset{\text{def}}{=} a \text{ if } a \neq \Delta \text{ and } \delta \text{ otherwise}$$

We write $\mathfrak{R}^{-1}$ to denote the inverse of $\mathfrak{R}$. The mapping $\mathfrak{R}$ and its inverse extend readily over sets of actions. The extension of the mapping $\mathfrak{R}$ (and its inverse) over (sets of) traces, denoted by the mapping $\mathfrak{R}^*$ (resp. $\mathfrak{R}^{-1*}$), is defined in the obvious way.

Property 1. Let $F_1||F_2$ be the intersection of F_1 and F_2, and let s_1 be a state of F_1, s_2 be a state of F_2, (q_1, q_2) be a state of $F_1||F_2$ and $\sigma \in Act_\delta^*$. Then:

1. $F_1||F_2$ is deterministic,
2. $[([s_1]_\sigma, [s_2]_\sigma)]_a \neq \emptyset$ implies $([s_1]_{\sigma\mathfrak{R}(a)}, [s_2]_{\sigma\mathfrak{R}(a)}) \in [([s_1]_\sigma, [s_2]_\sigma)]_a$,
3. $\mathbf{out}(([q_1]_\epsilon, [q_2]_\epsilon)) \setminus \{\delta\} = \mathfrak{R}^{-1}(\mathbf{out}([q_1]_\epsilon) \cap \mathbf{out}([q_2]_\epsilon))$.

Some of the above properties should not come as a surprise: at the basis of the intersection operator is the *Suspension Automata* transformation of [13], which codes a non-deterministic specification into a deterministic LTS with explicit suspension transitions. That transformation is known to retain the exact same **ioco** testing power as the original specification, albeit on different domains of specification models.

Strong Distinguishability. Recall that the intersection $F_1||F_2$ codes the behaviours that are shared among the LTSs F_1 and F_2. This means that in states of $F_1||F_2$ that have no output transitions, both LTSs disagree on the outputs that should occur, providing the opportunity to eliminate at least one of the two fault models. We say that such a state is *discriminating*. If a tester always has a finite "winning strategy" for steering an implementation to such a discriminating state, the fault models are *strongly distinguishable*. Recall that testing is sometimes portrayed as a (mathematical) game in which the tester is in control of the inputs and the system is in control of the outputs. We next formalise the notion of strong distinguishability.

Definition 5. *The intersection $F_1||F_2 = \langle S, Act_I, Act_U^\Delta, \to, \bar{s} \rangle$ is said to be root-discriminating if there exists a natural number k, such that $\bar{s} \in \mathcal{D}_{F_1||F_2}(k)$, where $\mathcal{D}_{F_1||F_2} : \mathbb{N} \to 2^S$ is inductively defined by:*

$$\begin{cases} \mathcal{D}_{F_1||F_2}(0) & = \{t \in S \mid \mathbf{out}([t]_\epsilon) = \{\delta\}\} \\ \\ \mathcal{D}_{F_1||F_2}(n+1) & = \bigcap_{a \in Act_U^\Delta} \{t \in S \mid [t]_a \subseteq \mathcal{D}_{F_1||F_2}(n)\} \\ & \quad \cup \bigcup_{a \in Act_I} \{t \in S \mid \emptyset \neq [t]_a \subseteq \mathcal{D}_{F_1||F_2}(n)\} \end{cases}$$

A state $s \in \mathcal{D}_{F_1||F_2}(k)$ is referred to as a k-discriminating state. If it is clear from the context, we drop the subscript $F_1||F_2$ from the mapping $\mathcal{D}_{F_1||F_2}$. We say that fault models F_1 and F_2 are strongly distinguishable *iff $F_1||F_2$ is root-discriminating.*

Property 2. For all intersections $F_1||F_2$ and all $k \geq 0$, we have $\mathcal{D}(k+1) \supseteq \mathcal{D}(k)$.

Note that a state s is allowed to be $(k+1)$-discriminating if there is a strategy to move from state s to a state which is k-discriminating via some input, even though there are some outputs that would not lead to a k-discriminating state. This is justified by the fact that the implementations that we consider are input enabled. This means that they have to be able to accept inputs at all times, and input may therefore pre-empt possible output of a system. Strong distinguishability is preserved under **ioco**-conformance which means that if two fault models are strongly distinguishable, then also the implementations/refinements they model behave observably differently.

Property 3. Let F_1, F_2 be fault models, and let I_1, I_2 be implementations. If I_1 **ioco** F_1 *and I_2* **ioco** F_2 *and F_1 and F_2 are strongly distinguishable, then so are I_1 and I_2.*

Strong distinguishability can be characterised by means of a modal μ-calculus formula. The formal connection is established by the following theorem.

Theorem 1. *Let F_1, F_2 be two fault models. Then F_1 and F_2 are strongly distinguishable iff $F_1||F_2 \models \phi_{sd}$, where*

$$\phi_{sd} \stackrel{\text{def}}{=} \mu X. \, [\mathcal{A}ct_U^\Delta]X \vee \langle \mathcal{A}ct_I \rangle X$$

Weak Distinguishability. Strong distinguishability as a property is quite powerful, as it ensures that there is a testing strategy that inevitably leads to a verdict about one of the two fault models. However, it is often the case that there is no such fail-safe strategy, even though reachable discriminating states are present in the intersection. We therefore introduce the notion of *weak distinguishability*.

Definition 6. *Two fault models F_1, F_2 are said to be* weakly distinguishable *if and only if* $\mathbf{der}(F_1||F_2) \cap \mathcal{D}(0) \neq \emptyset$.

The problem of deciding whether two fault models are weakly distinguishable is a standard reachability property as testified by the following correspondence.

Theorem 2. *Let F_1, F_2 be two fault models. Then F_1 and F_2 are weakly distinguishable iff $F_1||F_2 \models \phi_{wd}$, where*

$$\phi_{wd} \stackrel{\text{def}}{=} \mu X. \, \langle \mathcal{A}ct^\Delta \rangle X \vee [\mathcal{A}ct_U^\Delta]\text{ff}$$

Unlike strong distinguishability, weak distinguishability is not preserved under **ioco**. This is illustrated by the following example:

Example 2. Let F_1 and F_2 be two fault models and let I be an implementation (see Fig. 1). Clearly, I **ioco** F_1 *and I* **ioco** F_2. *Moreover, F_1 and F_2 are weakly distinguishable, as illustrated by the trace $?b.!e$. However, I is clearly not weakly distinguishable from itself, as distinguishability is irreflexive.*

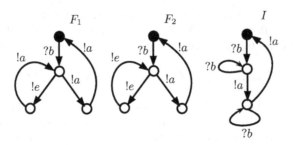

Fig. 1. Fault models F_1 and F_2 and implementation I

3.2 Orthogonality

Whereas distinguishability focuses on the differences in output for two given fault models, it is equally well possible that there is a difference in the specified inputs. Note that this is allowed in **ioco**-testing theory: a specification does not have to be input complete; this partiality with respect to inputs supports a useful form of underspecification. In practice, a fault hypothesis can often be tested by focusing testing effort on particular aspects. Exploiting the differences in underspecifications of the fault models gives rise to a second heuristic, called *orthogonality*, which we describe in this section. We start by extending the intersection operator of Def. 4.

Definition 7. *Let* $F_i = \langle S_i, Act_I, Act_U, \rightarrow_i, \bar{s}_i \rangle$*, for* $i = 1, 2$ *be two fault models. Assume* $\Theta = \{ \Theta_a \mid a \in Act_I \}$ *is a set of fresh constants disjoint from* Act^Δ*. We denote* $Act \cup \Theta$ *by* Act_Θ^Δ*. The orthogonality-aware intersection of* F_1 *and* F_2*, denoted* $F_1 \|_\Theta F_2$*, is an LTS defined by* $\langle (2^{S_1} \setminus \emptyset) \times (2^{S_2} \setminus \emptyset), Act_I^\Theta, Act_U^\Delta, \rightarrow, ([\bar{s}_1]_\epsilon, [\bar{s}_2]_\epsilon) \rangle$*, where* $\rightarrow$ *is defined by the two rules of Def. 4 in addition to the following two rules:*

$$\frac{\emptyset \neq q_1 \subseteq S_1 \quad \emptyset \neq q_2 \subseteq S_2 \quad [q_1]_a \neq \emptyset \quad [q_2]_a = \emptyset \quad a \in Act_I}{(q_1, q_2) \xrightarrow{\Theta_a} (q_1, q_2)}$$

$$\frac{\emptyset \neq q_1 \subseteq S_1 \quad \emptyset \neq q_2 \subseteq S_2 \quad [q_2]_a \neq \emptyset \quad [q_1]_a = \emptyset \quad a \in Act_I}{(q_1, q_2) \xrightarrow{\Theta_a} (q_1, q_2)}$$

Property 4. Let $F_1 \|_\Theta F_2$ be the orthogonality-aware intersection of F_1 and F_2, and let (q_1, q_2) be a state of $F_1 \|_\Theta F_2$. Then:

1. $F_1 \|_\Theta F_2$ is deterministic,
2. For all inputs $a \in Act_I$, $(q_1, q_2) \xrightarrow{a}$ implies $(q_1, q_2) \not\xrightarrow{\Theta_a}$.

Note that the reverse of Property 4, item 2 does not hold exactly because of the input incompleteness of fault models in general. Intuitively, the occurrence of a label Θ_a in the orthogonality-aware intersection models the fact that input a is specified by only one of the two LTSs and is left unspecified by the other LTS. The presence of such labels in the orthogonality-aware intersection are therefore pointers to the orthogonality of two systems. Once an experiment arrives in a state with an orthogonality label Θ_a, testing can focus on one of the two fault models exclusively. Any test failure that is subsequently found is due to the incorrectness of the selected fault model. We next formalise the notions of strong and weak orthogonality, analogously to distinguishability.

Definition 8. *Let* $F_1||_\Theta F_2 = \langle S, Act_I^\Theta, Act_U^\Delta, \rightarrow, \overline{s} \rangle$. F_1 *and* F_2 *are said to be* strongly orthogonal *if there is a natural number* k *such that* $\overline{s} \in \mathcal{O}_{F_1||_\Theta F_2}(k)$, *where* $\mathcal{O}_{F_1||_\Theta F_2}$: $\mathbb{N} \rightarrow 2^S$ *is inductively defined by:*

$$
\begin{cases}
\mathcal{O}_{F_1||_\Theta F_2}(0) &= \{t \in S \mid \exists a \in Act_I : t \xrightarrow{\Theta_a}\} \\
\mathcal{O}_{F_1||_\Theta F_2}(n+1) &= \bigcap_{a \in Act_U^\Delta} \{t \mid [t]_a \subseteq \mathcal{O}_{F_1||_\Theta F_2}(n) \wedge \exists a' \in Act_U : [t]_{a'} \neq \emptyset\} \\
&\quad \cup \bigcup_{a \in Act_I} \{t \mid \emptyset \neq [t]_a \subseteq \mathcal{O}_{F_1||_\Theta F_2}(n) \vee t \xrightarrow{\Theta_a}\}
\end{cases}
$$

The following theorem recasts strong orthogonality as a modal property.

Theorem 3. *Fault models* F_1 *and* F_2 *are strongly orthogonal iff* $F_1||_\Theta F_2 \models \phi_{so}$, *where*
$$
\phi_{so} \stackrel{\text{def}}{=} \mu X. (\langle Act_U^\Delta \rangle \text{tt} \wedge [Act_U^\Delta]X) \vee \langle Act_I \rangle X \vee \langle \Theta \rangle \text{tt}
$$

Analogously to distinguishability, we define a weak variation of strong orthogonality, which states that it is possible to reach a state in which an orthogonal label Θ_a for some a is enabled.

Definition 9. *Given* $F_1||_\Theta F_2 = \langle S, Act_I^\Theta, Act_U^\Delta, \rightarrow, \overline{s} \rangle$. F_1 *and* F_2 *are said to be* weakly orthogonal *iff* $\mathbf{der}(F_1||_\Theta F_2) \cap \mathcal{O}(0) \neq \emptyset$.

A recast of weak orthogonality into the μ-calculus is as follows.

Theorem 4. *Fault models* F_1 *and* F_2 *are weakly orthogonal iff* $F_1||_\Theta F_2 \models \phi_{wo}$, *where*
$$
\phi_{wo} \stackrel{\text{def}}{=} \mu X. \langle Act^\Delta \rangle X \vee \langle \Theta \rangle \text{tt}
$$

Orthogonality is not preserved under **ioco** conformance, which is illustrated by the following example.

Example 3. Let F_1 and F_2 be two fault models and let I_1 and I_2 be two implementations, depicted in Fig. 2. Clearly, I_1 **ioco** F_1 and I_2 **ioco** F_2. Moreover, F_1 and F_2 are (strongly and weakly) orthogonal, as illustrated by the trace $?b.?b$ which is applicable for F_1, but not applicable for F_2. However, I_1 and I_2 are not orthogonal. Note that by repeatedly executing experiment $?b.?b$ and subsequently observing output confidence in the correctness of (aspects of) F_1 can increase.

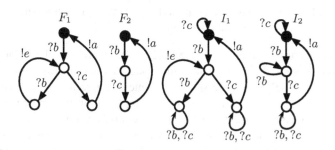

Fig. 2. Fault models F_1 and F_2 and implementations I_1 and I_2

4   Automating Diagnostic Testing

In the previous section we formalised the notions of *distinguishability* and *orthogonality*, both in terms of set-theory and modal logic. In this section, we rely on the latter results for defining provably correct algorithms for eliminating fault models and for isolating behaviours of fault models for further scrutiny.

First, we introduce the basic tools that we rely on for defining our on-the-fly diagnostic testing algorithms and semi-decision procedures. Then, in Section 4.2 we define the algorithms for strong distinguishability and orthogonality, and in Section 4.3, the semi-decision procedures for weak distinguishability and orthogonality are given.

4.1   Preliminaries

For the remainder of these sections, we assume that I is an implementation that we wish to subject to diagnostic testing, and $F_i = \langle S_i, Act_I, Act_U, \rightarrow_i, \bar{s}_i \rangle$, for $i = 1, 2$ are two given fault models. $F_1 ||_{(\Theta)} F_2 = \langle S, Act_I^{(\Theta)}, Act_U^{\Delta}, \rightarrow, \bar{s} \rangle$ is the (orthogonality-aware) intersection of F_1 and F_2. From this time forth, we assume to have the following four methods at our disposal:

1. Apply(a): send input action $a \in Act_I$ to an implementation,
2. Observe(): observe some output $a \in Act_U \cup \{\delta\}$ from an implementation,
3. Counterexample(L, ϕ): returns an arbitrary counterexample for $L \models \phi$ if one exists, and returns $\perp$ otherwise.
4. Counterexample$_s(L, \phi)$: returns one among possibly many shortest counterexamples for $L \models \phi$ if a counterexample exists, and returns $\perp$ otherwise.

We refer to [9] for an explanation of the computation of counterexamples for the modal μ-calculus. In our ordeals we assume that $\perp$ is a special character that we can concatenate to sequences of actions.

4.2   Strong Distinguishability and Strong Orthogonality

Suppose F_1 and F_2 are strongly distinguishable or orthogonal. Algorithm 1 derives and executes (on-the-fly) an experiment that (see also Theorem 5), depending on the input:

- allows to eliminate at least one fault model from a set of fault models, or
- isolates a fault model for further testing.

Recall that $\bar{\phi}$ denotes the dual of ϕ (see Section 2). Informally, the algorithm works as follows for strongly distinguishable fault models F_1 and F_2 (likewise for strongly orthogonal fault models): η is the shortest counterexample for F_1 and F_2 *not* being strongly distinguishable. The algorithm tries to replay η on the implementation, and recomputes a new counterexample when an output produced by the system-under-test does not agree with the output specified in the counterexample. When the counterexample has length 0, we can be sure to have reached a discriminating state, and observing output in this state eliminates at least one of the two considered fault models.

Algorithm 1. Algorithm for exploiting strong distinguishability/orthogonality

---

Require: $P \subseteq S$, $|P| = 1$, η is a shortest counterexample for $P \models \overline{\phi}_x$, $\phi_x \in \{\phi_{sd}, \phi_{so}\}$
Ensure: Returns a sequence executed on I.
  1: **function** $\mathcal{A}_1(P, \eta, \phi_x)$
  2:    **if** $\eta = \epsilon$ **then**
  3:        **if** $\phi_x = \phi_{sd}$ **then return** Observe();
  4:        **else choose** a **from** $\{y \in Act_I \mid [P]_{\ominus_y} \neq \emptyset\}$; **return** a;
  5:    **end if**
  6:   **else**   ▷ Assume $\eta \equiv e\,\eta'$ for some action e and sequence η'
  7:       **if** $e \in Act_I$ **then** Apply(e); **return** $e\,\mathcal{A}_1([P]_e, \eta', \phi_x)$;
  8:       **else** $a := $ Observe();
  9:          **if** $a = e$ **then return** $e\,\mathcal{A}_1([P]_e, \eta', \phi_x)$;
 10:          **else if** $\mathfrak{R}^{-1}(a) \in \mathbf{out}(P)$ **then**
 11:              **return** $a\,\mathcal{A}_1([P]_a, \mathfrak{R}^*(\texttt{Counterexample}_s([P]_a, \overline{\phi}_x)), \phi_x)$;
 12:          **else return** a;
 13:          **end if**
 14:       **end if**
 15:   **end if**
 16: **end function**

Theorem 5. *Let F_1 and F_2 be strongly orthogonal or strongly distinguishable fault models. Let $\phi = \phi_{sd}$ when F_1 and F_2 are distinguishable and let $\phi = \phi_{so}$ when F_1 and F_2 are orthogonal. Then algorithm $\mathcal{A}_1(\{\overline{s}\}, \texttt{Counterexample}_s(F_1\|_{\ominus}F_2, \overline{\phi}), \phi)$ terminates and the sequence $\sigma \equiv \sigma'\,a$ it returns satisfies:*

1. $a \in Act_\delta \backslash Act_I$ implies $\mathbf{out}([I]_{\sigma'}) \not\subseteq \mathbf{out}([F_1]_{\sigma'})$ or $\mathbf{out}([I]_{\sigma'}) \not\subseteq \mathbf{out}([F_2]_{\sigma'})$,
2. $a \in Act_I$ implies $\phi = \phi_{so}$ and $[F_1]_\sigma = \emptyset$ or $[F_2]_\sigma = \emptyset$.

The sequence that is returned by the algorithm can be used straightforwardly for checking which fault model(s) can be eliminated, or which fault model is selected for further scrutiny (see also Section 4.5). Such "verdicts" are easily added to our algorithms, but are left out for readability.

4.3   Weak Distinguishability and Weak Orthogonality

In case F_1 and F_2 are not strongly but weakly distinguishable (resp. weakly orthogonal), there is no guarantee that a discriminating (resp. orthogonal) state is reached. By conducting sufficiently many tests, however, chances are that one of such states is eventually reached, unless the experiment has run off to a part of the state space in which no discriminating (resp. orthogonal) states are reachable. Semi-decision procedure 2 conducts experiments on implementation I, and terminates in the following three cases:

1. if a sequence has been executed that led to a discriminating/orthogonal state,
2. if an output was observed that conflicts at least one of the fault models,
3. if discriminating/orthogonal states are no longer reachable.

So long as neither of these cases are met, the procedure does not terminate. The semi-decision procedure works in roughly the same manner as the algorithm of the previous section. The main differences are in the termination conditions (and the result it returns), and, secondly the use of arbitrary counterexamples, as shorter counterexamples are not necessarily more promising for reaching a discriminating/orthogonal state.

Algorithm 2. Procedure for exploiting weak distinguishability/orthogonality

---

Require: $P \subseteq S$, $|P| = 1$, η is any counterexample for $P \models \overline{\phi}_x$, $\phi_x \in \{\phi_{wo}, \phi_{wd}\}$
Ensure: Returns a sequence executed on I upon termination
1: **function** $\mathcal{A}_2(P, \eta, \phi_x)$
2:     **if** $\eta = \epsilon$ **then**
3:         **if** $\phi_x = \phi_{wd}$ **then return** Observe();
4:         **else choose** a **from** $\{y \in Act_I \mid [P]_{\ominus_y} \neq \emptyset\}$; **return** a;
5:     **end if**
6:     **else**   ▷ Assume $\eta \equiv e\,\eta'$ for some action e and sequence η'
7:         **if** $e \in Act_I$ **then** Apply(e); **return** $e\,\mathcal{A}_2([P]_e, \eta', \phi_x)$;
8:         **else** $a := $ Observe();
9:             **if** $a = e$ **then return** $e\,\mathcal{A}_2([P]_e, \eta', \phi_x)$;
10:            **else if** $\mathfrak{R}^{-1}(a) \in \mathbf{out}(P) \wedge$ Counterexample$([P]_a, \overline{\phi}_x) \neq \bot$ **then**
11:                **return** $a\,\mathcal{A}_2([P]_a, \mathfrak{R}^*($Counterexample$([P]_a, \overline{\phi}_x)), \phi_x)$;
12:            **else if** $\mathfrak{R}^{-1}(a) \in \mathbf{out}(P) \wedge$ Counterexample$([P]_a, \overline{\phi}_x) = \bot$ **then**
13:                **return** $\bot$;
14:            **else return** a;
15:            **end if**
16:        **end if**
17:    **end if**
18: **end function**

Theorem 6. *Let F_1 and F_2 be weakly orthogonal or weakly distinguishable fault models. Let $\phi = \phi_{wd}$ when F_1 and F_2 are distinguishable and let $\phi = \phi_{wo}$ when F_1 and F_2 are orthogonal. If algorithm $\mathcal{A}_2(\{\overline{s}\}, $ Counterexample$(F_1\|_\ominus F_2, \overline{\phi}), \phi)$ terminates it returns a sequence $\sigma \equiv \sigma'\,a$ satisfying:*

1. $a \in Act_\delta \backslash Act_I$ *implies* $\mathbf{out}([I]_{\sigma'}) \not\subseteq \mathbf{out}([F_1]_{\sigma'})$, *or* $\mathbf{out}([I]_{\sigma'}) \not\subseteq \mathbf{out}([F_2]_{\sigma'})$,
2. $a \in Act_I$ *implies* $\phi = \phi_{wo}$ *and* $[F_1]_\sigma = \emptyset$ *or* $[F_2]_\sigma = \emptyset$.
3. $a = \bot$ *implies either* $\phi = \phi_{wo}$ *and* $der([\overline{s}]_{\sigma'}) \cap \mathcal{O}(0) = \emptyset$, *or* $\phi = \phi_{so}$ *and* $der([\overline{s}]_{\sigma'}) \cap \mathcal{D}(0) = \emptyset$.

The following example illustrates that the semi-decision procedure does not necessarily terminate.

Example 4. Suppose the intersection of two fault models is given by $F_1\|F_2$ and the malfunctioning implementation is given by I (see Fig. 3). Clearly, F_1 and F_2 are weakly distinguishable, which means semi-decision procedure 2 is applicable. A counterexample to non-weak distinguishability is e.g. *?b!e?b?b!a*, so the procedure might try to execute this sequence. However, termination is not guaranteed, as the implementation may never execute action *!a*, but output *!e* instead, making the semi-decision procedure recompute new counterexamples.

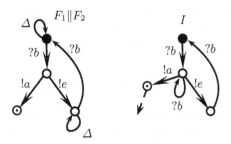

Fig. 3. No termination guaranteed for semi-decision procedure 2

4.4 Optimisations

The algorithms for strong distinguishability (resp. strong orthogonality) in the previous section can be further optimised in a number of ways. First, one can include a minor addition to the standard model-checking algorithm, marking each k-discriminating (resp. k-orthogonal) state in the LTS that is checked with its depth k. While this has a negligible negative impact on the time complexity of the model checking algorithm, the state markings allow for replacing the method Counterexample$_s$() with a constant-time operation. Secondly, upon reaching a node in $\mathcal{D}(k)$ ($\mathcal{O}(k)$, respectively), the semi-decision procedure for weak distinguishability/orthogonality could continue to behave as algorithm 1. Furthermore, the orthogonality aware intersection is an extension of the plain intersection. Computing both is therefore unnecessary: only the former is needed; in that case, the formulae for strong and weak distinguishability need to be altered to take the extended set of input actions into account.

4.5 Diagnostic Testing Methodology

Distinguishability and orthogonality, and their associated algorithms, help in reducing the effort that is required for diagnostic testing. Thus far, we presented these techniques without addressing the issue of when a particular technique is worth investigating. In this section, we discuss a methodology for employing these techniques in diagnostic testing. For the remainder of this section, we assume a faulty implementation I and a given set of fault models $F = \{F_1, \ldots, F_n\}$.

We propose a stepwise refinement of the diagnostic testing problem using distinguishability and orthogonality. The first step in our methodology is to identify the largest non-symmetric set of pairs of strongly distinguishable fault models G. We next employ the following strategy: so long as $G \neq \emptyset$, select a pair $(F_i, F_j) \in G$ and provide this pair as input to algorithm 1. Upon termination of the algorithm, an experiment $\sigma \equiv \sigma' a$ is returned, eliminating F_k from F iff $a \notin \mathbf{out}([F_k]_{\sigma'})$ ($k = i, j$). Moreover, remove all fault models F_l for which $[F_l]_{\sigma'} \neq \emptyset$ and $a \notin \mathbf{out}([F_l]_{\sigma'})$ and recompute G. A worst case scenario requires at most $|G|$ iterations to reach $G = \emptyset$. The process can be further optimised by ordering fault models w.r.t. **ioco**-testing power, but it is beyond the scope of this paper to elaborate on this.

When G is empty, no strongly distinguishable pair can be found in F. The set of fault models can be further reduced using the weak distinguishability and strong orthogonality heuristics, in no particular order, as neither allows for a fail-safe strategy to a conclusive verdict. As a last resort, weak orthogonality is used before conducting naive testing using the remaining fault models.

5   Example

As an illustration of some of the techniques that we presented in this paper, we consider a toy example concerning the prototypical coffee machine. The black-box behaviour of the coffee-machine is defined by specification S in Fig. 4, where action $?c$ and $!c$ represent a coffee request and production, $?t$ and $!t$ represent a tea request and production, and $?m$ and $!m$ represent a coffee-cream request and production. Among the set of fault

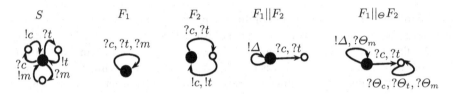

Fig. 4. Specification S and fault models F_1, F_2 and F_3 of a coffee machine

models for a misbehaving implementation of S are fault models F_1 (modelling e.g. a broken keypad in the machine) and F_2 (modelling e.g. a broken recipe book). Computing their intersection and their orthogonal-aware intersection, we find that F_1 and F_2 are strongly distinguishing and strongly orthogonal. The preferred choice here would be to run algorithm 1 with arguments setting it to check for strong distinguishability using e.g. $?t$ as input for the shortest counterexample. Algorithm 1 would first offer $?t$ to the implementation (which is accepted by assumption that implementations are *input-enabled*). Since then the shortest counterexample to non-strong distinguishability would be the empty string ϵ, the algorithm queries the output of the implementation and terminates. Any output the implementation produces either violates F_1 or F_2, or both. In case one would insist on using strong orthogonality, algorithm 1 would be used with the emtpy string ϵ as the shortest counterexample to non-strong orthogonality. The algorithm would return the sequence $?m$, indicating that isolated aspects of F_1 can be tested by experiments starting with input $?m$.

6   Concluding Remarks

We considered the problem of diagnosis for reactive systems, the problem of finding an explanation for a detected malfunction of a system. As an input to the diagnosis problem, we assumed a set of fault models. Each fault model provides a formal explanation of the behaviour of an implementation in terms of an LTS model. From this set of fault models, those models that do not correctly describe (aspects of) the implementation must be eliminated. As may be clear, this can be done naively by testing the implementation against each fault model separately, but this is quite costly. We have introduced several methods, based on model-based testing and model checking techniques, to make this selection process more effective.

Concerning issues for future research, we feel that the techniques that we have described in this paper can be further improved upon by casting our techniques in a quantitative framework. By quantifying the differences and overlap between the outputs described by two fault models, a more effective strategy may be found. The resulting quantitative approach can be seen as a generalisation of our notion of weak distinguishability. Such a quantitative approach may very likely employ techniques developed in model checking with costs (or rewards). A second issue that we intend to investigate is the transfer of our results to the setting of real-time, in particular for fault models given by Timed Automata. In our discussions, we restricted our attention to the problem of selecting the right fault models from a set of explicit fault models by assuming this set was obtained manually, thereby side-stepping the problem of obtaining such fault models in the first place. Clearly, identifying techniques for automating this process is required for a full treatment of diagnosis for LTSs. Lastly, and most importantly, the efficacy of the techniques that we have developed in this paper must be assessed on real-life case-studies. There is already some compelling evidence of their effectiveness in [5] where a notion of distinguishability is successfully exploited in the setting of communicating FSM nets.

Acknowledgement. The authors would like to thank Vlad Rusu, Jan Tretmans and René de Vries for stimulating discussions and advice on the subjects of diagnosis and testing.

References

1. Belinfante, A., Feenstra, J., de Vries, R.G., Tretmans, J., Goga, N., Feijs, L., Mauw, S., Heerink, L.: Formal test automation: A simple experiment. In: Csopaki, G., Dibuz, S., Tarnay, K. (eds.) Testcom '99, pp. 179–196. Kluwer, Dordrecht (1999)
2. Bradfield, J.C., Stirling, C.P.: Modal logics and mu-calculi: an introduction. In: Bergstra, J., Ponse, A., Smolka, S. (eds.) Handbook of Process Algebra, ch. 4, pp. 293–330. Elsevier, Amsterdam (2001)
3. El-Fakih, K., Prokopenko, S., Yevtushenko, N., von Bochmann, G.: Fault diagnosis in extended finite state machines. In: Hogrefe, D., Wiles, A. (eds.) TestCom 2003. LNCS, vol. 2644, pp. 197–210. Springer, Heidelberg (2003)
4. El-Fakih, K., Yevtushenko, N., von Bochmann, G.: Diagnosing multiple faults in communicating finite state machines. In: Proc. FORTE'01, pp. 85–100. Kluwer, Dordrecht (2001)
5. Gromov, M., Kolomeetz, A., Yevtushenko, N.: Synthesis of diagnostic tests for fsm nets. Vestnik of TSU 9(1), 204–209 (2004)
6. Guo, Q., Hierons, R.M., Harman, M., Derderian, K.: Heuristics for fault diagnosis when testing from finite state machines. Softw. Test. Verif. Reliab. 17, 41–57 (2007)
7. Jard, C., Jéron, T.: Tgv: theory, principles and algorithms. STTT 7(4), 297–315 (2005)
8. Jéron, T., Marchand, H., Pinchinat, S., Cordier, M.-O.: Supervision patterns in discrete event systems diagnosis. In: Proc. WODES 2006, IEEE, New York (2006)
9. Kick, A.: Generation of Counterexamples and Witnesses for Model Checking. PhD thesis, Fakultät für Informatik, Universität Karlsruhe, Germany (July 1996)

10. Lamperti, G., Zanella, M., Pogliano, P.: Diagnosis of active systems by automata-based reasoning techniques. Applied Intelligence 12(3), 217–237 (2000)
11. Petrenko, A., Yevtushenko, N.: Testing from partial deterministic fsm specifications. IEEE Trans. Comput. 54(9), 1154–1165 (2005)
12. Pietersma, J., van Gemund, A.J.C., Bos, A.: A model-based approach to sequential fault diagnosis. In: Proceedings IEEE AUTOTESTCON 2005 (2005)
13. Tretmans, J.: Test generation with inputs, outputs and repetitive quiescence. Software—Concepts and Tools 17(3), 103–120 (1996)

Model-Based Testing of
Service Infrastructure Components[*]

László Gönczy[1], Reiko Heckel[2], and Dániel Varró[1]

[1] Department of Measurement and Information Systems
Budapest University of Technology and Economics
Budapest, Magyar tudósok krt. 2. H-1117, Budapest- Hungary
{gonczy,varro}@mit.bme.hu
[2] Department of Computer Science
University of Leicester
University Road, LE1 7RH, Leicester - United Kingdom
reiko@mcs.le.ac.uk

Abstract. We present a methodology for testing service infrastructure components described in a high-level (UML-like) language. The technique of graph transformation is used to precisely capture the dynamic aspect of the protocols which is the basis of state space generation.

Then we use model checking techniques to find adequate test sequences for a given requirement. To illustrate our approach, we present the case study of a fault tolerant service broker which implements a well-known dependability pattern at the level of services. Finally, a compact Petri Net representation is derived by workflow mining techniques to generate faithful test cases in a non-deterministic, distributed environment.

Note that our methodology is applicable at the architectural level rather than for testing individual service instances only.

Keywords: Model-based testing, Graph Transformation, Model Checking, Fault-Tolerant Services.

1 Introduction

Beyond the usual characteristics of distributed systems, like asynchrony and communication over potentially lossy channels, service-oriented systems are characterised by a high degree of dynamic reconfiguration. Middleware protocols for such systems, therefore, have to specify not only the communication behaviour exhibited by the components involved, but also the potential creation and deletion of their connections, or indeed the components themselves.

This additional focus on structural changes requires an approach to protocol specification which allows to describe (i) the class of configurations the components involved can assume, (ii) their interaction, and (iii) changes to the configuration through actions of either the components under consideration or the

---

[*] This work was partially supported by European Research Training Network *SegraVis* (on *Syntactic and Semantic Integration of Visual Modelling Techniques*) and the *SENSORIA* European FP6 project (IST-3-016004).

environment. In general, such a model (however it is specified) will have an infinite state space, which makes full automatic verification more problematic. What is more, to verify the implementation of such protocols, the components implementing them will have to be tested against their specifications.

In this paper we address the testing of service infrastructure components against their specifications. By service infrastructure components we refer to services that are not part a specific application, but play a dedicated role in the service middleware. A typical example are services acting as proxies for implementing fault-tolerance mechanisms, e.g., by managing redundancy, forwarding requests to one of a number of available services chosen based on their performance. In our approach, service infrastructure reconfiguration protocols are specified by *dynamic metamodelling* [9], a combination of static metamodelling for describing structures as instances of class diagrams, with graph transformation rules for modelling changes to these structures. Using a UML-inspired notation for rules, specifications can be understood at an intuitive level while, at the same time, a formal semantics, theory, and tools are available to support verification.

We make use of that background through the state space generation tool Groove [25] for deriving (a bounded subset of) the transition system described by the metamodel and graph transformation rules. This transition system is employed to generate test sequences (i.e. desirable actions and their ordering) by model checking for a specific test requirement expressed as a reachability property. Unfortunately, the direct adaptation of derived test sequences as test cases in a SOA environment is problematic due to (i) internal (non-observable) steps in a test sequence, (ii) the distributed test environment where the interleaving of independent actions is possible, and (iii) not all steps of test execution are controllable.

For this purpose, we propose a technique to synthesize a compact Petri net representation for the possible interactions between the service under test and the test environment by using workflow mining techniques [1]. Concrete test cases can be defined by a sequence of controllable actions in this Petri net, while the test oracle accepts an observable action if the corresponding step can be executed in the Petri net.

The paper is organised as follows. Section 2 presents the structural metamodel for our case study and its extension by graph transformation rules. Section 3 discusses the specification of requirements for test cases and the generation of test sequences using model checking. Section 4 presents the architecture of the test environment and the derivation of test cases. Section 5 describes related work while Section 6 concludes the paper and discussed future research.

2   Modelling a Solution for Fault-Tolerant Service Infrastructure

As running example, we first introduce the dynamic metamodel of a service broker implementing a pattern for fault-tolerant services. The broker acts as a *proxy* for service clients, maintaining a list of available services and forwarding clients' requests to individual service *variants*. The replies of these variants will be validated by a separate *checker* before being forwarded to the caller.

During the broker's lifetime services may be created or disappear, or may be temporarily unavailable due to loss of connection. Traditionally in the field of fault-tolerant systems, reconfiguration is captured by semi-formal design patterns. To provide a foundation for test generation, we propose a formal model which generalises these patterns while retaining the intuitive nature of semi-formal descriptions.

2.1 Structural Model

Our approach is based on metamodelling. A *metamodel* describes the ontology of a domain in the form of UML class diagrams. Domain concepts are denoted by *classes* with *attributes* defining their properties. *Inheritance* specifies subclass relations while *associations* define binary relations between classes. Multiplicities of association ends (at-most-one by default, or arbitrary denoted by an asterisk) restrict the connectivity along -instances of- an association. The formal interpretation of such a metamodel is a *type graph* [7], i.e., a graph whose nodes and edges represent types. Instances of the metamodel are formalised as corresponding instance graphs.

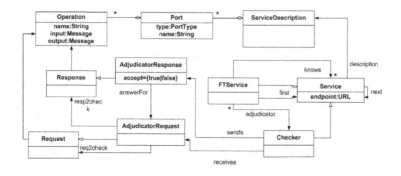

Fig. 1. Metamodel for Fault-Tolerant Services

The metamodel for fault-tolerant services is shown in Fig. 1. The FTService (also known as *broker* or *dispatcher*) is realised as a service, too. It is responsible for forwarding incoming requests to service components with the required functionality, designated by the knows association. The number of service components receiving the same request is determined by the fault-tolerance strategy applied. Responses to a particular request are sent back to the FTService, which sends them to a Checker service (also known as *adjudicator*).

The Checker service can be provided by a third-party component or by a local service running on the same machine. An AdjudicatorRequest sent to this service is composed of the original request of the client and the response of the variant service. The Checker evaluates the incoming request and decides about its acceptance. As this step is highly application- or domain-dependent, we do not intend to give a general description here. Usually a simple comparison between the expected result, an approximate value, often determined offline, and the response of the variant service is sufficient. If there are multiple answers, another possibility is to compare the different values. The answer of the checker is wrapped in an AdjudicatorResponse and sent back to the FTService. If the answer is acceptable, it is forwarded to the client. In case of an erroneous answer, the next action is chosen according to the applied fault-tolerance algorithm and the number of available variants.

The metamodel presents an overview of the structure of the fault-tolerance pattern, but it does not specify the protocol executed by the component. This is described in the following section by graph transformation rules. In particular, we will model the *Recovery Block* pattern [24], where requests are sent to one variant at a time: the "best" one available according to some metrics gathered over time. This requires to maintain a list of components in the order of preference. More sophisticated strategies (such as load balancing between components by permuting the available components, etc.) are also possible. To mention other FT modeling solutions, OMG introduced an UML profile for QoS and FT [27], however, our solution uses a metamodel specially tailored to needs of SOA and patterns are modeled in more details (which is obviuosly needed for test generation).

2.2 Behavioural Rules

This section describes how the dynamic behaviour of service infrastructure component is specified in a formally verifiable way by graph transformation rules. The theory of graph transformation is described in detail e.g. in [7]. Here we only summarise the background.

A graph transformation rule consists of a *Left Hand Side (LHS)*, a *Right Hand Side (RHS)* and (optionally) a *Negative Application Condition (NAC)*, defined as instances of the type graph representing the metamodel. The *LHS* is a graph pattern consisting of the *mandatory* elements which prescribes a precondition for the application of the rule. The *RHS* is a graph pattern containing all elements which should be present after the application. Elements in $RHS \cap LHS$ are left unchanged by the execution of the transformation, elements in $LHS \setminus RHS$ are deleted while elements in $RHS \setminus LHS$ are newly created by the rule. The negative condition prevents the rule from being applied in a situation where undesirable elements are present in the graph. Formally, we follow the Single Pushout (SPO) Approach with negative application conditions.

A *graph grammar (GG)* consists of a start graph and a set of graph transformation rules. A *graph transition system* is a labelled transition system whose states are all the graphs reachable from the start graph by the application of rules, and whose transitions are given by rule applications labelled rule names.

We use the tool Groove [25] for creating graph transformation systems and generating their transition systems [25]. The "traditional" representation of rules separates LHS from RHS and shows the negative condition as part of the left-hand side (crossed out). The compact representation of Groove, on the other hand, uses a single graph only, with tags on the nodes and arcs to distinguish *new*ly created elements *del*eted elements and elements that must *not* be present, forming part of a negative application condition.

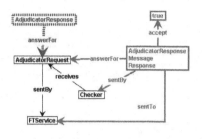

Fig. 2. Compact representation

Fig. 2 shows the compact representation of a sample transformation rule. This rule expresses the behaviour of the proxy when a decision has arrived from the adjudicator, reporting that the response of a particular service variant has passed the acceptance check. In this case, the proxy will send the response of the variant back to the client. In this simplified model, we abstract from changes to the original messages, however, in a real life system, technical changes can be performed on the reply (e.g., the sender of the message can be substituted).

Altogether we have identified four classes of transformation rules according to the nature of the behaviour they describe:

Reconfiguration rules determine the behaviour of the service under test. In our case, these are the rules which identify the actions of the FT proxy (forwarding requests to variant, register a new variant, etc.).

Environmental rules describe the dynamic behaviour of the other components in the infrastructure, still at a high level of abstraction (ignoring implementation-dependent steps). In our case, rules for service variant and checker components will belong to this set. The main difference between these and the reconfiguration rules is that these are possible "intervention points" where the concrete test case may affect the system since they relate to the behaviour of the tester component(s). However, if the *System Under Test* (SUT) changes (e.g., the checker component is the subject of testing), then the classification of rules may change.

Platform-dependent implement low-level operations, such as sending a SOAP message. They are needed to create a connection between the behaviour of different infrastructure components, e.g. to model that a message can be received by the target component only after the source has sent it. They also provide flexibility as other middleware-related aspects can easily be integrated. For instance, if one would like to extend the model by logging or reliable messaging features (e.g. creating acknowledgements for each messages), these extensions can be separated from the main component's high level logic. An example for such an extension was described in [13].

Test-related rules express actions which influence the tests but happen outside the system, including fault injection rules. In the case study, rules describing actions of the client are considered to be clearly test-related. In our fault model we consider *external service faults* representing an incorrect response which will fail the acceptance check. The checker component is considered to be always correct, but the model is extendable to deal with a possibly unavailable/wrong checker.

The rules of the example are listed in Fig. 3 with their classification and a brief description. The rule classification has an impact on the test case generation, as rules of the above classes will affect test cases in different ways, as described in Sect. 3.

Rule name	Description
Reconfiguration rules	
callAdjudicator	The proxy calls the adjudicator when it receives an answer from a service variant.
callFirstVariant	The proxy forwards the request of the client to the first service variant in the list.
callNextVariant	The proxy forwards the request of the client to the next available service variant.
createFailureMessage	Since there are no further variants, and no acceptable response was given to the client request, the proxy indicates a failure to the client.
createProxyResponse	As the response of the variant was correct, the proxy forwards it to the client.
createNoServiceFailureMessage	As there are no available service variant for a particular request type, the proxy returns with a special failure.
registerVariant	The proxy registers a variant to the service list.
registerFirstVariant	The proxy registers the first variant to the service list.
Environmental rules	
createResponse	A service variant creates a response.
newSubscription	A service variant send a subscription request to the proxy.
makeNegativeDecision	The checker component rejects a variant response.
makePositiveDecision	The checker component accepts a variant response.
Platform-dependent rules	
sendMessage	A message is sent by the middleware.
receiveMessage	A message is received by the middleware.
Test-related rules	
newRequest	The client creates a request.
receiveAnswer	The client receives the answer of the proxy.

Fig. 3. Rules of the fault-tolerant proxy case study

3   Generation of Execution Sequences by Model Checking

This section describes the use of state space generation and model checking to derive executable test cases for the service broker. An architectural overview of our approach is presented in Fig. 4. Conceptually, we follow the principles of [4] to generate test cases as counterexamples for a given property using model checking. The properties are derived from test requirements specifying sequences of transformation steps to be used as test cases.

Given the counterexamples in form of rule sequences, we build a structure representing the possible combination of test sequences. This way non-determinism introduced by distributed computing is included in our model, and the test oracle will be able to treat all possible branches (with the restriction that we will manage only test cases given as a result of the model checking).

For representing test cases we use the formalism of Petri Nets. We use *critical pair analysis* of the GT rules to find non-determinism in the system. Once we have the rule dependencies and test cases, the α-algorithm of [1] is performed to synthesize a complex Petri Net. Finally we reduce this Petri Net by filtering the rules which are neither observable nor controllable and therefore are not needed for the test oracle. Rules corresponding to middleware behavior and internal operations of the SUT are typically erased from the net.

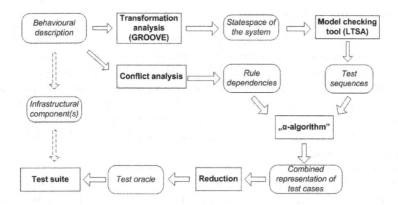

Fig. 4. High-level architectural view of the testing framework

3.1   Test Requirements

The test requirements we express can prescribe a desired action, following a specified sequence as a precondition. More formally in EBNF, our simple property language is defined as follows.

```
<requirement> ::= <sequence> => <rule>
<sequence> ::= <rule> | <rule>.<sequence>
```

Here, arrow (=>) means implication, dot (.) concatenation, while | and ::= are the usual EBNF (meta) operators. Note that although the conclusion of a requirement consists of the application of a single rule, a choice between multiple actions can be modelled by describing requirements for several test cases.

To illustrate our approach, we describe test case generation for a sample requirement: *If a variant response passed the acceptance test, the proxy will forward it to the client.* In terms of graph transformation, this translates to the following rule sequence, using the rule names of Fig. 3.

```
callAdjudicator.makePositiveDecision => createProxyResponse
```

Typical requirements correspond to forbidden behavior (such as, *"If there is a variant which has not been asked, no failure message can be sent to the client"*) and required actions, e.g., *"If a checker accepts a result, the proxy must forward it to the client".*

The rules used in requirements will typically belong to the classes of reconfiguration, environmental, or test-related rules, expressing high-level functionality observable at the application level. Platform-dependent steps are normally neglected, which results in reusable requirement patterns. For example, message-based communication could be replaced by remote procedure calls without affecting these requirements.

3.2 State Space Generation and Model Checking

Given the transformation rules described in Sect. 2.2 and an initial configuration (e.g., a proxy, a number of service variants not registered with the proxy and a client) one can generate the entire state space of the graph transition system using the GROOVE tool [25]. Groove performs a bounded state space generation by applying graph transformation rules in all possible ways to the start graph, up to a given search depth. Unfortunately, the implementation of model checking of temporal logic formulae is still in an early stage for GROOVE, therefore we use a separate model checking tool.

We transfer both the graph transition system generated by Groove and the requirements into the Labelled Transition System Analyzer (LTSA) tool [18]. For model checking, LTSA composes an automaton from the LTS of the original system and the property automaton of the requirement. The analysis will find a violation trace if the property automaton reaches an error state.

Thus, a requirement has to be translated into a property automaton with the obvious modification that the "required action" is considered as an error state. Moreover, a separate automaton is derived from the state space of the graph transformation system generated by Groove. In the process, all information about the internal structure of states and transformation steps is lost. Therefore, we have to restrict our execution path retrieved by the LTSA analysis to handle *one request at a time*. However, this does not prevent to apply our test generation technique of Sec. 4 to be used for concurrent messages.

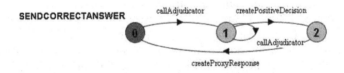

Fig. 5. Requirement expressing the behavior of the proxy

Given such an input, the LTSA tool is able to find low-level rule sequences in the state space of the system which "violate" the property automaton derived from the original requirement by negating the required action. These sequences will serve as the basis for deriving the actual test cases.

As an example, we examine the rule set (of Fig. 3) for a sample configuration consisting of one client, one proxy, one checker and three service variants. The formulation of our sample requirement as a property automaton is given in Fig. 5. The sequence which is found as a "counterexample" for this property is shown in the following example. (We modified the output format of LTSA to make the sequence more readable.)

```
newSubscription => sendMessage => receiveMessage =>
registerFirstVariant => newRequest => sendMessage =>
receiveMessage => callFirstVariant => sendMessage =>
receiveMessage => createResponse => sendMessage =>
receiveMessage => callAdjudicator => sendMessage =>
receiveMessage => createPositiveDecision => sendMessage
=> receiveMessage => createProxyResponse
```

This corresponds to a sequence describing the desired functionality of the system, and it will serve as the basis for a test cases for this requirement. This sequence is one of the shortest possible rule sequences since the execution of some of the steps can be carried out in parallel (e.g. the subscription of a variant and the creation of a client request). That means, although the test case could contain concurrent steps, the model checker will return only a sequence.

Note that although we focus on the generation of test sequences, the same technique can also be used to verify the dynamic behaviour of the model as described in [13]. In this case the output of the model checker represents a decision about whether the system meets a particular requirement. If not, a sequence of events is provided that violates the requirement.

In our case, if the analysis results in a positive decision about the negated requirement, the original requirement is not satisfied by the rules of the model. This provides, almost as a side effect, with a verification of the model (e.g. as described in [13]) possibly leading to a re-design of the rules according to the results of the test case generation.

4 Derivation of Test Cases

At this point, execution sequences derived by the LTSA model checker are available. However, their direct adaptation for test cases in a SOA environment is not at all straightforward as (i) certain steps in the execution sequence are internal to the proxy thus they are not observable, (ii) the tests need to be executed in a distributed environment where the interleaving of independent actions is possible, and (iii) we cannot deterministically control all steps of test execution (non-deterministic testing [22]).

For the first problem, many existing approaches [22,5] typically use an abstract representation of the test case which only includes controllable and observable actions. For the second problem, one may ask the model checker to derive all possible execution paths which satisfies a given requirement [14]. However, this results in a huge set of test sequences, i.e. a different sequence for each interleaving, which can be infeasible in practice. For the third problem, a test oracle needs to be derived which identifies if one of the correct execution paths were executed by the FT proxy (i.e. service under test, SUT) for a given input.

After discussing the architecture of the test environment, we propose a technique to synthesize a compact Petri net representation for the possible interaction between the FT proxy and the test environment by using workflow mining techniques [1]. Concrete test cases can be defined by a sequence of controllable actions in this Petri net, while the test oracle accepts an observable action if the corresponding step can be executed in the Petri net.

4.1   A Distributed Test Architecture

The component architecture of the test framework is shown in Fig. 6 with the interfaces of messages arriving to each component and potential interactions between the components.

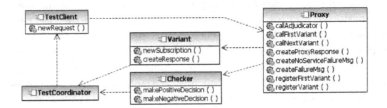

Fig. 6. Architecture of test components

Since in a service-oriented architecture the server will be unaware of the client implementation and communicate only via messages, thus the TestClient does not have to implement any method of the FT proxy (SUT).

Methods of Variant and Checker will be forwarded to the TestCoordinator, thus the test coordinator implements the interfaces of all the other components in the test environment. Operations on these interfaces will be interaction points or controllable and observable actions (see later in Sec. 4.2). This way, no modifications are made to the SUT (the FT proxy) for testing purposes as services implementing other infrastructural elements can replace the interfaces of the coordinator.

The execution of a test case requires to make certain decisions available as test configuration parameters, which are application-dependent during normal operation. For instance, decisions like the result of an acceptance test or the availability of a variant will be influenced by pre-defined test parameters for each decision in the test. For instance, if multiple variants will be asked by the proxy, each of them will ask the coordinator whether to answer the request.

In the paper, we assume that tests components are stand-alone services in a distributed SOA environment, but no further implementation details are provided to better concentrate on presenting the test generation approach itself.

4.2   Creating the Test Suite

In the field of model based testing, generating executable test sets from abstract test sequences is a well-known problem. Actions in a sequence can be *controllable*, *observable* or *hidden*. These categories respectively correspond to decision points to set up a certain test case (controllable), automatically executed actions within the test framework (observable) and actions inside the SUT (hidden).

In our case, a rule sequence produced by the model checker may contain many rules to be executed automatically, without any test-case specific intervention. *Reconfiguration* rules (like registerVariant) are obviously part of the SUT, i.e., the FT proxy. *Platform-dependent rules* can be observable or hidden, depending on whether they are executed in the SUT or in the tester. However, as the middleware rules are not directly affected during testing, we consider them hidden. Hence, only controllable *test* and *environmental* rules will included in a test case.

Our goal is to build a combined representation of multiple test cases and test oracles in the form of Petri nets. Petri Nets (PN) are a special class of bipartite graphs used to formally model and analyze concurrent systems with a wide range of available tool support. The reader is referred, for instance, to [23] for the theory and application of PN.

For this purpose, we combine *critical pair analysis* of graph transformation rules with the α-algorithm used for *workflow mining* in [1]. The former technique aims at statically detecting conflicts and causal dependency between graph transformation rules, while the latter method is used for building instances of a special class of Petri Nets (called Workflow Nets) from workflow logs.

Step 1: Partial ordering of an execution path. First, we build a Petri Net of each individual test case which makes concurrent behaviour explicit by de-sequencing (totally ordered) actions in the execution path derived by the model checker into a partially ordered transitions in the PN.

In order to detect concurrent actions, we use basic concepts of graph transformation theory. Two rules are in *conflict* with each other, if the execution of a rule does disables the execution of the other rule (otherwise, they are *parallel independent*). A rule is causally dependent on another rule, if the first rule enables the other (e.g. by creating new matchings for it), otherwise, they are called *sequentially independent*.

A partial ordering between actions can be derived by performing *critical pair analysis*[19] of our rules, which is a well-known static analysis technique in the field of graph transformation to detect potential conflicts and causalities of graph transformation rules. Critical pair analysis is able to show minimal conflicting (or causal dependent) situations between two graph transformation rules. These suspicious situations can be inspected by the user to decide if they arise in a certain application or not.

In our case, a critical pair analysis will detect some trivial conflicts due to the semantics of Groove which always explicitly requires a NAC to prevent the system from getting to an infinite loop. After eliminating such trivial dependencies, the result of the analysis for an execution sequence will reveal those parts of the sequence which can be executed concurrently. These will correspond to the behavior of distributed components and the middleware, the order of which cannot be determined. Fig. 7 shows a partially ordered version of rule sequence in Sect. 3.2 as a PN where controllable and observable actions are highlighted.

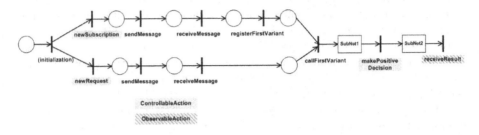

Fig. 7. Petri Net representing a test case

Step 2: Constructing workflow nets from partially ordered paths. Using the method of [1] the system model is reconstructed by workflow mining techniques from individual *observations (cases)*. Our problem is very similar: we have to create an abstract model of observable and controllable actions, which explicitly contains concurrent behavior and potential non-determinism.

This workflow mining technique groups relations between pairs of actions into the following categories: *potential parallelism*, *direct causality*, *precedence* and *concurrency*. These relations can be derived from the critical pair analysis in the previous step. We also have the restriction that the net will be of class Free Choice Net. The only difference is that we do not expect the final model to be a valid WF-Net. The result of the algorithm is shown in Fig. 8.

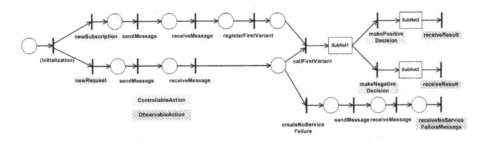

Fig. 8. The Petri Net created after the combination of test sequences

Step 3: Reduction of an observable Petri net. The net is then reduced using standard PN reduction rules described for instance in [23]. The main principle of the reduction is that we erase all sequences and parallel constructs which do not contain any controllable or observable actions, since these correspond to internal behavior of SUT (this case, the proxy) and therefore will not affect the

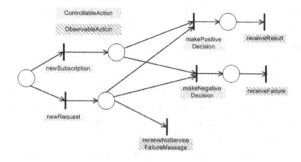

Fig. 9. The Abstract Petri Net created after reduction

tester. The result of the reduction is shown in Fig. 9. The resulting abstract PN can be used as a combined test suite and test oracle for a set of given requirements.

4.3 Discussion

Our approach relies on a combination of various formal techniques. Now we discuss the role of each individual technique in the overall approach.

Why to combine two model checkers? At this point, the official Groove release does not yet support model checking facilities, only state space generation. However, this feature of the tool is strong enough because dynamically changing models are supported. Therefore, we project the state space generated by Groove into the LTSA tool to derive actual execution sequences for a given test criterion (this is practically renaming an LTS structure).

Why to use a Petri Net representation for a test case? The final Petri net representation offers two advantages: (i) a compact representation of test case with explicit concurrency and without interleavings (which is not the case of the original GTS state space) (ii) mining techniques are available to derive the PN.

Direct bridging of graph transformation and Petri nets. There are existing approaches to generate a Petri net representation of a graph transformation system on various levels of abstraction [2,28]. In the future, we plan to investigate more on their applicability in a testing environment. However, we believe that the Petri net representation of a test case is more simple compared to them.

5   Related Work

The modelling technique in our paper is conceptually derived from [3] where SOA-specific reconfigurations where first defined by a combination of graph transformation and metamodelling techniques.

Graph transformation is used as a specification technique for dynamic architectural reconfigurations in [10] using the algebraic framework CommUnity. Executable visual contracts derived from graph transformation rules are facilitated in [20] where JML code is generated for the run-time checking of (manual) service implementations. In [6] the same contracts are used for specification matching. Graph transformation rules guided the model-based discovery of web services in [15].

The specification and analysis of fault behaviors have been carried out in [8] using graph grammars. While this approach is not directly related to SOA, it uses similar techniques for modeling the behaviour of the system, and also applies model checking techniques for verifying the behavioural specification. However, the behaviour of SOA components typically induces an infinite state space, such a full verification is problematic.

In [16], one of the authors applies graph transformation-based modelling for conformance testing. The novel contribution of the current paper is that (i) we use model checking to generate test sequences, which leads to a higher level of automation (ii) our models focus on changes at the architectural level rather than on the data state transformation with a single service.

The work presented in [17] aims at test generation for processes described in OWL-S. Our work is different as our test cases are derived from high-level formal specification of the dynamic behaviour, rather than being abstracted from its implementation. The same applies to [12] where the SPIN model checker is used to generate test cases for BPEL workflows. Authors of [21] also use SPIN to create test sequences to meet coverage criteria. Categories of actions and formalisms for describing test cases are defined among others in [22] and [5]. However, synthesis of test cases is still an open issue. We already discussed the work described in [22] and [5].

LTSA [18] has already been applied successfully in a SOA context for the formal analysis of business processes given in the form of BPEL specifications in [11]. However, the direct adaptation of this approach is problematic, since the inherent dynamism in the reconfiguration behaviour of the service infrastructure is difficult to be captured in BPEL.

A de-facto industrial standard in the telecommunications domain for a highly available service middleware is the Application Interface Specification (AIS) of SA Forum [26]. Our future work includes the application of our approach to testing of components of the AIS infrastructure.

6   Conclusions and Future Work

We proposed a model-based approach for generating test cases for service infrastructure components exemplified by testing a fault-tolerant proxy. The reconfiguration behaviour of the service infrastructure was captured by a combination of static metamodels and graph transformation rules. The (bounded) state space of the service infrastructure was derived by the Groove tool [25], and post-processed by the LTSA model checker to derive an execution sequence for a given requirement. In order to generated faithful test cases to be executed in a distributed service environment, a compact Petri net representation was derived by workflow mining techniques. At the final step, this Petri net was reduced by abstracting from internal actions.

The scalability of our method is at the moment mainly limited by the statespace generation feature of Groove which is the range of 100 thousand states; however, these states represent a dynamically changing structure (vs. a BDD with predefined state variables). The quality of our generated test cases strongly corresponds to the requirements which are under investigation (as usual in requirement-based testing).

In the paper, we limited our tests to configurations with one proxy and one service type only. That means, all variants implement the same service. This, however, is only a limitation for illustration purposes, since the rules can easily be extended to model a proxy maintaining multiple variant lists, one for each type of service. Multiple requests can also be tested by starting a corresponding Petri net for observing each request. On the other hand, in case of more sophisticated requirements information about the structure of the graphs needs to be expressed. This is currently is not supported by the state space generation and modelchecking tools we use. Future developments in model checking for graph transformation systems are likely to ameliorate this problem. Our long-term purpose is to develop a methodology for testing automatically generated components, modelled by a visual notation that enables (semi-) automatic code generation.

References

1. van der Aalst, W., Weijters, T., Maruster, L.: Workflow mining: discovering process models from event logs. In: IEEE Trans. on Knowledge and Data Engineering, vol.16(9) (2004)
2. Baldan, P.B., König, B., Stürmer, I.: Generating Test Cases for Code Generators by Unfolding Graph Transformation Systems. In: Ehrig, H., Engels, G., Parisi-Presicce, F., Rozenberg, G. (eds.) ICGT 2004. LNCS, vol. 3256, pp. 194–209. Springer, Heidelberg (2004)
3. Baresi, L.R., Heckel, S., Thöne, S., Varró, D.: Style-Based Modeling and Refinement of Service-Oriented Architectures. Journal of Software and Systems Modelling 5(2), 187–207 (2006)
4. Beyer, D., Chlipala, A.J., Majumadr, R.: Generating Tests from Counterexamples. In: Proc. 26th Intern. Conf. on Software Engineering, pp. 326–335 (2004)
5. Campbell, C., Grieskamp, W., Nachmanson, L.: Model-Based Testing of Object-Oriented Reactive Systems with Spec Explorer. Technical Report MSR-TR-2005-59, Microsoft Research (2005)
6. Cherchago, A., Heckel, R.: Specification Matching of Web Services Using Conditional Graph Transformation Rules. In: Ehrig, H., Engels, G., Parisi-Presicce, F., Rozenberg, G. (eds.) ICGT 2004. LNCS, vol. 3256, pp. 304–318. Springer, Heidelberg (2004)
7. Corradini, A., Montanari, U., Rossi, F.: Graph Processes. Special Issue of Fundamenta Informaticae 26(3-4), 241–266 (1996)
8. Dotti, L., Ribeiro, L., dos Santos, O.M.: Specification and analysis of fault behaviours using graph grammars. In: Pfaltz, J.L., Nagl, M., Böhlen, B. (eds.) AGTIVE 2003. LNCS, vol. 3062, pp. 120–133. Springer, Heidelberg (2004)
9. Engels, G.J., Hausmann, J., Heckel, R., Sauer, S.: Dynamic meta modeling: A graphical approach to the operational semantics of behavioral diagrams in UML. In: Evans, A., Kent, S., Selic, B. (eds.) UML 2000. LNCS, vol. 1939, pp. 323–337. Springer, Heidelberg (2000)

10. Wermelinger, M., Fiadeiro, J.L.: A graph transformation approach to software architecture reconfiguration. Science of Comp. Progr. 44(2), 133–155 (2002)
11. Foster, H., Uchitel, S., Magee, J., Kramer, J.: Model-based verification of web service compositions. In: 18th IEEE Intern. Conf. on Automated Software Engineering (ASE 2003), Montreal, Canada, pp. 152–163. IEEE, New York (2003)
12. Garca-Fanjul, J., Tuya, J., de la Riva, C.: Generating Test Cases Specifications for BPEL Compositions of Web Services Using SPIN. In: Proc. Intern. Workshop on Web Service Modeling and Testing (WS-MATE 2006), pp. 83–85 (2006)
13. Gönczy, L., Kovács, M., Varró, D.: Modeling and verification of reliable messaging by graph transformation systems. In: Proc. of the Workshop on Graph Transformation for Verification and Concurrency (GT-VC 2006), Elsevier, Amsterdam (2006)
14. Hamon, G., de Moura, L., Rushby, J.: Generating Efficient Test Sets with a Model Checker. In: Proc. of SEFM 04, Beijing, China (September 2004)
15. Hausmann, J.H., Heckel, R., Lohmann, M.: Model-based Discovery of Web Services. In: IEEE Intern. Conf. on Web Services (ICWS), USA (June 6-9, 2004)
16. Heckel, R., Mariani, L.: Automated Conformance Testing of Web Services. In: Cerioli, M. (ed.) FASE 2005. LNCS, vol. 3442, pp. 34–48. Springer, Heidelberg (2005)
17. Huang, H., Tsai, W.-T., Paul, R., Chen, Y.: Automated Model Checking and Testing for Composite Web Services. In: Proc. of 8th IEEE Intern. Symp. on Object-Oriented Real-Time System Computing (ISORC'05) pp. 300–307 (2005)
18. Labelled   Transition   System   Analyser   (Version   2.2)
    http://www-dse.doc.ic.ac.uk/concurrency/ltsa-v2/index.html
19. Lambers, L., Ehrig, H., Orejas, F.: Conflict Detection for Graph Transformation with Negative Application Conditions. In: Corradini, A., Ehrig, H., Montanari, U., Ribeiro, L., Rozenberg, G. (eds.) ICGT 2006. LNCS, vol. 4178, pp. 61–76. Springer, Heidelberg (2006)
20. Lohmann, M., Sauer, S., Engels, G.: Executable Visual Contracts. In: Proc. IEEE Symposium on Visual Languages and Human Centric Computing (VL/HCC 05), pp. 63–70 (2005)
21. Micskei, Z., Majzik, I.: Model-based Automatic Test Generation for Event-Driven Embedded Systems using Model Checkers. In: Proc. of lnt'l Conf. on Dependability of Computer Systems (DEPCOS-RELCOMEX'06), pp. 191–198 (2006)
22. Muccini, H.: Software Architecture for Testing, Coordination and Views Model Checking. PhD Thesis (2002)
23. Murata, T.: Petri Nets: Properties, Analysis and Applications. In: Proc. of IEEE, vol. 77(4) (1989)
24. Randell, B., Xu, J.: The Evolution of the Recovery Block Concept, in Software Fault Tolerance. In: Lyu, M. (ed.) Trends in Software, pp. 1–22. J. Wiley, New York (1994)
25. Rensink, A.: The GROOVE simulator: A tool for state space generation. In: Pfaltz, J.L., Nagl, M., Böhlen, B. (eds.) AGTIVE 2003. LNCS, vol. 3062, pp. 479–485. Springer, Heidelberg (2004)
26. S.A. Forum: Application Interface Specification. http://www.saforum.org
27. UML Profile for Modeling Quality of Service and Fault Tolerance Characteristics and mechanisms.
28. Varró, D., Varró-Gyapay, S., Ehrig, H., Prange, U., Taentzer, G.: Termination Analysis of Model Transformations by Petri Nets. In: Corradini, A., Ehrig, H., Montanari, U., Ribeiro, L., Rozenberg, G. (eds.) ICGT 2006. LNCS, vol. 4178, pp. 260–274. Springer, Heidelberg (2006)

Testing Input/Output Partial Order Automata

Stefan Haar[1], Claude Jard[2], and Guy-Vincent Jourdan[3]

[1] IRISA/INRIA
Rennes, France
Stefan.Haar@irisa.fr
[2] IRISA, ENS Cachan Bretagne
Campus de Ker-Lann, F-35170 Bruz, France
Claude.Jard@bretagne.ens-cachan.fr
[3] School of Information Technology and Engineering (SITE)
University of Ottawa
800 King Edward Avenue
Ottawa, Ontario, Canada, K1N 6N5
gvj@site.uottawa.ca

Abstract. We propose an extension of the Finite State Machine framework in distributed systems, using *input/output partial order automata (IOPOA)*. In this model, transitions can be executed non-atomically, reacting to asynchronous inputs on several ports, and producing asynchronous output on those ports. We develop the formal framework for distributed testing in this architecture and compare with the synchronous I/O automaton setting. The advantage of the compact modelling by IOPOA combines with low complexity : the number of tests required for concurrent input in our model is polynomial in the number of inputs.

1 Introduction

Finite State Machines (FSMs) have been used to model many types of sequential systems. However, it is distributed applications over networks that become increasingly important; they do not fit into this sequential model, because inputs may be applied simultaneously and events are not necessarily totally ordered. In the context of testing, distributed testing models use *multi-port* automata in which each transition is guarded by a required vector of inputs (possibly $\perp$, i.e. no input on some channels) on a collection of channels, and produces a vector of outputs (possibly $\perp$) on those channels. This model, often called *Multiports Deterministic FSM* in the literature, but that we call *sequential input automata* in this paper, has been widely studied from a distributed system testing perspective; emphasis is given in that work to the coordination messages, between testers at different ports, that are necessary to avoid *controllability* and *observability* problems in distributed systems testing [1, 2, 3, 4, 5, 6, 7, 8, 9, 10, 11, 12, 13, 14, 15, 16]. However, this model is intrinsically sequential regarding the inputs, which must be specified one at a time (although one such single input can generate several, concurrent outputs on different ports). In order to specify that from a given state, two concurrent inputs a and b are required, one has to specify either 'a

A. Petrenko et al. (Eds.): TestCom/FATES 2007, LNCS 4581, pp. 171–185, 2007.
© IFIP- International Federation for Information Processing 2007

then b' or 'b then a'. Consider the more detailed example in Figure 1. In that context, we need to specify that in order to go from state $\mathbf{s}_i$ to state $\mathbf{s}_f$, we need to input i_1, i_2 and i_3 on ports 1, 2 and 3 respectively. On port 1, the output o_1 should be produced after i_1 was input. On port 3, output o_3 should be produced after i_3 was input, and on port 2, o_2 should be produced after i_1, i_2 and i_3 have all been input. In general, when n inputs must be provided concurrently, the only option is to enumerate all $n!$ ordering for the inputs, leading to a specification that is large and difficult to create, hard to interpret and thus to understand, and whose size makes it difficult to test. Another approach would be to arbitrarily impose a given ordering for the inputs, which seems a poor option and which adds needless constraints at the implementation level.

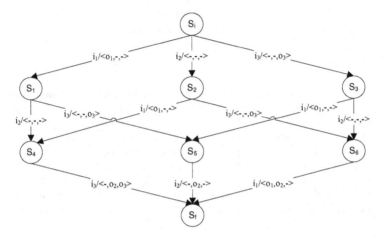

Fig. 1. (Partial) Multiports Deterministic FSM

We therefore endeavour to explore a model that allows specifications to relax synchronization constraints: equipping partial order automata with input/output capabilities. We define a class of *IO-PO-automata (IOPOA)* in which

- inputs can arrive asynchronously, and
- transitions may occur partially, and in several steps, reacting to inputs as they arrive and producing outputs as soon as they are ready, without dedicated synchronization.

The important additional feature (in addition to state transition and output production) of transitions is then a *causal order*: for p channels, we have a bipartite graph of (p inputs) * (p outputs) such that input on channel i precedes output on channel i produced by that transition. Cross-channel dependencies may persist between input on some channel j and output on channel $i \neq j$; at most, the input on j can trigger a broadcast to all channels. However, inputs are not ordered among one another, neither are outputs.

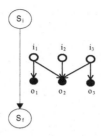

Fig. 2. The IOPOA corresponding to the multiports deterministic FSM of Figure 1

Figure 2 shows the IOPOA corresponding to the set of transitions of Figure 1. Clearly, the result is a much simpler model, with two states and one transition. The role of states for testing in this model will be redefined, and a new role emerges for the partially ordered patterns ; the theoretical toolbox of distinguishing sequences etc. needs to be adapted, yet keeps its importance. Concerning the complexity of checking, one might have expected that the new model were just a concise way of specifying the same behavior, and thus that testing would be the same in both cases (that is, that all combinations of concurrent inputs would have to be tested anyways). It turns out not to be the case; in fact, the number of tests required for concurrent input in our model is *polynomial* in the number of inputs.

The rest of the paper is structured as follows. In section 2, the IOPOA Framework is introduced, section 3 focuses on the differences between the partial order model and the classical one regarding conformance testing. Finally, section 4 discusses future extensions to more general IOPOA classes, and shows that IOPOA can have homing and synchronizing sequences, but may not have state identification or state verification sequences. Section 5 concludes.

2   IOPOA Framework

We introduce the model of IOPO Automaton with I/O vector sequences. The definition of *conformance*, given in 2.3 in this framework needs the notions of well-behavedness and of completion, which we discuss in 2.4.

2.1   IOPO Automata

Definition 1. *An* Input/Output Partial Order Automaton (or IOPO Automaton, IOPOA) *is a tuple* $\mathcal{M} = (S, \mathbf{s}^{\mathbf{in}}, Chn, \mathcal{I}, \mathcal{O}, \delta, \lambda, \omega)$, *where*

1. S *is a finite set of* states *and* $s_1 = \mathbf{s}^{\mathbf{in}} \in S$ *is the* initial state; *the number of states of* $\mathcal{M}$ *is denoted* $n \triangleq |S|$ *and the states of* $\mathcal{M}$ *are enumerated, giving* $S = \{s_1, \ldots, s_n\}$;
2. $Chn = \pi_1, \ldots, \pi_p$ *is the set of I/O channels (ports),*

3. $\mathcal{I}$ *is the common input alphabet, and* $\mathcal{O}$ *the common output alphabet for all channels. Note that the literature often notes different alphabets* $\mathcal{I}_1, \ldots, \mathcal{I}_p$ *for different channels ; the above implies no loss of generality provided that the port to which an input is applied is uniquely identifiable. Taking*

$$\overline{\mathcal{I}} \triangleq \bigcup_{i=1}^{p} \mathcal{I}_i \; ; \mathcal{I} \triangleq \overline{\mathcal{I}} \times Chn,$$

such that (a, i) *denotes input* a *on port* i, *one can switch from one representation to the other. We require a special symbol* $\perp \in \mathcal{I} \cap \mathcal{O}$ *to represent empty input/output. Let* Θ *be the p-tuple* $\Theta \triangleq (\perp, \ldots, \perp)$, *and*

$$\mathcal{X} \triangleq \mathcal{I}^p \backslash \{\Theta\}, \; \mathcal{X}_\Theta \triangleq \mathcal{X} \cup \{\Theta\}$$
$$\mathcal{Y} \triangleq \mathcal{O}^p$$

be the sets of input/output p-vectors, respectively.
4. $\delta : S \times \mathcal{X} \to S$ *is a* **(partial) next state function:** $s' = \delta(s, \mathbf{x})$ *for states* $s, s' \in S$ *and* $\mathbf{x} = (\mathbf{x}_1, \mathbf{x}_2, \ldots, \mathbf{x}_p) \in \mathcal{X}$ *means that if* $\mathcal{M}$ *is in state* s, *and inputs* $\mathbf{x}_1, \mathbf{x}_2, \ldots, \mathbf{x}_p$ *are applied to ports* $1, 2, \ldots, p$, *respectively, then* $\mathcal{M}$ *will enter state* s';
5. $\lambda : S \times \mathcal{X} \to \mathcal{Y}$ *is the* **output function;** *if* $\mathcal{M}$ *is in state* s, *and input* $\mathbf{x} = (\mathbf{x}_1, \mathbf{x}_2, \ldots, \mathbf{x}_p) \in \mathcal{X}$ *is applied, then the output* $\lambda(s, \mathbf{x}) = (\mathbf{y}_1, \mathbf{y}_2, \ldots, \mathbf{y}_p)$ *is observed; write* $\lambda_i(s, \mathbf{x}) = \mathbf{y}_i$ *to indicate that* $\mathbf{y}_i$ *is observed at port* i;
6. ω *is a* **PO transition label function:** *For any* $(s, \mathbf{x}) \in S \times \mathcal{X}$ *such that* $\delta(s, \mathbf{x}) = s'$ *and* $\lambda(s, \mathbf{x}) = \mathbf{y} \in \mathcal{Y}$, $\omega(s, \mathbf{x}) \subseteq (\{\mathbf{x}_1, \ldots, \mathbf{x}_p\} \times \{\mathbf{y}_1, \ldots, \mathbf{y}_p\})$ *is a partial order that satisfies*
   (a) $\mathbf{x}_i < \mathbf{y}_i$ *for all* $i \in \{1, \ldots, p\}$ *such that* $\mathbf{x}_i \neq \perp$ *and* $\mathbf{y}_i \neq \perp$, *and*
   (b) *if* $\mathbf{x}_i = \perp$, *then* $\mathbf{x}_i \not\leq \mathbf{y}_j$ *for all* $j \in Chn$.

We assume throughout this paper that the underlying transition graph is strongly connected for all IOPOA considered. δ and λ extend to sequence-valued functions $S \times \mathcal{X}^* \to S^*$ and $S \times \mathcal{X}^* \to \mathcal{Y}^*$, which we denote by the same function names.

2.2   I/O Vector Sequences

We allow I/O with restricted concurrency. That is, in each round, one input may be given and one output be received on each channel, and I/O on different channels in that round are pairwise concurrent; in particular, inputs can be made in any order. By contrast, I/Os in different rounds are never concurrent: earlier rounds strictly precede all subsequent ones, for all channels.

For $\mathbf{x}, \mathbf{x}' \in \mathcal{X}_\Theta$, say that $\mathbf{x} \leq \mathbf{x}'$ iff for all $i \in \{1, \ldots, p\}$, $\mathbf{x}_i \neq \mathbf{x}'_i$ implies $\mathbf{x}_i = \perp$. Write $\mathbf{x} < \mathbf{x}'$ iff $\mathbf{x} \leq \mathbf{x}'$ and $\mathbf{x} \neq \mathbf{x}'$. Intuitively, if $\mathbf{x} < \mathbf{x}'$, $\mathbf{x}$ can be seen as an *incomplete input* of $\mathbf{x}'$; one may "enter $\mathbf{x}$ first, and later add the rest of $\mathbf{x}'$". This is in fact a key to our technique for transition identification, see below. For vector sequences $\alpha, \beta \in \mathcal{X}^*$, write $\alpha \sqsubseteq \beta$ iff

1. $\alpha_1 \ldots \alpha_{|\alpha|-1}$ is a prefix of β, and
2. $\alpha_{|\alpha|} \leq \beta_{|\alpha|}$.

Note that this is more restrictive than the general partial order prefix relation.

Subtraction:

- For vectors $\mathbf{x} \leq \mathbf{x}'$, let $\mathbf{x}' \ominus \mathbf{x}$ be the vector w such that $w_i = \mathbf{x}'_i$ iff $\mathbf{x}_i = \bot$, and $w_i = \bot$ otherwise.
- For vector sequences $\alpha \sqsubseteq \beta$, let

$$\beta \ominus \alpha \triangleq (\beta_{|\alpha|} \ominus \alpha_{|\alpha|}) \circ \beta_{|\alpha|+1} \cdots \beta_{|\beta|}.$$

2.3   Completion of an IOPOA

Intermediate states: Suppose states s, s' and vectors $\mathbf{x}, \mathbf{x}' \in \mathcal{X}$ such that $\delta(s, \mathbf{x}) = s'$ and $\Theta < \mathbf{x}' < \mathbf{x}$. In general, $\delta(s, \mathbf{x}')$ may be undefined; remedy this by using an extended state space, with an intermediate state $s^{\mathbf{x}'} \notin S$ such that input $\mathbf{x}'$ leads from s to $s^{\mathbf{x}'}$, and input $\mathbf{x} \ominus \mathbf{x}'$ leads from $s^{\mathbf{x}'}$ to s'. Formally, we extend S to a superset $\overline{S}$ and assume δ, λ, ω extend to partial functions $\overline{\delta} : (\overline{S} \times \mathcal{X}) \to \overline{S}$, $\overline{\lambda} : (\overline{S} \times \mathcal{X}) \to \mathcal{Y}$ and $\overline{\omega} : (\overline{S} \times \mathcal{X}) \to 2^{(\mathcal{X} \times \mathcal{Y})}$ such that the following properties hold:

1. $\overline{\delta}_{|(S \times \mathcal{X})} \equiv \delta$, $\overline{\lambda}_{|(S \times \mathcal{X})} \equiv \lambda$, and $\overline{\omega}_{|(S \times \mathcal{X})} \equiv \omega$;
2. **Monotonicity:** Changing the order in which inputs are received must not alter the behavior of $\overline{\delta}$, $\overline{\lambda}$ and $\overline{\omega}$. Formally, $\alpha \sqsubseteq \beta$ must imply for all $s \in S$ ($\circ$ denotes concatenation):
   (a) $\overline{\delta}(s, \beta) = \overline{\delta}(\overline{\delta}(s, \alpha), \beta \ominus \alpha)$;
   (b) $\overline{\lambda}(s, \beta) = \overline{\lambda}(s, \alpha) \circ \overline{\lambda}(\overline{\delta}(s, \alpha), \beta \ominus \alpha)$;
   (c) $\overline{\omega}(s, \beta) = \overline{\omega}(s, \alpha) \circ \overline{\lambda}(\overline{\delta}(s, \alpha), \beta \ominus \alpha)$;

If the above are satisfied by $\mathcal{M}$, we say that $\mathcal{M}$ is **well-behaved**. If $\mathcal{M} \triangleq (S, \mathbf{s}^{\text{in}}, \mathcal{I}, \mathcal{O}, Chn, \delta, \lambda, \omega)$ is well-behaved, call $\overline{\mathcal{M}} \triangleq (\overline{S}, \mathbf{s}^{\text{in}}, \mathcal{I}, \mathcal{O}, Chn, \overline{\delta}, \overline{\lambda}, \overline{\omega})$ its **completion**.

Well-behavedness captures the *strong input determinism* of a transition in a IOPOA. If one transition specifies several inputs, then necessarily these inputs are concurrent, and thus can be input in the system in any order without impact on the state reached at the end of the transition. This is a reasonable assumption since if the state reached was different for different orderings of the input, it would imply that the inputs were in fact causally related, and therefore the specification should not have treated them as concurrent.

Thus, in the following, we require all IOPOAs to be well-behaved, thus we are dealing with strongly deterministic IOPOAs for which no order needs to be enforced for concurrent inputs.

2.4   Morphisms and Conformance

Let $\mathcal{M}$ and $\mathcal{M}'$ be two IOPO automata over the same in/output alphabets:

$$\mathcal{M} = (S, s_1, \mathcal{I}, \mathcal{O}, Chn, \delta, \lambda, \omega)$$
$$and \ \mathcal{M}' = (S', s_2, \mathcal{I}, \mathcal{O}, Chn, \delta', \lambda', \omega').$$

A **morphism** from $\mathcal{M}$ to $\mathcal{M}'$ is a total mapping $\Phi : S \to S'$ with the property that for all $(s, \mathbf{x}) \in S \times \mathcal{X}$ such that $\delta(s, \mathbf{x})$ is defined,

1. $\delta'(\Phi(s), \mathbf{x})$ is defined, and $\delta'(\Phi(s), \mathbf{x}) = \Phi(\delta(s, \mathbf{x}))$;
2. $\lambda'(\Phi(s), \mathbf{x}) = \lambda(s, \mathbf{x})$;
3. Φ induces a partial order isomorphism $\omega(s, \mathbf{x}) \to \omega'(\Phi(s), \mathbf{x})$.

We say that $\mathcal{M}'$ **conforms** to $\mathcal{M}$ iff there exists a *bijective* morphism $\Phi : S \to S'$, called a *conformal map*. Φ is an **isomorphism** iff (i) it is bijective and (ii) Φ^{-1} is a morphism from $\mathcal{M}'$ to $\mathcal{M}$. Note that conformance is not a symmetric relation, and strictly weaker than isomorphism. We note that:

Lemma 1. *The composition of conformal maps yields a conformal map, i.e. conformance is transitive.*

Theorem 1. *Let $\mathcal{M}_1$ and $\mathcal{M}_2$ be well-behaved IOPO automata. If $\mathcal{M}_2$ conforms to $\mathcal{M}_1$ under $\Phi : S_2 \to S_1$, then $\overline{\mathcal{M}}_2$ conforms to $\overline{\mathcal{M}}_1$.*

Proof. Suppose $\mathcal{M}_2$ conforms to $\mathcal{M}_1$ under $\Phi : S_2 \to S_1$. Let u_1 be an intermediate state of $\overline{\mathcal{M}}$, and $(s_1, \alpha) \in S_1 \times \mathcal{X}^*$ such that $\overline{\delta}_1(s_1, \alpha) = u_1$. By construction of $\overline{\mathcal{M}}_1$, there exists $s_1' \in S_1$ and $\alpha' \in \mathcal{X}^*$ such that $\alpha \sqsubseteq \alpha'$ and

$$\overline{\delta}_1(s_1, \alpha') = \delta_1(s_1, \alpha') = u_1'. \tag{1}$$

Isomorphism of $\mathcal{M}_1$ and $\mathcal{M}_2$ implies that

$$\overline{\delta}_2(s_2, \alpha') = \delta_2(s_2, \alpha') = u_2', \tag{2}$$

where $s_2 \triangleq \Phi(s_1)$, $s_2' \triangleq \Phi(s_1')$, and $u_2' \triangleq \Phi(u_1')$. By construction of $\overline{\mathcal{M}}_2$, there exists an intermediate state u' of $\overline{\mathcal{M}}'$ such that $\overline{\delta}_2(s_2, \alpha) = u_2$. Input determinism implies that u_2 is unique with this property. Set $\overline{\Phi}(u_1) \triangleq u_2$. One obtains an extension $\overline{\Phi} : \overline{S}_1 \to \overline{S}_2$ of $\Phi : S_1 \to S_2$, and checks that $\overline{\Phi}$ is bijective and defines a morphism $\overline{\mathcal{M}}_1 \to \overline{\mathcal{M}}_2$.

3 Conformance Testing for Automata with Distinguishing Sequences

The utility of the theorem 1 lies in the following application: Suppose we are given an implementation $\mathcal{M} = (S, \mathbf{s}^{\mathbf{in}}, Chn, \mathcal{I}, \mathcal{O}, \overline{\delta}, \overline{\lambda}, \overline{\omega})$ and a specification $\mathcal{M}_1 = (S_1, \mathbf{s}_1^{\mathbf{in}}, Chn, \mathcal{I}, \mathcal{O}, \delta_1, \lambda, \omega)$. Let $\mathcal{L}_1 \subseteq \mathcal{X}^*$ be the set of all input vector sequences α such that $\delta_1(\mathbf{s}_1^{\mathbf{in}}, \alpha)$ is defined, i.e. application of α in $\mathbf{s}^{\mathbf{in}}$ takes $\mathcal{M}_1$ to some specification state $s^\alpha = \delta(\mathbf{s}_1^{\mathbf{in}}, \alpha) \in S_1$. Let $\mathcal{M}_2$ be the IOPO automaton obtained by applying $\mathcal{L}_1$ in $\mathcal{M}$, i.e. let

$$\mathcal{M}_2 \triangleq (S_2, \mathbf{s}_1^{\mathbf{in}}, \mathcal{I}, \mathcal{O}, Chn, \delta_2, \lambda_2, \omega_2),$$

$$where: \quad S_2 \triangleq \left\{ s \in S \mid \exists\, \alpha \in \mathcal{L}_1 : \delta(\mathbf{s}^{\mathbf{in}}, \alpha) = s \right\},$$

$$\delta_2 \triangleq \delta_{|S_2 \times \mathcal{L}_1},$$

$$\lambda_2 \triangleq \lambda_{|S_2 \times \mathcal{L}_1},$$

$$\omega_2 \triangleq \omega_{|S_2 \times \mathcal{L}_1}.$$

Here, $\overline{\mathcal{L}}_1$ denotes the closure of $\mathcal{L}_1$ under subtraction of prefixes. By construction, $\mathcal{M}$ conforms to $\overline{\mathcal{M}}_2$. Using well-known techniques [17], conformance of $\mathcal{M}_2$ to $\mathcal{M}_1$ can be tested. If the test is passed, we know by Theorem 1 that $\overline{\mathcal{M}}_2$ conforms to $\overline{\mathcal{M}}_1$; thus Lemma 1 yields that $\mathcal{M}$ conforms to $\mathcal{M}_1$. Hence the task of testing conformance for IOPO automata is indeed completed.

In order to actually perform a test of conformance, we use a *checking sequence*. Let $C(\mathcal{M})$ be the set of IOPOA having no more states than $\mathcal{M}$, the same number of ports and the same input and output alphabet.

Definition 2 (Checking Sequence). *Let $\mathcal{M}_1 = (S_1, \mathbf{s}_1^{\mathsf{in}}, Chn, \mathcal{I}, \mathcal{O}, \delta_1, \lambda_1, \omega_1)$ be an IOPOA. A checking sequence of $\mathcal{M}_1$ is an input sequence I which distinguishes $\mathcal{M}_1$ from any IOPOA $\mathcal{M}_2 = (S_2, \mathbf{s}_2^{\mathsf{in}}, \mathcal{I}, \mathcal{O}, Chn, \delta_2, \lambda_2, \omega_2)$ in $C(\mathcal{M}_1)$ that does not conform to $\mathcal{M}_1$, i.e. such that $\forall \mathbf{s} \in S_2, \overline{\lambda_1}(\mathbf{s}_1^{\mathsf{in}}, I) \neq \overline{\lambda_2}(\mathbf{s}, I)$ or $\overline{\omega_1}(\mathbf{s}_1^{\mathsf{in}}, I) \neq \overline{\omega_2}(\mathbf{s}, I)$.*

Distinguishing sequences are usually defined as a sequence of inputs that will produce a different output for every state [17]. In the case of IOPOAs, we need to expand this definition to include the possibility of having the same output but different partial order labels.

Definition 3 (Distinguishing Sequence). *An IOPOA $\mathcal{M}$ admits an adaptive distinguishing sequence if there is a set of n input sequences $\{\xi_1, \ldots, \xi_n\}$, one per state of S, such that for all $i, j \in [1, \ldots, n], i \neq j$, ξ_i and ξ_j have a non-empty common prefix ξ_{ij} and $\lambda(s_i, \xi_{ij}) \neq \lambda(s_j, \xi_{ij})$ or $\omega(s_i, \xi_{ij}) \neq \omega(s_j, \xi_{ij})$.*

The automaton has a preset distinguishing sequence *if there is an adaptive one such that for all $i, j \in [1, \ldots, n], \xi_i = \xi_j$; in that case, ξ_i distinguishes state s_i.*

Not all automata have adaptive distinguishing sequences, but by definition, if an automaton has a preset checking sequence, it has an adaptive one.

3.1 Assumptions

In the following, we assume that the number q of states in the implementation does not exceed the number of states in the specification, i.e. $q \leq n$. We also assume that the directed graph induced by δ on S in strongly connected (and thus, by construction, the directed graph induced by $\bar{\delta}$ on $\bar{S}$ is also strongly connected). We finally assume that the IOPOA has an adaptive distinguishing sequence.

3.2 Sequential Input Automata

Since sequential input automata form a special case of IOPOA, it is instructive to look at that class first. It is known that we can construct a checking sequences of polynomial length [18, 19, 17], using polynomial time algorithms [20]. One example of such an algorithm is the following [19]. We call a *transfer sequence* $\tau(\mathbf{s}_i, \mathbf{s}_j)$ a sequence taking the machine from state $\mathbf{s}_i$ to state $\mathbf{s}_j$. Such a sequence always exists, since the state graph is strongly connected. In order to prove the morphism between the specification and the implementation, we need to show that every state on the specification exists in the implementation, and that every transition of the specification is in the implementation as well, going from the correct state to the correct state and generating the correct output when given the correct input.

Assuming that the machine starts in its initial state $\mathbf{s^{in}} = \mathbf{s}_1$ and that we have a distinguishing sequence ξ_i for every state s_i, the following test sequence checks that the implementation has n states, each of which reacts correctly when input the distinguishing sequence for that state:

$$\xi_1 \circ \tau(\delta(\mathbf{s}_1, \xi_1), \mathbf{s}_2) \circ \xi_2 \circ \tau(\delta(\mathbf{s}_2, \xi_2), \mathbf{s}_3) \circ \ldots \circ \xi_n \circ \tau(\delta(\mathbf{s}_n, \xi_n), \mathbf{s}_1) \circ \xi_1 \quad (3)$$

In order to test a transition a/b going from state $\mathbf{s}_i$ to $\mathbf{s}_j$, assuming the implementation is currently in a state $\mathbf{s}_k$, we can use the following test sequence:

$$\tau(\mathbf{s}_k, \mathbf{s}_{i-1}) \circ \xi_{i-1} \circ \tau(\delta(\mathbf{s}_{i-1}, \xi_{i-1}), \mathbf{s}_i) \circ a \circ \xi_j \quad (4)$$

Applying the test sequence 3, then applying the test sequence 4 for each transition provides a checking sequence. Unfortunately, this simple approach will not directly work with IOPOAs, because causal relationships between inputs and outputs between processes are not directly observable. In order to overcome this issue, we need to create longer test sequences that check causal relationships as well.

In order to explain our solution, we first illustrate our technique on a single transition, assuming that the implementation is correct.

3.3   Complete Transition Identification

We will test transitions by *delaying* input on only one channel, i; let us formalize this as input in the **i-test mode**: Let $1 \le i \le p$, and suppose an input vector $\mathbf{x} \in \mathcal{X}$ given. Then define input vector $\check{\mathbf{x}}^i$ as

$$\check{\mathbf{x}}^i_j \triangleq \begin{cases} \bot & : i = j \\ \mathbf{x}_j & : i \ne j, \end{cases}$$

and let $\hat{\mathbf{x}}^i \triangleq \mathbf{x} \ominus \check{\mathbf{x}}^i$; i.e.

$$\hat{\mathbf{x}}^i_j \triangleq \begin{cases} \mathbf{x}_i & : i = j \\ \bot & : i \ne j. \end{cases}$$

Let $\mathbf{x}$ be an input vector occurring in some input sequence $\alpha = \alpha_1 \ldots$, such that $\alpha_m = \mathbf{x}$ for some m. Applying $\mathbf{x}$ in **i-test mode** in α means applying, instead of α, the sequence $\alpha' \triangleq \alpha_1 \ldots \alpha_{m-1} \check{\mathbf{x}}^i \hat{\mathbf{x}}^i \alpha_{m+1} \ldots$.

Denote as $\Delta_i(\alpha)$ the sequence obtained from $\alpha = \alpha_1 \ldots$ by replacing *each* α_k by the pair $\check{\alpha}^i_k \hat{\alpha}^i_k$, i.e. in $\Delta_i(\alpha)$, input i is delayed in *all* rounds. It is important to note that delaying creates *equivalent* sequences, in the sense that for all α and i,

$$\lambda\left(\Delta_i(\alpha)\right) = \lambda(\alpha),$$
$$\delta\left(\Delta_i(\alpha)\right) = \delta(\alpha),$$
$$\text{and } \omega\left(\Delta_i(\alpha)\right) = \omega(\alpha).$$

Fix some input vector $\mathbf{x}$ and state $\mathbf{s}$, and set $\mathbf{y} \triangleq \lambda(\mathbf{s}, \mathbf{x})$. Assume we are interested in the label $\omega(\mathbf{s}, \mathbf{x})$; more precisely, since input and output are given, look for the partial order $<_\omega \subseteq (\mathbf{x} \times \mathbf{y})$. Denote as $\tau_{\mathbf{s}}^{\mathbf{x}} \triangleq \tau(\delta(\mathbf{s}, \mathbf{x}), \mathbf{s})$ a sequence that brings the machine back to state $\mathbf{s}$ after having input $\mathbf{x}$ from state $\mathbf{s}$. The test is now performed by inputting

$$\sigma \triangleq \check{\mathbf{x}}^1 \hat{\mathbf{x}}^1 \tau_{\mathbf{s}}^{\mathbf{x}} \check{\mathbf{x}}^2 \hat{\mathbf{x}}^2 \tau_{\mathbf{s}}^{\mathbf{x}} \ldots \tau_{\mathbf{s}}^{\mathbf{x}} \check{\mathbf{x}}^p \hat{\mathbf{x}}^p \tau_{\mathbf{s}}^{\mathbf{x}}, \tag{5}$$

that is, return to state $\mathbf{s}$ and test the same input vector $\mathbf{x}$, delaying a different channel in each round. Call $\mathbf{s}_i \triangleq \delta(\check{\mathbf{x}}^i, \mathbf{s})$ and $\check{\mathbf{y}}^i \triangleq \lambda(\check{\mathbf{x}}^i, \mathbf{s})$. Now, exactly those outputs that are generated only after input $\mathbf{x}_i$ are causal consequences of $\mathbf{x}_i$. That is, we obtain $<_\omega$ as follows:

$$<_\omega \triangleq \{(\mathbf{x}_i, \mathbf{y}_i) \mid i \in Chn, \mathbf{x}_i \neq \perp \text{ and } \mathbf{y}_i \neq \perp\} \tag{6}$$
$$\cup \left\{(\mathbf{x}_i, \mathbf{y}_j) \mid j \in Chn - \{i\} \wedge \check{\mathbf{y}}_j^i = \perp \wedge \mathbf{y}_j \neq \perp\right\}. \tag{7}$$

In fact, consider $i \neq j$ and $\mathbf{x}_i \neq \perp$ and $\mathbf{y}_j \neq \perp$.

- If $\mathbf{x}_i <_\omega \mathbf{y}_j$, then output $\mathbf{y}_j$ cannot be produced before input $\mathbf{x}_i$ arrives, hence $\check{\mathbf{y}}_j^i = \perp$; and
- conversely, if $\mathbf{x}_i \not<_\omega \mathbf{y}_j$, then $\mathbf{y}_j = \perp$.

Note that we assume here that all enabled outputs are produced and observed immediately, that is, we can actually decide whether or not output has been produced; reading $\perp$ means that no output was produced, we do not consider delayed outputs (where $\perp$ could mean 'no output *yet*').

3.4 Algorithm for IOPOA Conformance Testing

Single State Identifying Sequence: The implementation can be said to have implemented a state $\mathbf{s}_k$ if it can be shown that there is a state in the implementation that behaves like $\mathbf{s}_k$ when the input ξ_k is entered. The state $\mathbf{s}_k$ has been *identified* in the implementation. As already pointed out, the difficulty lies in the inter-channels causal relationships: We can easily observe that $\lambda(\mathbf{s}_k, \xi_k)$ is produced by the implementation, but checking that $\omega(\mathbf{s}_k, \xi_k)$ is correct requires more work.

Theorem 2. *An implementation of an IOPOA, assumed to be in a state $\mathbf{s}_k$ for which ξ_k is a distinguishing sequence, can be verified to have implemented $\mathbf{s}_k$ with the following test sequence:*

$$\left[\Delta_1(\xi_k) \circ \tau_{\mathbf{s}_k}^{\xi_k}\right]^n \circ \left[\Delta_2(\xi_k) \circ \tau_{\mathbf{s}_k}^{\xi_k}\right]^n \circ \ldots \circ \left[\Delta_p(\xi_k) \circ \tau_{\mathbf{s}_k}^{\xi_k}\right]^n, \tag{8}$$

where $[I]^n$ stands for the application of input sequence I n times.

Proof. By assumption, the IOPOA is deterministic and the implementation has at most n states. Thus, after entering the same input n times, the implementation is necessarily "locked" in a cycle of states and will not leave that cycle while the same input is entered. The input sequence will thus clearly loop between states that output $\lambda(\mathbf{s}_k, \xi_k)$ when input ξ_k. There are between 1 and n such states. By entering $[\Delta_i(\xi_k).\tau(\delta(\mathbf{s}_k, \xi_k), \mathbf{s}_k)]^n$ for some $i \in [1, \ldots, p]$, we can verify that the (at most n) states we are looping through do exhibit the correct causal relationships on port i. Since we test all ports, at the end of the test sequence we have identified in the implementation between 1 and n states that produce $\lambda(\mathbf{s}_k, \xi_k)$ and $\omega(\mathbf{s}_k, \xi_k)$ when input ξ_k.

Denote by $\Gamma(\mathbf{s}_i)$ the input sequence (8) for state $\mathbf{s}_i$. When adaptive distinguishing sequences exist, it is possible to find one of size $O(n^2)$ [21]. Moreover, transfer sequences have size $O(n)$, so the entire test sequence is of size $O(pn^3)$ when using adaptive distinguishing sequence.

3.5   Checking Sequence Construction

As a direct consequence of Theorem 2, it is easy to see that, assuming that the machine starts in its initial state $\mathbf{s}^{\mathbf{in}} = \mathbf{s}_1$ and that $\{\xi_1, \ldots, \xi_n\}$ is a set of adaptive distinguishing sequences, the following test sequence checks that the implementation has n states, each of which reacts correctly when input the corresponding distinguishing sequence:

$$\Gamma(\mathbf{s}_1) \circ \tau(\mathbf{s}_1, \mathbf{s}_2) \circ \Gamma(\mathbf{s}_2) \circ \tau(\mathbf{s}_2, \mathbf{s}_3) \circ \ldots \circ \Gamma(\mathbf{s}_n) \circ \tau(\mathbf{s}_n, \mathbf{s}_1) \circ \Gamma(\mathbf{s}_1) \quad (9)$$

When using adaptive checking sequences, this test sequence is of size $O(pn^4)$, since we have seen that a state identification sequence $\Gamma(\mathbf{s}_i)$ can be executed in size $O(pn^3)$ and a transfer sequence in $O(n)$, and since we have n states to verify.

In order to check the transitions, let us assume that in the IOPOA we have $\mathbf{x} \in \mathcal{X}, \mathbf{s}_i, \mathbf{s}_j \in S$ such that $\mathbf{s}_j = \delta(\mathbf{s}_i, \mathbf{x})$. We need to test that when the implementation is in a state identified as $\mathbf{s}_i$, if $\mathbf{x}$ is input the implementation outputs $\lambda(\mathbf{s}_i, \mathbf{x})$ to move into a state identified as $\mathbf{s}_j$, while respecting $\omega(\mathbf{s}_i, \mathbf{x})$.

The test sequence $\Gamma(\mathbf{s}_i).x.\Gamma(\mathbf{s}_j)$ can be used to check the transition's end states and $\lambda(\mathbf{s}_i, \mathbf{x})$. In order to verify $\omega(\mathbf{s}_i, \mathbf{x})$, the test sequence (5) can be used, but we have to ensure that $\tau(\delta(\mathbf{s}_i, \mathbf{x}), \mathbf{s}_i)$ brings indeed the implementation back to a state identified as $\mathbf{s}_i$, which can be achieved by the test sequence $\Gamma(\mathbf{s}_i).x.\tau(\delta(\mathbf{s}_i, \mathbf{x}), \mathbf{s}_i).\Gamma(\mathbf{s}_i)$. So, writing $\tau_{\mathbf{s}_i}^{\mathbf{x}} = \tau(\delta(\mathbf{s}_i, \mathbf{x}), \mathbf{s}_i)$, the entire test of the transition $\mathbf{x}$ can be done with the test sequence:

$$\Gamma(\mathbf{s}_i) \circ x \circ \Gamma(\mathbf{s}_j) \circ \tau_{\mathbf{s}_i}^{\mathbf{x}} \circ \Gamma(\mathbf{s}_i) \circ \Delta_1(\mathbf{x}) \circ \tau_{\mathbf{s}_i}^{\mathbf{x}} \circ \Delta_2(\mathbf{x}) \circ \tau_{\mathbf{s}_i}^{\mathbf{x}} \circ \ldots \circ \Delta_p(\mathbf{x}) \circ \tau_{\mathbf{s}_i}^{\mathbf{x}} \quad (10)$$

When using adaptive checking sequences, this test sequence is of size $O(pn^3)$. It must be done for every transition. If we assume t transitions, and since it is possible to go from any state to any other state in $O(n)$, testing every transition can be done in $O(tpn^3)$. The following result immediately follows:

Theorem 3. *Given an IOPOA of n states and t transitions having an adaptive checking sequence, assuming that the implementation is in the initial state, the following test sequence is a checking sequence of size $O(tpn^3 + pn^4)$:*

1. *Check all states with the test sequence (9)*
2. *For all transitions do:*
   (a) transfer to the starting state of the transition
   (b) check the transition with the test sequence (10)

Note that given that the IOPOA modeling leads to an exponential reduction of the size of the model compared to the multiport deterministic model, the checking sequence constructed with theorem 3 is also considerably shorter than one for a multiport deterministic model of the same system when dealing with sufficiently large and concurrent systegms.

4   Extensions and Outlook

4.1   Conformance Testing for Automata Without Distinguishing Sequences

Not every automaton has distinguishing sequences. In the absence of distinguishing sequences, one can always create checking sequences based on *separating families of sequences*. Adapting the definition of [17] to IOPOAs, a separating family of sequences for an IOPOA is a collection of n sets $Z_1, Z_2, \ldots, Z_n$, one collection per state. Each collection is made of up to $n-1$ sequences, such that for every pair of states $\mathbf{s}_i, \mathbf{s}_j \in S$, there is an input string α such that α is a prefix of some sequence of Z_i and of some sequence of Z_j and $\lambda(\mathbf{s}_i, \alpha) \neq \lambda(\mathbf{s}_j, \alpha)$ or $\omega(\mathbf{s}_i, \alpha) \neq \omega(\mathbf{s}_j, \alpha)$.

Separating families always exist for minimized automata, and can be used to create checking sequences based on *identifying sequences*; these checking sequences are in the worst case of exponential length in the number of separating sequences.

The construction carries over to the IOPOA case; the details will be given in an extended version of the present paper.

4.2   State Identification and State Verification

The *state identification* and *state verification* problems are two common and well known questions: find out the state the implementation is in (state identification) and verify that the implementation is indeed in a given state (state verification). With sequential input automata, the former can be answered if the automata has a distinguishing sequence, while the latter can be answered if the state has a *unique input output (UIO)*.

Unfortunately, neither questions can be answered with IOPOAs, even with a distinguishing sequence. The problem lies again in the inter-channel causal relationships that cannot be directly observed, and yet can be the only difference between two states. In order to uncover these differences, several tests of the states can be necessary, which is simply impossible when the state is unknown or unsure.

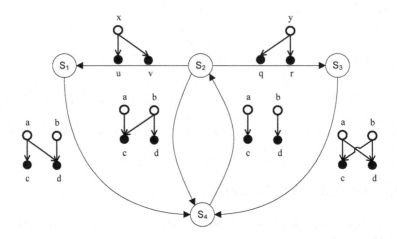

Fig. 3. An IOPOA for which states can neither be identified nor verified

The figure 3 illustrate the problem. In this IOPOA, the simple input (a, b) is a distinguishing sequence. Yet, the only strategies, a then b or b then a cannot distinguish between states s_2 and s_3 or s_1 and s_3 respectively. And since, whatever strategy, the implementation should be in state s_4 afterward, it is not possible to extend the test any further to clarify the situation, and thus state identification is not possible. For the same reason, it is not possible to ensure that the implementation is currently in state s_3.

4.3   Homing and Synchronizing Sequences

As opposed to the state identification and verification problems outlined in Section 4.2, homing sequences and synchronizing sequences are not difficult with IOPOAs.

A synchronizing sequence is a sequence that always takes the implementation to a particular state regardless of the state it was in when the sequence was entered. Clearly, not every automaton has such a synchronizing sequence. On the other hand, a synchronizing sequence does not involve any observation of the outputs of the implementation. Thus, if such a sequence exists, it can be used even with an IOPOA.

A *homing* sequence has the weaker property of taking the implementation to some known state, although not necessarily the same state depending on the unknown initial state. If the automaton is reduced, then such a homing sequence necessarily exists. In this case, the output plays a key role, since allows us to know the ending state. Yet, the classical way of constructing such an homing sequence is to pick two random states and build a sequence that tells them apart (such a sequence always exists in a reduced machine), and keep going until we have told all states pairwise apart. This can be easily achieved with IOPOAs, even if the two states differ only by non directly observable inter channels causal relationships, since we know what we are trying to uncover, and we can thus test for it. As an example, consider the IOPOA of Figure 4. Initially, we do not know the current steate, so it could be $\{s_1, s_2, s_3, s_4\}$. Say we want to separate s_1 from s_2; this can been done by delaying input a and observe whether d is output. Thus, the input sequence $< \bot, b >, < a, \bot >$ will generate either $< \bot, \bot >< c, d >$ or $< \bot, d >< c, \bot >$. In the first case, we were on s_1 or s_4, and we are now on $\{s_2, s_1\}$, and in the other case we were on s_2 or s_3, and we are now on $\{s_3, s_4\}$. The very same input again will tell apart the elements of these two sets.

So, the homing sequence is $< \bot, b >, < a, \bot >, < \bot, b >, < a, \bot >$, and the interpretation of the observation is, for the final state:

$$< \bot, \bot >< c, d >< \bot, \bot >< c, d > \Rightarrow s_2$$
$$< \bot, \bot >< c, d >< \bot, d >< c, \bot > \Rightarrow s_3$$
$$< \bot, d >< c, \bot >< \bot, \bot >< c, d > \Rightarrow s_1$$
$$< \bot, d >< c, \bot >< \bot, d >< c, \bot > \Rightarrow s_4$$

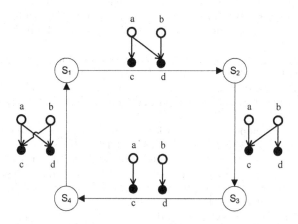

Fig. 4. Homing sequences can be found for IOPOA

5   Conclusion

We have introduced a generalized testing framework that includes and generalizes the classical I/O automaton setup. Using I/O partial order automata, asynchrony in inputs can be easily and concisely specified. Where a listing of all possible combinations of concurrent inputs is required with the Multiports Deterministic FSM model usually seen in the literature, a single transition is necessary with I/O partial order automata, leading to a model that can be exponentially smaller. I/O partial order automata allow also to specify the causal order between inputs and outputs, including unobservable interprocess causal relationships.

We have provided a test method to check the correctness of an implementation for a specification provided with an I/O partial order automata that has an adaptive distinguishing sequence. We show that in this case, we can produce a checking sequence of polynomial size in the number of transitions and the number of ports, thus we are not "paying back" the exponential reduction achieved by the model. This non intuitive result shows that I/O partial order automata are a powerful model when it comes to specifying and testing concurrency in distributed systems.

References

1. Chen, J., Hierons, R., Ural, H.: Conditions for resolving observability problems in distributed testing. In: de Frutos-Escrig, D., Núñez, M. (eds.) FORTE 2004. LNCS, vol. 3235, pp. 229–242. Springer, Heidelberg (2004)
2. Chen, X.J., Hierons, R.M., Ural, H.: Resolving observability problems in distributed test architecture. In: Wang, F. (ed.) FORTE 2005. LNCS, vol. 3731, pp. 219–232. Springer, Heidelberg (2005)
3. Sarikaya, B., Bochmann, G.v.: Synchronization and specification issues in protocol testing. IEEE Transactions on Communications 32, 389–395 (1984)
4. Luo, G., Dssouli, R., Bochmann, G.V., Venkataram, P., Ghedamsi, A.: Test generation with respect to distributed interfaces. Comput. Stand. Interfaces 16, 119–132 (1994)
5. Tai, K., Young, Y.: Synchronizable test sequences of finite state machines. Computer Networks and ISDN Systems 30, 1111–1134 (1998)
6. Hierons, R.M.: Testing a distributed system: Generating minimal synchronised test sequences that detect output-shifting faults. Information and Software Technology 43, 551–560 (2001)
7. Khoumsi, A.: A temporal approach for testing distributed systems. Software Engineering, IEEE Transactions on 28, 1085–1103 (2002)
8. Wu, W.J., Chen, W.H., Tang, C.Y.: Synchronizable for multi-party protocol conformance testing. Computer Communications 21, 1177–1183 (1998)
9. Cacciari, L., Rafiq, O.: Controllability and observability in distributed testing. Inform. Software Technol. 41, 767–780 (1999)
10. Boyd, S., Ural, H.: The synchronization problem in protocol testing and its complexity. Information Processing Letters 40, 131–136 (1991)
11. Dssouli, R., von Bochmann, G.: Error detection with multiple observers. In: Protocol Specification, Testing and Verification, vol. V, pp. 483–494. Elsevier, North Holland (1985)
12. Dssouli, R., von Bochmann, G.: Conformance testing with multiple observers. In: Protocol Specification, Testing and Verification, vol. VI, pp. 217–229. Elsevier, North Holland (1986)
13. Rafiq, O., Cacciari, L.: Coordination algorithm for distributed testing. The. Journal of Supercomputing 24, 203–211 (2003)
14. Hierons, R.M., Ural, H.: Uio sequence based checking sequence for distributed test architectures. Information and Software Technology 45, 798–803 (2003)
15. Chen, J., abd, H.U., R.M.H.: Overcoming observability problems in distributed test architectures (Information Processing Letters) (to appear)

16. Jourdan, G.V., Ural, H., Yenigün, H.: Minimizing coordination channels in distributed testing. In: Najm, E., Pradat-Peyre, J.F., Donzeau-Gouge, V.V. (eds.) FORTE 2006. LNCS, vol. 4229, pp. 451–466. Springer, Heidelberg (2006)
17. Lee, D., Yannakakis, M.: Principles and methods of testing finite–state machines – a survey. In: Proceedings of the IEEE, vol. 84, pp. 1089–1123 (1996)
18. Gill, A.: Introduction to The Theory of Finite State Machines. McGraw Hill, New York (1962)
19. Hennie, F.C.: Fault–detecting experiments for sequential circuits. In: Proceedings of Fifth Annual Symposium on Switching Circuit Theory and Logical Design, Princeton, New Jersey, pp. 95–110 (1964)
20. Lee, D., Yannakakis, M.: Testing finite state machines: state identification and verification. IEEE Trans. Computers 43, 306–320 (1994)
21. Sokolovskii, M.N.: Diagnostic experiments with automata. Journal Cybernetics and Systems Analysis 7, 988–994 (1971)

A Framework for Testing AIS Implementations

Tamás Horváth and Tibor Sulyán

Dept. of Control Engineering and Information Technology, Budapest University of
Technology and Economics, Budapest, Hungary
{tom, stibi}@iit.bme.hu

Abstract. Service availability has become one of the most crucial parameter of
telecommunications infrastructure and other IT applications. Service
Availability Forum (SAF) is a leading organization in publishing open
specifications for Highly Available (HA) systems. Its Application Interface
Specification (AIS) is a widely accepted standard for application developers.
Conformance to the standard is one of the most important quality metrics of
AIS implementations. However, implementers of the standard usually perform
testing on proprietary test suites, which makes difficult to compare the quality
of various AIS middleware. This paper presents a testing environment which
can be used to perform both conformance and functional tests on AIS
implementations. The results and experiences of testing a particular AIS
middleware are also summarized. Finally we show how to integrate our testing
environment to be part of a comprehensive TTCN-3 based AIS implementation
testing framework.

Keywords: Application Interface Specification (AIS), Conformance Testing,
Functional Testing, Service Availability.

1 Introduction

Service Availability Forum's Application Interface Specification (SAF AIS) [1]
defines a standard for a distributed middleware which can be used to implement
highly available carrier-grade services. Several implementations of the specification,
both commercial and open-source, are available to developers. To select the most
appropriate product, thorough testing is required. One of the most important quality
parameter of AIS implementations is standard compliance, but performance
characteristics have to be taken into consideration when choosing the appropriate
middleware.

Due to the complexity and distributed nature of AIS services, testing of the
implementations of the standard requires specialized test systems. Implementers of
AIS usually perform functional and performance testing on proprietary environments.
However, these systems cannot test other implementations, thus they are unable to
perform comparative examinations.

In this paper we present a new test suite which can be used to perform
conformance, functional and performance tests on various AIS implementations.
First, we give a short summary of the specification and the services it provides. Next,

A. Petrenko et al. (Eds.): TestCom/FATES 2007, LNCS 4581, pp. 186–198, 2007.

we evaluate some public test systems designed for AIS implementation testing. In the second part of the paper, the high level design of our proposed framework is introduced, followed by the test experiences and results of a particular AIS middleware product. Finally, we sketch the development direction of the system to become part of a TTCN-3-based testing framework.

2 Application Interface Specification Overview

Application Interface Specification defines Availability Management Framework (AMF) and a set of services offering functionality which supports the development of highly available applications. AMF provides a logical view of the cluster and supports the management of redundant resources and services on it. To build highly available services, AMF functionality is extended by a set of basic services, grouped into service areas:

– *Cluster Membership Service (CLM)* maintains information about the logical cluster and dynamically keeps track of the cluster membership status as nodes join or leave. CLM can be notify the application process when the status changes.
– *Event Service (EVT)* offers a publish-subscribe communication mechanism based on event channels which provide multipoint-to-multipoint event delivery.
– *Message Service (MSG)* is a reliable messaging infrastructure based on message queues. MSG service enables multipoint-to-point communication.
– *Checkpoint Service (CKPT)* supports the creation of distributed checkpoints and the incremental recording of checkpoint data. Application failure impact can be minimized by resuming to a state recorded before the failure.
– *Lock service (LCK)* provides lock entities in the cluster to synchronize access to shared resources.

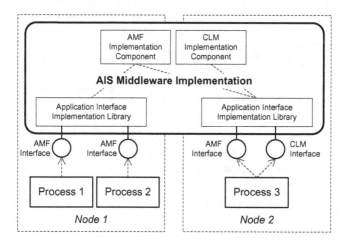

Fig. 1. Interaction between the AIS middleware implementation and the application processes on a two-node cluster

The detailed description of AIS services is out of the scope of this paper, an in-depth overview can be found in [1]. Nevertheless, the interfaces defined by the specification needs to be discussed here, because these interfaces can be considered as the only points of control and observation (PCO) of the AIS implementation. The relation of the middleware and its clients is shown on Figure 1.

Services provided by the middleware implementation are used by application *processes*. The term *process* can be considered equivalent to that defined in the POSIX standard. Communication between the application processes and AIS implementation is managed by *Application Interface Implementation Library (AIIL)*. Interfaces and implementations of the service areas are separated. Moreover, *Implementation Components* are not covered by the standard; the internal design of the middleware is unknown to the middleware tester. Service area interfaces (*AMF and CLM interface*) represent a logical communication channel between *processes* and the AIS implementation. The logical nature of interface connections is emphasized on Figure 2 by displaying two *AMF interface* objects. The standard provides both synchronous and asynchronous programming models for the communication. Moreover, certain requests can be performed either ways.

In general the synchronous API is much easier to use. Synchronous communication is based on blocking API function calls. The user invokes a request by calling an API function. The request is considered performed by the time the function has been returned. Data exchange between the application process and the AIS middleware is realized by the parameters and the return value of the API function. Although the synchronous model greatly simplifies the programming tasks, certain services cannot be used this way. For example, cluster membership change notifications require a

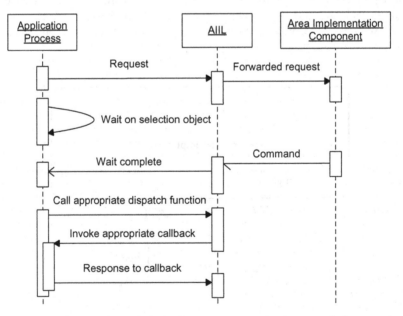

Fig. 2. Sequence diagram of a typical asynchronous communication scenario between the AIS implementation and the application process

mechanism that permits the middleware to send information to the application asynchronously. Long-running requests are another example where synchronous requests are not recommended.

To support asynchronous communication, the standard employs a callback mechanism. The request API function returns immediately to the caller, and a standard-defined *callback function* is called when the request has been completed. Callback functions are standard-defined, but implemented by the process. Since the middleware cannot invoke a function directly in the application process, a notification is sent first on a *selection object* by the AIIL. In response, the process invokes an area-specific dispatcher function which finally invokes the appropriate callback function of the application process. The body of callback functions usually concludes in a response call carrying status information to the middleware. An illustration of a typical asynchronous communication scenario is presented on Figure 2.

This communication model has high importance considering AIS implementation testing, because all kinds of control and observation tasks can be derived to a series of the following extensions to the model:

- Addition of control operations, such as AIS service requests;
- Inspection of callback function parameters;
- Addition of administrative code (result evaluation, logging, communication with other test components).

3   Current AIS Implementation Testing Systems

Our research covered a survey of currently available AIS testing frameworks. We examined two open source systems considering the following quality parameters:

- **Executable test types.** The most natural expectation from a test system is that it should support multiple test types. We distinguish four classes of tests when testing AIS implementations. Conformance tests address the verification of the API implemented in the middleware product. Functional tests verify that the behavior of the middleware conforms to the standard. Performance tests mean any performance measurements. Robustness tests examine the operation of the system under extreme conditions, for example operation system crashes or hardware failures. In our research, we primarily focus on conformance, functional and performance tests.
- **Capability of automated testing.** This criterion is also a common expectation from a testing tool. Automated testing means not only automatic test case execution, but also automatic evaluation of results and run-time reconfiguration of the IUT. AIS doesn't define the configuration methods of the cluster, so it can be different in various middleware products.
- **Availability of test results.** The test system should report not only the test verdict, but also additional information about the test case execution. Performance tests for example may require timing information. Additional information is also needed to determine the cause of a particular test execution failure.
- **Adaptability.** This is a requirement specific to AIS standard. The standard evolves constantly, and has multiple versions. Different products realize different versions;

even the same middleware may implement different versions of particular service areas. A universal test framework must support this diversity to be able to test multiple implementations.

Based on the criteria above, we shortly describe and evaluate the test suites examined.

3.1 SAFtest

SAFtest [2] is a test execution framework for running AIS and HPI (Hardware Platform Interface) conformance tests. SAFtest test cases are small C programs that implement the procedure shown on Figure 2. In addition, this sequence is extended by calls of AIS API functions with correct and incorrect parameters and execution order. The main purpose of the test cases is to verify that particular API functions exist and yield the expected result when called with different parameters. The framework itself is a collection of makefiles and shell scripts which can configure the IUT, run test cases and collect results. Test cases can be extended with meta-information such as test case description or reference to the specification fragment tested. This meta-information is used by SAFtest to create test coverage reports automatically.

Example test case metadata containing the name of the function under test, assertion description and specification coverage information.

```
<assertions spec="AIS-B.01.01"
 function="saClmSelectionObjectGet">
   <assertion id="1-1" line="P21-38: P22-1">
      Call saClmSelectionObjectGet(), before
      saClmFinalize() is invoked, check if the returned
      selection object is valid.
   </assertion>
   <assertion id="1-2" line="P21-38: P22-1">
      Call saClmSelectionObjectGet(), then invoke
      saClmFinalize(), check if the returned selection
      object is invalid.
   </assertion>
</assertions>
```

This snippet also shows that SAFtest is primarily designed for AIS API function testing as assertions are grouped by the AIS function under test. The range of executable test cases is limited to API conformance tests. The most important flaw of this framework is that test cases of SAFtest ignore the distributed nature of the middleware. Test cases run on a single computer, not on a cluster. This way the majority of the AIS functionality cannot be tested properly.

To summarize, SAFtest is a compact test framework well suited for testing API conformance. It can run tests and generate test result summary automatically. Automatic configuration is not necessary in this case since tests run on a single host. However, this means that most of the AIS functionality cannot be tested with this framework. In addition, testing a different specification version requires a different test case set.

3.2 SAFtest Next Generation

SAFtest-NG [3] is a recent test suite which tries to eliminate most of the limitations of the SAFtest system. Figure 3 shows the main system components of the test suite. The main objectives of this framework are the following:

- To offer a general-purpose AIS test framework which supports not only API conformance but functional and performance tests as well.
- To support fully automatic test execution including automated test environment configuration.
- To be able to test multiple AIS implementations with a minimum amount of reconfiguration overhead.

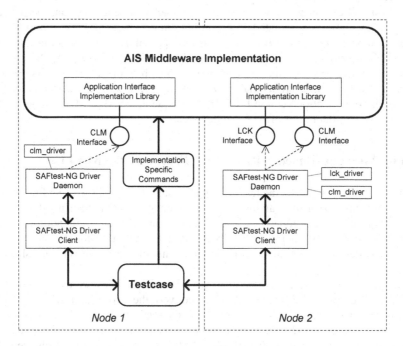

Fig. 3. A sample SAFtest-NG test suite configuration on a two-node cluster. It illustrates relationship between the test case and the various driver components.

Test cases in SAFtest-NG are written in Ruby, a high-level object-oriented script-like language. The abstraction level Ruby enables very clean and straightforward test case implementation. Test cases do not run directly on the AIS middleware, they control *drivers*. Drivers are the main components of the SAFtest-NG architecture. A driver consists of three parts. *Driver clients* or short-lived drivers are accepting high-level test case instructions, converting and relaying them to *Driver daemons*. By using this indirection, a single test case is able to drive the whole cluster which AIS implementation manages. *Driver daemons* or long-lived drivers communicate with the AIS middleware via one or more service area interface. Driver daemons implement the communication process shown on Figure 2, and execute AIS function calls. The actual API calling is implemented in separate shared libraries (*clm_driver, lck_driver*). This decomposition enables the testing of special implementations, where the service areas are realized according to different versions of the standard. Moreover, this design enhances the reusability of driver libraries when adapting to a new version of the specification.

SAFtest-NG offers a solution to test environment setup as well by using implementation hooks. These hooks describe the *implementation specific commands* for each particular AIS middleware implementation to perform any operation which is not defined by the standard. These operations include cluster management (addition or removal of nodes), and information retrieval commands (for example, gathering information about the current cluster membership status). AIS middleware vendors only need to provide these hooks to test their product.

SAFtest-NG enhances the executable test range with functional and performance tests. It also supports automated IUT configuration, which SAFtest supported only partially. Support for test result collection is only partial, so test evaluation (especially in case of performance tests) requires log analyzer tools. Unfortunately, SAFtest-NG is an incomplete system, and it seems to be an abandoned project. As Figure 3 suggests, driver libraries are available only for the CLM and LCK service areas, and only for the specification version B.01.01. Consequently, SAFtest-NG cannot be used for a complete in-depth test of AIS middleware.

4   The Message-Based AIS Testing Framework (MATF)

In this chapter we introduce the test system (Message-based AIS Testing Framework - MATF) we developed to examine AIS implementations. Our primary design goals were to meet the requirements we enumerated in chapter 3. In addition, future integration of the system into a TTCN-3 based framework was also an important design consideration. The architectural components of the framework are shown on Figure 4.

The architecture enables remote testing of the AIS middleware as defined in ISO 9646 [4]. The main idea of MATF is to convert the procedural AIS API into a message-based interface. *Test Coordinator (TC)* sends test control messages to the *Local Test Components (LTC)*. LTC then interprets the message with *Message Parser* and either communicates with the middleware (via *Dispatcher*) or controls the cluster node itself (via *LC*). Middleware responses are forwarded to the *Remote Log Server* and potentially to other log servers. Test Coordinator evaluates the results based on the entries of the remote log. In the following chapters we summarize the roles of the components of MATF.

4.1   Test Coordination

Test case messages are sent by the Test Coordinator component. Messages can be transmitted through any reliable communications channel, for example TCP/IP. The TC component implements a log server, which collects incoming data from all Local Test Components. Test case verdict is evaluated based on this data. The format of messages is analogous to function signatures. A message has an identifier and zero or

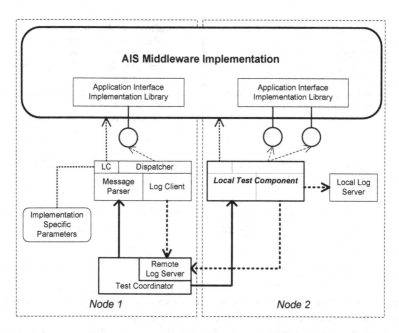

Fig. 4. Components of the proposed Message-based AIS Testing Framework. *Local Test Component* consists of four modules: the *Local Controller (LC)*, the *Dispatcher*, the *Message Parser* and the *Log Client*.

more parameters. This allows the direct mapping of the AIS functions into messages. Although the use of messages introduces an indirection between the tester and the implementation under test, message-based testing has several advantages over the direct use of AIS API.

The most important among them is the capability of abstraction. Common tasks which require multiple AIS function calls can be encoded in a single message. These tasks include for example connection initiation between the application process and the middleware. Another aspect of abstraction is detail hiding. Messages can hide details of API functions by using default message parameters analogous to C++ default function parameters. When adapting to a new version of the specification, only incremental changes are needed to be performed on the previous Message Parser module. This incremental nature applies also to the test cases. The format of a specific message can be the same for different versions of the specification. Consider two versions of the message queue opening API call [5] [6]:

These function specifications have three differences. The return type has changed, the passing form of the *msgHandle* parameter has been altered, and the two last parameters have been swapped. These changes can be hidden from the test case developer. Different versions of the Message Parser modules may translate them to the appropriate AIS function calls.

Another important advantage of the message-based testing is that messages can transparently extend test control instructions by operations that are not covered by the standard, but are necessary to perform successful testing. For example the details of

```
SaErrorT saMsgQueueOpen(              SaAisErrorT saMsgQueueOpen(
    const SaMsgHandleT *msgHandle,        SaMsgHandleT msgHandle,
    const SaNameT *queueName,             const SaNameT *queueName,
    const SaMsgQueueCreationAttributesT   const SaMsgQueueCreationAttributesT
        *creationAttributes,                  *creationAttributes,
    SaMsgQueueOpenFlagsT openFlags,       SaMsgQueueOpenFlagsT openFlags,
    SaMsgQueueHandleT *queueHandle,       SaTimeT timeout,
    SaTimeT timeout                       SaMsgQueueHandleT *queueHandle
);                                    );
```

Fig. 5. The same API functions from version A.01.01 (left) and version B.01.01 (right) of the specification

adding a new node to a cluster or completely shutting down a node are not defined in the AIS standard. Specific messages can be implemented in MATF to these operations.

4.2 Message Processing

Messages sent by the Test Coordinator are processed by the Message Parser. MP interprets messages and forwards them to the appropriate communication module (Local Controller or Dispatcher). The Message Parser is an object-oriented recursive-descent parser, which provides high reusability of the parser components, since parsing of different message entities are encapsulated in different parser objects.

Messages specific to local node control are not translated to API functions; rather they are passed to the Local Controller component. The LC will execute operating system commands to control the cluster or the middleware implementation itself. The concrete effect of control messages can be configured by *Implementation Specific Parameters*, a configuration mechanism similar to *implementation hooks* in SAFtest-NG. This way the test suite can be adapted to test multiple IUTs.

4.3 Controlling and Observing the IUT

Messages that drive the AIS implementation are handled by the Dispatcher component. The Dispatcher performs two main tasks.

Primarily the component provides synchronous and asynchronous interfaces for the Message Parser to enable communication with the AIS middleware. This is implemented by running a *dispatch loop*, which is a generalized version of the communication sequence shown on Figure 2. By default, all requests run on a single thread. However, to test the multi-threaded operation of the middleware,

Dispatcher also has to maintain all session information required to the communication. Session information includes handles, identifiers, or any specification-defined object that persists between multiple API calls. For example, message handle, message queue name and handle parameters on Figure 5 are session information.

After a synchronous operation, the results of the request are immediately available, so dispatcher can forward the results to the *Log Client*. Logging of the results of asynchronous operations is performed in the callback functions. According to its configuration, the Log Client sends the messages to multiple *Log Servers*. A Log

Server can be a local file or a process, either local or remote, which collects log data and maintains correct order between log entries. The *Remote Log Server* collects all incoming information from all local test components. Overall test results can be evaluated based on the data collected by the *Remote Log Server.*

5 Testing Experiments

To verify the operability of the architecture above, we executed a set of test cases on OpenAIS [7], an open-source AIS middleware implementation. This chapter summarizes the test results and the experiments we gained during the testing process.

5.1 Test Suite Configuration

The structure of OpenAIS is a straightforward mapping of the standard. Each service area is implemented in a separate process, interconnected by a private communication protocol. The Application Interface Implementation Library (see Figure 1) is implemented by a process called *AIS executive* or *aisexec*. The middleware can be configured by two configuration files, *openais.conf* and *amf.conf*. The former contains operational settings such as network setting or node authorization information. Since this data is implementation-specific, we don't need to alter its contents during testing. As the name suggests, *amf.conf* is used by the Availability Management Framework. The file stores the actual redundancy model of the cluster. To configure OpenAIS, the behavior of Control Component of MATF has been defined as:

- Adding or removing a node is equivalent to starting or shutting down *aisexec* on that particular node;
- AMF Redundancy model setting is equivalent to the replacement of *amf.conf* on all nodes, followed by a cluster restart. The latter step is needed because OpenAIS doesn't support dynamic redundancy model modification.

To examine OpenAIS, we set up a test suite consisting of three cluster nodes. *Test Coordinator* and *Remote Log Server* relied on a separate controller host. Clocks on all nodes were synchronized from a Network Time Protocol server.

5.2 Test Execution

We have tested OpenAIS version 0.70.1, the latest production ready version available at the time. This release implements version A.01.01 of the Availability Management Framework, and version B.01.01 of the CLM, EVT and CKPT service areas. Distributed Locks (LCK) and Message Service (MSG) are not implemented at all. To test OpenAIS, we established 8 possible configurations of the cluster. The configuration included the number of clusters, the number of local test components and middleware configuration, such as AMF redundancy model.

A total number of 113 test cases had been elaborated based on the specification versions OpenAIS implements. Although the test cases don't provide an exhaustive evaluation of the IUT, they inspect all functionality of the service areas implemented.

Passed test cases row requires no explanation. *Passed with conformance issue* means that although the functionality under test is correctly implemented, some

Table 1. Summary of the test results

Total number of test cases	113
Test cases passed	53
Test cases passed but conformance issues encountered	6
Functionality not implemented	31
Test cases failed	21
Test verdict cannot be determined	2

output function parameters or return values were unexpected. Failed test cases include incorrectly implemented functionality and critical errors of test case execution. Incorrect functionality manifested in invalid data or missing callback invocations. Critical error means unexpected *aisexec* termination which is equivalent to node shutdown. Finally, in two cases the information gathered after the test case execution were insufficient to evaluate the result.

6 Future Work

The test system we developed is far from being a complete framework. We implemented only a prototype version of the architectural elements described above. This prototype system is not capable to perform automated test execution, because of the rudimentary Test Controller and the Log components. The actual purpose of these components is to provide a primitive front-end to the Message Parser and Dispatcher components and to perform actual testing with it.

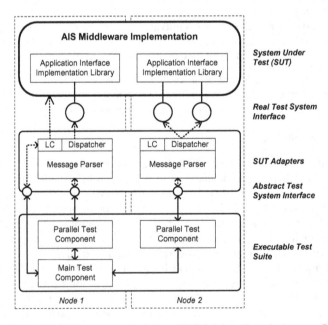

Fig. 6. Integration of MATF components into a TTCN-3 based environment. *Circles* denote TTCN-3 test ports.

Nevertheless, the prototypical implementation of the front-end is intentional. The next step of our development is the design and implementation of a TTCN-3 [8] based front end which will replace the Test Controller and the logging components. The new components of the framework are shown on Figure 6.

The rightmost column of the figure denotes the corresponding elements of the TTCN-3 runtime system architecture [9]. TTCN-3 provides highly sophisticated test coordination and evaluation mechanisms. The *Executable Test Suite (ETS)* can be considered as an advanced replacement of the prototype front-end described in the previous chapter. ETS supports the automation of test execution, test result collection and evaluation. The main modules of MATF (LC, Dispatcher and Message Parser) can be integrated with minor modification into this architecture as *SUT adapters*. *Abstract Test System Interface* is the interface defined by Message Parser and Local Controller. Although this interface already exists, TTCN-3 requires its adaptation to function as TTCN-3 test ports. By the introduction of test ports, the test configuration messages and the actual test messages can be separated. The middleware configuration messages are sent through a configuration port, while the test messages are sent to Message Parser via a separate port. This is possible because *Main Test Component* not only coordinates *Parallel Test Components*, but can directly send messages to SUT adapters via the Abstract Test System Interface.

7 Conclusion

In this paper we examined two currently available AIS implementation testing frameworks. We found that both systems can be used for particular testing tasks. Nevertheless both systems have certain flaws that prevent them from being general purpose test frameworks. We presented the architecture of a new framework which can be used for comprehensive testing of AIS middleware. To test the usability of the new system we implemented a prototype of the framework and a set of functional test cases. We executed these tests on an open source AIS implementation and summarized the results. The success of the testing process showed that MATF can be used to test AIS middleware. The next step of development is the TTCN-3 integration of the framework. We presented the architectural design of the future test system which highly reuses the actual components of MATF.

References

1. Service Availability Forum, Application Interface Specification, vol. 1, Overview and Models, SAI-AIS-B.01.01
2. SAFTest, http://www.saf-test.org/
3. SAFTest Next Generation, http://saftest.berlios.de/
4. International Organization for Standardization, Information technology – Open Systems Interconnection – Conformance testing methodology and framework – Part 1: General concepts, ISO/IEC 9646-1:1994
5. Service Availability Forum, Application Interface Specification, SAI-AIS-A.01.01
6. Service Availability Forum, Application Interface Specification, vol. 6: Message Service SAI-AIS-MSG-B.01.01

7. OpenAIS: Standards-Based Cluster Framework. http://developer.osdl.org/dev/openais/
8. European Telecommunications Standards Institute, Methods for Testing and Specification (MTS); The Testing and Test Control Notation version 3; Part 1: TTCN-3 Core Language, ETSI ES 201 873-1 (v3.1.1), Sophia Antipolis (June 2005)
9. European Telecommunications Standards Institute, Methods for Testing and Specification (MTS); The Testing and Test Control Notation version 3; Part 5: TTCN-3 Runtime Interface, ETSI ES 201 873-1 (v3.1.1), Sophia Antipolis (June 2005)

An Object-Oriented Framework for Improving Software Reuse on Automated Testing of Mobile Phones

Luiz Kawakami[1], André Knabben[1], Douglas Rechia[2], Denise Bastos[2], Otavio Pereira[2], Ricardo Pereira e Silva[2], and Luiz C.V. dos Santos[2]

[1] Motorola - Brasil Test Center
{wlk023,wak023}@motorola.com
[2] Computer Science Department - Federal University of Santa Catarina, Brazil
{rechia,denise,otavio,ricardo,santos}@labsoft.ufsc.br

Abstract. To be cost effective, the decision to automate tests that are usually hand-executed has to rely on a tradeoff between the time consumed to build the automation infrastructure and the time actually saved by the automated tests. Techniques which improve software reuse not only reduce the cost of automation, but the resulting productivity gain speeds up development. Such issues are specially relevant to the software development for mobile phones, where the time-to-market pressure asks for faster design and requires quicker deployment of new products. This paper presents a novel object-oriented framework tailored to support the automation of user-level test cases so as to improve the rate of deployment of mobile phones. Despite inherent test automation limitations, experimental results show that, with automation, the overall testing effort is about three times less than the manual effort, when measured within a one-year interval.

Keywords: Software verification, Software reusability, Software metrics.

1 Introduction

Many mobile phone models are released to the market every year with improved or brand new features. Examples of common phone features are messaging (*short messages* – SMS, *multimedia messages* – MMS and E-mail), phone and appointment books, alarm, embedded camera and so on. These functionalities are largely implemented in software. Every feature of each new phone model must be tested prior to its release to the end users. Also, the interaction among features must be checked, so as to ensure their proper integration. Such user-level functional tests are crucial to reduce both customer dissatisfaction and technical assistance costs.

Functional testing relies on checking many use-case scenarios. Each *test case* (TC) is a sequence of steps that performs a specific task or a group of tasks. TCs not only specify the steps, but also the expected results.

A. Petrenko et al. (Eds.): TestCom/FATES 2007, LNCS 4581, pp. 199–211, 2007.

Test engineers usually execute TCs manually. A TC may be repeated several times during successive software development life cycles. Manual test execution is both time-consuming and error prone, especially because exact TC reproduction cannot be guaranteed through different phone software versions. Such inadequacy leads to the use of software to automate TC execution. An *automated test case* (ATC) can automatically reproduce the steps that would be performed manually by the test engineer.

This paper presents a novel object-oriented framework tailored to support ATC creation for user-level functional testing of mobile phones. Our *test automation framework*, from now on called TAF, was designed to allow as much ATC reuse as possible across distinct phone models of a given product family. Essentially, TAF allows the automation of functional TCs and the efficient retargeting of pre-existing ATCs to distinct phone models. Such a retargeting within a product family is the very key to achieving productivity gain. Once a TC is automated, it is ready to be executed as many times as needed, through all phone development life cycles, reducing the time-to-market. The main contribution of this paper consists in the analysis of solid experimental results which show the actual impact of software reuse on test automation. The reuse achieved with our framework makes it worthy to be employed in the corporate environment, given certain conditions described later in this paper.

The remainder of this paper is organized as follows: Section 2 reviews related work; the structure of the automation framework is described in Section 3; Section 4 summarizes experimental results and finally our conclusions are drawn in Section 5.

2   Related Work

2.1   Practices in the Corporate Environment

While TC manual execution is still current practice, many companies are widely adopting test automation for unit and regression testing, since they are recurrent during software development life cycles, despite the inherent limitations imposed upon automation (for instance, 50% of the TCs within typical user-level test suites are not suitable for automation).

Although the approaches vary from one company to another, test automation has been progressively adopted at the user-level as a way of reducing the required effort for test execution. At this level, the main approaches employ either in-house developed test ware, such as PTF [1], or third party test systems, such as TestQuest Pro (R) [2].

Motorola relies on in-house test automation infrastructure. TAF, which will be described in Section 3, is the keystone of such infrastructure.

2.2   Related Research Topics

Related approaches on test automation address two basic goals: test case generation, and test execution and analysis.

Test suite generation focuses on finding ways of capturing test cases from code or design specifications. As an example, model-based testing is an approach in which the behavior of the software under test is described by means of formal models (such as Petri nets, state charts and so forth) as a starting point to automatic or semi-automatic TC generation [3] [4] [5] [6]. Other approach relies on algorithms able to create test cases which cover the interaction among different applications in rich-feature communicating systems [7].

The automation of test execution and result analysis aims to produce software artifacts able to execute test suites (automatically generated or not) and to compare the obtained results to the expected ones [8].

Recent work seems to indicate that there is not so far an ultimate solution for software test automation challenges [9]. On the contrary, distinct successful approaches are reported [8], [10], [11], [12], [13].

To assess the economic viability of test automation, a preliminary trade-off analysis [14] should be performed. Since high frequency of invocation is a prerequisite for automating a TC, common pitfalls should be avoided, such as overestimating the required effort for manual execution or underestimating the percentage of tests that are actually suitable for automation [15] [16].

As test automation often consists in producing software to test software, an alternative approach to achieve a better trade-off is to promote software reuse when constructing *testware*. Object-oriented frameworks are reusable software artifacts able to support testware development [17]. JUnit is a well-known example of framework applied to the domain of test development [18].

Since there is a trade-off between generality and effectiveness of reuse, domain-specific frameworks (such as JUnit) are expected to lead to a lower percentage of reuse than application-specific ones. That was the motivation to the development of a novel application-specific framework tailored to mobile phones.

There is lack of evidence in the literature quantifying the impact of software reuse on test automation. That's why the main contribution of this paper is to report the quantitative impact of an application-specific framework on real-life state-of-the-art product deployment.

3   TAF Design Description

TAF is an object-oriented framework tailored to automate functional user-level test-case *execution* for mobile phones. TAF provides the proper infrastructure to automate a test, but this process is essentially manual. In other words, TAF addresses the automation of test *execution*, not the automatic *generation* of tests.

TAF enables reuse by raising the abstraction level so as to make ATCs largely independent of model-specific phone properties. Therefore, it has to rely on a lower-level infrastructure, as described in the following subsection.

TAF was developed by Brasil Test Center (BTC), an R&D network of research institutes under Motorola's leadership[1].

3.1 Low-Level Implementation Infrastructure

In order to interface with the phone, TAF relies on a Motorola proprietary artifact, the so-called *phone test framework* (PTF) [1]. PTF provides an application-programming interface (API) that allows the user to simulate events from the phone's input/output behavior, like key pressing and display capture. Since most API methods are encoded at low abstraction levels, PTF leads to test scripts that are hard to read, difficult to maintain and inefficient to port to other phones. However, PTF represents a highly appropriate basis for test automation implementation.

3.2 High-Level ATC Encoding

The key to raising the abstraction level is to encapsulate lower-level test input actions (such as sequence of key pressings) and test output analysis (such as checking the phone display contents) into a so-called *utility function* (UF). UFs are primitive entities that hierarchically isolate functionality from implementation, leading to high-level ATCs. An ATC tells "what" to test, but not "how" to perform some input action or output analysis. As a result, UFs must rely on PTF components for actual test implementation.

Fig. 1 shows an example of a high-level ATC using utility functions. This ATC fragment performs the following sequence of steps: first, it takes a picture and stores it as a file (Steps 1 to 3); then, it checks some attributes and deletes the file (Steps 4 to 8). Note that seven UFs are employed: *LaunchApp* (it launches the camera application), *CapturePictureFromCamera* (it takes the picture), *storeMultimediaFileAs* (it stores the picture into the phone file system), *scrollToAndSelectMultimediaFile* (it scrolls through a list and opens a specific multimedia file), *openCurrentFileDetails* (it opens the screen which displays file attributes such as type, size, etc), *verifyAllMultimediaFileDetails* (it checks whether the picture file has the expected properties) and *DeleteFile* (it simply deletes the file from the phone file system).

Although different phones exhibit distinct input/output behavior, a same high-level ATC is applicable to several phone models of a given product family, since they basically implement the same features. Therefore, a TC is automated only once for a product family and the resulting high-level ATC must be retargeted to every distinct phone model within the family. This retargeting process is called *porting*.

TAF was designed to allow efficient porting of high-level ATCs, as will be described in the next subsection.

---

[1] TAF's initial design and development involved the Computer Science Department of Federal University of Santa Catarina (INE–UFSC) and the Center for Informatics of Federal University of Pernambuco (CIn–UFPE).

```
...
// Step 1: launch Camera application
   navigationTk.launchApp(PhoneApplication.CAMERA);

// Step 2: take the picture
   multimediaTk.capturePictureFromCamera();

// Step 3: store the picture and hold its file name in variable picture
   MultimediaFile picture =
       multimediaTk.storeMultimediaFileAs(MultimediaItem.STORE_ONLY);

// Step 4: take the phone to PICTURES_FILE_LIST screen
   navigationTk.launchApp(PhoneApplication.PICTURES);

// Step 5: open the picture
   multimediaTk.scrollToAndSelectMultimediaFile(picture);

// Step 6: open the file details screen
   multimediaTk.openCurrentFileDetails();

// Step 7: verify picture file attributes
   multimediaTk.verifyAllMultimediaFileDetails(picture);

// Step 8: return to PICTURE_VIEWER
   phoneTk.returnToPreviousScreen();

   multimediaTk.deleteFile(picture, true);
...
```

Fig. 1. An ATC fragment

3.3   TAF Organization

To enable the reuse of ATCs, TAF was designed to overcome the issues that are raised by PTF's low-level APIs, such as creating scripts that are hard to read and difficult to maintain.

Fig. 2 summarizes the organization of TAF in terms of class relations.

The interface *Step* lies at the top of the diagram. It provides a generic method (*execute*) allowing ATCs to invoke the functionality of distinct UFs.

The class *BaseTestCase* stores a collection of objects of type *Step*. It is extended to give rise to a test case (e.g. *Test1*).

On the one hand, the framework relies on key abstract classes that define distinct UF APIs that implement the interface *Step* (e.g. *LaunchApp* and *CapturePictureFromCamera*). They define additional methods to convey UF-specific information (e.g. *setApplication* and *setResolution*).

On the other hand, TAF employs concrete classes to extend UF APIs. UF implementations invoke PTF APIs, thereby enclosing the low-level input/output behavior of a specific phone (e.g. *LaunchAppImp* and *CapturePictureFromCameraImp*).

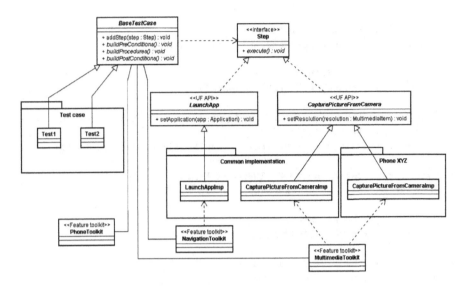

Fig. 2. A TAF class diagram

Target-independent and target-dependent classes are organized in distinct pack-ages (e.g. *Common Implementation* and *Phone XYZ*).

To allow proper instantiation of UF implementations for a given phone, TAF relies on the notion of Feature Toolkit (e.g. Phone Toolkit, Navigation Toolkit, Multimedia Toolkit), as illustrated in Fig. 2, at the bottom. Since TAF has potentially more than one implementation for each UF API, this class must know the appropriate UF implementation for the phone under test. This information is encoded within an XML file, which is maintained by TAF developers. Another role of a Feature Toolkit is to add the instantiated UF to the list of test case steps, and to launch their execution. As soon as the step list is created, the test case execution can be started. In brief, an ATC consists of several calls to methods encapsulated within Feature Toolkits. Fig. 3 summarizes the hierarchy of TAF layers from the highest to the lowest level.

3.4   Automating a Test Case with the Aid of TAF

The structure of TAF has facilities to create an ATC from a TC written in natu-ral language and conceived to be manually executed. First, a subclass *BaseTest-Case* has to be created as a template for the new ATC. Three of its abstract methods – *buildPreConditions()*, *buildProcedures()* and *buildPostConditions()* – must be overwritten. Such methods define the functional structure of a test: the phone configuration actions required for the test (e.g. date and hour settings, web browser set-up, e-mail accounts, etc.), the actual test steps and post-test clean-up procedures (e.g. the rollback of side effects that could possibly affect further tests). The next step consists in inserting calls to methods of Feature Toolkits. The code to be inserted within the overwritten methods ressembles the

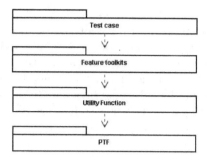

Fig. 3. TAF layer view

one shown in Fig. 1. Once the subclass is created (i.e. the new ATC), a preliminary checking is performed to verify if there are suitable implementations of the required UF APIs for the target-phone. If a proper implementation is found, it will be reused as it is; otherwise, a new one will be created.

3.5 Object-Oriented Framework: A Keystone for Worthy Automation

The object-oriented approach adopted by TAF achieves significant software reuse through three distinct mechanisms: inheritance-based creation of ATCs, reuse of available UFs and porting of pre-existent ATCs to other phones.

Inheritance-Based Creation of ATCs. Remember that the methods mentioned in Section 3.4 provide an interface between the ATC and the UF APIs. Since TAF currently has hundreds of distinct UF APIs available, suitable APIs are very likely to be found for a new test.

Note that, in the worst case, the implementation of a specific UF would require the creation of a subclass of the abstract class in which the UF API is defined, as illustrated in Fig. 2, where the specific implementation *CapturePictureFromCamera* is created when targeting phone XYZ. Note that, even in the worst case (no implementation reuse at all), the process of UF creation is still guided by TAF through the inheritance mechanism.

Reuse of Pre-existent UF Implementations. A new implementation is rarely created from scratch. Sometimes, it may be obtained through the creation of a subclass of a pre-existent UF by partially reusing its code. A non-negligible amount of reuse should be expected in this way, as explained in the following. Consider the classes within the *Common implementation* package (see Fig. 2). Their method *execute()* is a *template* that invokes several *hooks*[19]. Since an implementation inheriting another implementation must only re-implement the *hook* methods required by a specific phone, all the other methods already encoded in the superclass are reused.

In the best case, an untouched UF implementation is reused. Fortunately, the best case is dominant, as it will be seen in Section 4.1.

Porting ATCs to Other Phones. Given a set of pre-existing ATCs, let's analyze how TAF supports the porting of a test case to another phone. First, in the same way as described in the previous section, it should be checked if every UF in that ATC matches the expected behavior for the phone under porting. If not, a new low-level implementation for this UF must be created. Since UFs are extended only if no compatible UF could be found, TAF maximizes the amount of software reuse. In such a way, the whole code of the ATC is reused for different phone models.

4   Experimental Results

This section presents real-life values collected from BTC's Test Automation Project. Three classes of experiments were performed to assess how effective and sustainable is the impact of TAF on test automation. First, we provide a quantitative breakdown of software reuse induced by the framework (Section 4.1). Second, we quantify the impact of reuse in the process of porting new phones (Section 4.2). Later, we quantify the actual productivity gain by first adding the effort of automating TCs to the effort of executing ATCs and then comparing the overall effort with manual TC execution (Section 4.3).

4.1   Quantifying Reuse Upon TAF

Table 1 displays reuse figures measured for a set of 10 phones, each submitted to a same test suite containing 67 TCs. Such suite employs 246 distinct UFs. The second column shows the number of phone-specific UF implementations required to perform the porting to each phone (i.e. the number of new UF implementations created either extending the UF API or extending other UF implementation). The third column shows the number of UFs whose implementations were untouched when reused. The fourth column summarizes the percentage of untouched UFs with respect to the total number of invoked UFs.

   The high amount of achieved reuse (84.8% on average) contributes to attenuate the TC automation effort required as a result of the ATC-generation learning curve, as will be discussed in the next section.

4.2   The Impact of Software Reuse

Fig. 4 and Fig. 5 show the evolution of TC automation for two product families during a period of seven months.

   Fig. 4 shows the average number of hours required per ATC per developer, normalized to the number of hours required when automation was launched (Month 1). In practice, since the simplest TCs are automated first and the most complex ones later, the TC automation effort increases with time. The minimum effort corresponds to the simplest and shortest TCs. The maximum effort corresponds to the most complex TCs (requiring 1.8 times more effort than in Month 1). On average, the effort is about 1.4 times larger than at the

Table 1. UF reuse breakdown

Phone ID	# phone-specific UF Imps	# UFs reused	% of reuse
1	56	190	77.24%
2	48	198	80.49%
3	44	202	82.11%
4	44	202	82.11%
5	52	194	78.86%
6	54	192	78.05%
7	58	188	76.42%
8	6	240	97.56%
9	7	239	97.15%
10	5	241	97.97%

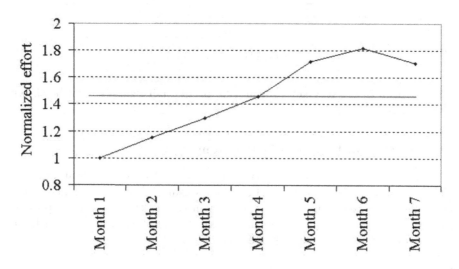

Fig. 4. TC automation effort

time automation was launched. In brief, Fig. 4 could be seen as a learning curve for ATC generation within typical product families. This is the price to pay to obtain a first ATC, which will be hopefully reused with the help of TAF.

Fig. 5 shows the average number of hours required per porting per developer, normalized to the same reference adopted in Fig. 4. Note that the ATC porting effort decreases with time as a consequence of software reuse. The minimum porting effort is approximately 1/4 of the effort required to automate the first TCs. On average, the time to port an ATC to a new phone is about 1/3 of the time to create a new ATC. This is a strong evidence that the TAF architecture effectively enables test reuse.

By correlating the average values extracted from Fig. 4 and Fig. 5, we conclude that porting is on average 4 times faster than building an ATC from scratch.

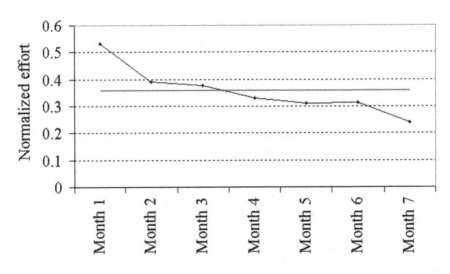

Fig. 5. ATC porting effort

This speed up is the very key to achieving productivity gain, as will be shown in the next subsection.

4.3 The Overall Impact of Test Automation

To be worth doing, the overall effort spent in all tasks involved in automation (TC automation, ATC porting, ATC execution and TAF maintenance) must be smaller than executing the same tests manually. In this section, we provide quantitative evidence that, despite the automation limitations at the user level, the adoption of a test framework does pay off.

Experimental Set up. The values summarized in next subsection were measured while testing 15 different phone models belonging to a same product family. Given a phone model, a test suite consisting of 60 TCs (suitable for automation) was selected. First, the TCs were automated gradually, giving rise to ATCs. While an ATC is not completed, its original TC is manually executed instead. The cumulative effort of testing with the aid of automation was measured along a fourteen-month interval. Since the selected TCs were invoked many times in distinct development life cycles, this procedure captured the overall effort spent with testing.

To assess the impact of automation as compared to purely manual text execution, we performed an estimation of how much time would be spent to run those TCs manually. An estimate of the effort required under purely manual test execution was obtained by multiplying the average manual execution time by the number of TCs that would be invoked for the same test suite during the same period. Such a manual test execution estimate will be used from now on as a reference for effort normalization.

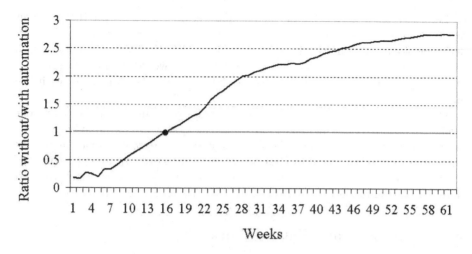

Fig. 6. The impact of automation on effort

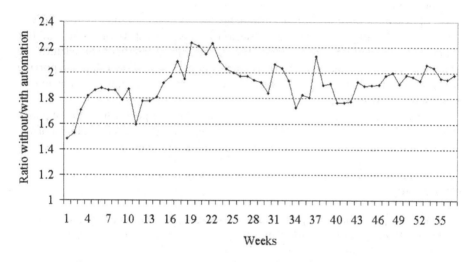

Fig. 7. Test execution speed-up with automation

Assessment of Productivity Gain. Fig. 6 depicts the overall testing effort along the monitored period, normalized to the manual test execution effort.

Note that the effort to manually execute the test suite for all phones under development within the product family would be 2.8 times larger, if automation was not employed. Therefore, a productivity gain of approximately 3 times was reached within about slightly more than one year. Note also that it took about three months to reach a breakeven point (that is, the time after which automation starts paying off). This indicates that, to deserve automation, TCs must be selected among the highest recurrent ones.

Let's now analyze the impact of automation on each software development cycle. Instead of reporting the cumulative value (as it was shown in Fig. 6), let's now focus only on actual test execution speed-up (i.e. the ratio between the estimated time that the manual test execution team would spend to execute the whole test suite and the actual time spent to execute the same suite with the aid of automation).

In Fig. 7, dots represent speed-up values obtained within distinct development cycles of various phone models. Note that, the speed-up is 2 on average. This means that, with automation, the whole test suite is executed twice as fast as compared to purely manual test execution. In other words, assuming a constant number of test engineers, manual test execution takes twice as much time to deliver test results.

5   Conclusions and Future Work

We reported an object-oriented framework granting significant test productivity gain by means of software reuse.

As opposed to most reported approaches, we present abundant quantitative evidence of the impact of test automation on real-life state-of-the-art mobile phones. Experimental results indicate that a productivity gain of three times can be achieved in about one year. To reach such a gain, it was shown that the porting of a pre-existing ATC should be around four times faster than automating an equivalent TC from scratch.

As future work, our framework will be extended to support other mobile phone platforms. We also intend to employ Aspect-Oriented Programming and Formal Verification so as to detect possible flaws in the test software.

References

1. Esipchuk, I., Validov, D.: PTF-based Test Automation for JAVA Applications on Mobile Phones. In: Proc. IEEE 10th International Symposium on Consumer Electronics (ISCE), pp. 1–3 (2006)
2. Test Quest, Test Quest Pro (2006) available at http://www.testquest.com
3. Pretschner, A., et al.: One Evaluation of Model-Based Testing and its Automation. In: Proc. International Conference on Software Engineering, pp. 392–401 (2005)
4. Dalal, S.R., et al.: Model-Based Testing in Practice. In: Proc. International Conference on Software Engineering, pp. 1–6 (1999)
5. Bredereke, J., Schlingloff, B.: An automated, Flexible Testing Environment for UMTS. In: Proc. 14th IFIP TC6/WG 6.1 International Conference on Testing of Communicating Systems, pp. 79–94 (2002)
6. Heikkilä, T., Tenno, P., Väänänen, J.: Testing Automation with Computer Aided Test Case Generation. In: Proc. 14th IFIP TC6/WG 6.1 International Conference on Testing of Communicating Systems, pp. 209–216 (2002)
7. Chi, C., Hao, R.: Test Generation for Interaction Detection in Feature-Rich Communication Systems. In: Proc. 17th IFIP TC6/WG 6.1 International Conference, TestCom, pp. 242–257 (2005)

8. Tkachuk, O., Rajan.: Application of automated environment generation to commercial software. In: Proc. International Symposium on Software Testing and Analysis, pp. 203–214 (2006)
9. Zhu, H., et al.: The first international workshop on automation of software test. In: Proc. International Conference on Software Engineering, pp. 1028–1029 (2006)
10. Gallagher, L., Offutt, J.: Automatically Testing Interacting Software Components. In: Proc. International Workshop on Automation of Software Test, pp. 57–63 (2006)
11. Okika, J.C., et al.: Developing a TTCN3 Test Harness for Legacy Software. In: Proc. International Workshop on Automation of Software Test, pp. 104–110 (2006)
12. Xia, S., et al.: Automated Test Generation for Engineering Applications. In: Proc. International Conference on Automated Software Engineering, pp. 283–286 (2005)
13. Kansomkeat, S., Rivepiboon, W.: Automated-Generating Test Case Using UML Statechart Diagrams. In: Proc. of the Annual Research Conference of the South African Institute of Computer Scientists and Information Technologists on Enablement Through Technology, pp. 296–300 (2003)
14. Ramler, R., Wolfmaier, K.: Economic Perspectives in Test Automation: Balancing Automated and Manual Testing with Opportunity Cost. In: Proc. International Workshop on Automation of Software Test, pp. 85–91 (2006)
15. Berner, S., et al.: Observations and Lessons Learned from Automated Testing. In: Proc. International Conference on Software Engineering, pp. 571–579 (2005)
16. Oliveira, J., et al.: Test Automation Viability Analysis Method. In: Proc. VII IEEE Latin-American Test WorkShop (LATW 2006) (2006)
17. Fayad, M.E., et al.: Building Application Frameworks: Object-Oriented Foundations of Framework Design. Prentice Hall, Englewood Cliffs (1999)
18. Gamma, E., Beck, K.: JUnit specification (2006) available at http://www.junit.org
19. Gamma, E., et al.: Design patterns: elements of reusable object-oriented software. Addison-Wesly, London (1994)

Model Based Testing of an Embedded Session and Transport Protocol

Vesa Luukkala and Ian Oliver

Nokia Research Center
Helsinki, Finland
{vesa.luukkala,ian.oliver}@nokia.com

Abstract. We describe an experience in applying model based testing in verifying especially the parallel behavior of a device level service and discovery protocol. Our approach is two phased: we first define a high level domain model in B and use cases in CSP that can be verified and then create a more detailed reference model that we use for testing the implementation on-the-fly. The use cases are used to drive both the B model and the reference model.

1 Introduction

This paper documents our experiences in applying formal methods and model-based testing approaches in an industrial, semi-formal environment. We were tasked with testing an embedded session and transport protocol for mobile devices based on SOA [1] principles. It is expected that subsystems connected by such a protocol would be provided by an external vendor and must be tested as black box implementations.

We were expected to construct the tester during development of the system, the implementation work had already started and there were no formal specifications for the system. Our main goal was to find bugs arising from concurrency. Experiences of the modeling work and some empirical evidence related to this is described in [2]. During the development the requirements and the the environment evolved forcing us to attempt a less formal and more pragmatic approach.

It is well known that parallel systems are hard to verify and test. Exhaustive verification and proving of correctness are options for systems that are constructed in well controlled environments and often with specialized languages. When testing parallel systems the above problems are augmented by the fact that it is not possible to force an executing parallel system to a certain state as timing and scheduling issues that cannot be influenced from outside of the implementation affect the behavior of the system under test.

We attempted a rigorous approach where we specified requirements at a high level, dividing them into use cases and system model and then refined that model strictly to a concrete model that could then be used as basis for automated testing which would partly alleviate the problems of testing a parallel system. Especially we planned to rely on on-the-fly testing technique to cope with parallelism.

A. Petrenko et al. (Eds.): TestCom/FATES 2007, LNCS 4581, pp. 212–227, 2007.

Our experience was that maintaining the refinement between a high level model and a more concrete model that was used for testing was too laborious, thus maintaining that was abandoned. Use cases remained the only link between those two models. To measure the coverage of the use cases we initially limited the model used for testing so that it would execute according to the use cases. We also post-processed traces of random execution of the model to recognize use case behaviors. Eventually tool support allowed us to drive the model directly with use cases.

We attempted to measure the quality of the testing by comparing function call counts from an instrumented implementation that was tested by our approach and with existing "traditional" style testers, but we noted that this kind of coverage values do not reflect quality of testing. Although on-the-fly testing typically gives higher values it is easy to create a traditional test suite that produces values of same or higher magnitude by blind repetition. For same call count values the model-based on-the-fly tester is able to order the sequence of calls in more different ways, which is useful especially in uncovering errors arising from concurrency. At the same time we note that it is usually not feasible to be able to exhaustively test all possible parallel combinations, thus we propose that a way to ensure uncovering parallel errors is to drive the model with a use case to a situation with high parallelism and then letting the tester tool execute unguided after that. The produced traces can be postprocessed to obtain a measure on how many of the possible parallel executions have been covered.

The rest of the paper is arranged as follows: section 2 describes our approach, section 3 describes the system that we are testing, section 4 elaborates on the construction of the tester, including modeling and other decisions during development, section 5 describes the results of testing and finally section 6 presents our conclusions.

2   The Approach

Our initial approach is outlined in figure 1: the system *requirements* (1) are split into *use cases* (2) and an *abstract system* (3). We have chosen to design the use cases as *CSP* (4) and the abstract *domain model* of the system in B [3] (5).The B model can be checked for consistency by theorem proving and further validated against the use cases by model checking. Another view of the same thing is that the CSP can drive the B model, which acts as an oracle.

Then we *ideally* construct a concrete *reference model* (6) based on the abstract B model using the B method and refinement. The CSP can be used to drive the reference model as well so in addition to the construction method the CSP is used as another mechanism for ensuring the validity of the reference model.

The reference model is concrete enough that it contains necessary information for testing an implementation of the defined system without adding behavior outside of the model. Note that the reference model is really a model of the system rather than model of the tests, we rely on tools to be able to infer the tests from the system model. We assumed that it would be less effort to operate

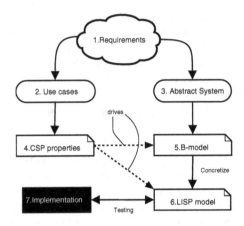

Fig. 1. The attempted process

on the same terms as the system design rather than having a completely separate tester. Also we are especially interested in testing of the parallel features and thus we wanted to perform *on-the-fly testing* [4] under the assumption that it would be easier to have the tester adapt to the parallel behavior of the system rather than to have test cases that would contain essentially the same functionality. These assumptions led us to use the commercial Conformiq Qtronic [5] tool for testing.

3   NoTA Architecture, Session and Transport Protocol Layer

The system under test is the "high interconnect" (H_IN) part of the Network on Terminal Architecture (NoTA) [6,7]. The driving force behind NoTA is to produce a device which supports "horizontalisation" of technology in the most extreme form currently available while adhering to the constraints of very small, embedded devices - namely, but not limited to, the mobile phone as we currently know today. The two goals that drive NoTA are modularity and service orientation.

Modularity is seen in the way devices are physically (and possibly logically) constructed while service orientation allows the functionality of the device to be abstracted away from that functionality's implementation and physical location. What this achieves is a complete separation of the functionality from the construction of the device. In other words the whole product line [8] based upon the Network on Terminal Architecture concepts becomes simply the choice of what functionality is required and then the choice of the most suitable implementation technologies; or even possibly vice versa.

A NoTA device is constructed in a modular fashion from components which are termed subsystems. A subsystem is simply a self-contained, independent

unit of the device which provides processing capabilities. Typically a subsystem manifests itself as a unit containing a processor, local memory and a set of local devices, for example, a camera, solid state storage and so on. A subsystem must also provide communication to the NoTA Interconnect allowing the services that run upon that subsystem to communicate with other services elsewhere in the whole NoTA device. Figure 2 shows this pictorially with a device containing two subsystems of various configurations.

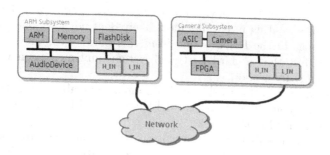

Fig. 2. Pictorial Representation of a NoTA Device

The Interconnect can be any suitable communication medium, although in a mobile device this means some kind of high speed bus. In the simplest design, the interconnect is either of star or bus topology, although any particular network topology is possible.

The interconnect is divided into three layers: High Interconnect (H_IN), Low Interconnect (L_IN) and Network (TCP/IP, MipiUnipro etc) layer.

The H_IN provides services with resource discovery and management, session and upper level transport facilities. It is into this layer that services and applications request communication channels to other services and applications, while services themselves announce and register their existence and provide functionality to the world. The communication mechanism provided by the H_IN is connection oriented and provides asynchronous message based passing and streaming data communication possibilities:

Typically the asynchronous message based type of communication is reserved for control commands and general interaction between services, while large quantities of data, eg: multimedia, are sent via the streaming connections. The L_IN is more of a device driver level which abstracts the underlying communication network away from the H_IN.

While devices are constructed out of subsystems and an interconnect, without services the device is not capable of providing any end-user functionality. The subsystems and interconnect serve to support the provision of services. The notion of service in NoTA is an abstraction of some logical grouping of functionality [9].

4   Modeling the System and Constructing the Tester

There are two requirements for testing of a NoTA system that we focus on: the parallel nature of the system and testing of third party provided subsystem implementations. Each subsystem must be able to communicate with entities (services or applications) on the H_IN network regardless of which subsystem they reside on. Each subsystem may be a separate computing entity and there is no global scheduling mechanism for the whole system, which means that even if there is a deterministic behavior for a single subsystem, the behavior of all the subsystems executing in parallel most likely does not have one. We wanted to flush out bugs in the H_IN arising from this parallel complexity. Secondly we wanted to test that a third party subsystem implementation would work properly in the H_IN network.

While not covered in this work, we also wanted to be able to test the services themselves as they also may have parallel behavior and hopefully use the same approach as described in this work.

We decided to model the H_IN layer itself rather than the protocol between the subsystems or the individual services and applications using the H_IN. The reason for this was that we wanted to concentrate our effort on the system itself and then derive the possible correct behaviors from the system model automatically to cope with the expected complexity. The downside of this decision is that the system model as such treats all of the entities using H_IN the same data-wise, that is, the model of the system (or the real implementation itself) does not contain any specific information about the content of the data that it passes. When needed, we planned to provide this data by use cases or as a more drastic measure, model the particular services in the same model as the H_IN.

Also, we are modeling the system (H_IN layer) and we want to test based on the same or refined model of the *system* rather than construct a separate model of the *tester* for the system. This is because after constructing a valid model of the parallel system we want to reuse this effort in automatically deriving the tests from this model. The difference between a tester model and a system model is that the tester exists outside of the system acting as a user whereas the system model describes the behavior from the system point of view. So when a tester model sends a message, the system model expects a message. One way of thinking about this is that we want to *compare* the system model with the implementation. Here the system model acts as a reference implementation. Another view is that the tester is an "inverse" of the system.

The model communicates with the outside world by sending messages to named ports and receiving messages from them. The number of ports that allow communication to the outside world is fixed, but easily parameterizable. Each of these ports corresponds to an entity offering a service or using some service on top of the H_IN layer, so the chosen number of ports determines the maximal parallelism for the model. When the model sends a message to the outside world, it means that we expect the system to produce that kind of message and when the model receives a message we expect the implementation to accept that kind of message.

For high level modeling of the H_IN layer we chose the B language because we had previous experience in building systems using that methodology [10] and using CSP for use cases followed from the available B tool support. Furthermore the B method has the notion of refinement which allows stepwise generation of less abstract models until B0 subset of B for which there exist mappings to imperative programming languages. We used the ProB animator and model checker [11] and the commercial ClearSy Atelier B for analyzing the B.

Also, since we expected to be testing a parallel system, we wanted to be able to perform testing on-the-fly, that is we wanted to "execute" the model in parallel with the implementation and adapt to the behavior of the implementation. It is possible to use the system model to generate a set of linear test cases, but using this mechanism to attempt testing of parallel systems is cumbersome. Firstly because attempting to generate all possible linear test cases covering all parallel interleavings becomes infeasible for nontrivial systems due to the state explosion problem. Secondly, a parallel system may execute correctly but differently than what a test case expects, which means that the test case signals "inconclusive" or "fail" and the effort spent for that test case is wasted. Finally, while typically testing languages have the possibility of branching based on replies from the implementation, constructing a test case that would attempt to adapt to the implementation would lead to implementing an on-the-fly tool using the testing language. We expect that on-the-fly testing alleviates these problems. There is existing work on generation of test cases from a model [12,13] and also on various *test selection* heuristics [14,15] which attempt to produce a subset of possible testing for best possible coverage of the model. We expect that the most successful of these heuristics will be applicable for test case generation as well as on-the-fly testing and that support for them will appear in tools.

The two particular requirements of the ability to base its testing on the model of the system and ability to execute testing on-the-fly led us to use the Conformiq Qtronic tool for testing. At the time of this decision Qtronic supported a variant of LISP to define the models. Qtronic executes the the LISP model symbolically in parallel with the executing system under test causing messages to be sent to the system under test. The feedack from the tested system is taken into account by the symbolic executor and influences the permissible behavior later on. The tool also has possiblity of guiding the testing by various coverage criteria: there is structural criteria, such as branch, condition and boundary value coverage over the LISP model and coverage over user specified checkpoints that are entries in the LISP model. In addition there is notion of coverage over use cases, which are also defined in LISP.

4.1   The Models

The B machine contains operations for each H_IN primitive and the CSP is then used to express the use cases that specify the desired behavior of the system over those primitives. The CSP can be then verified against the B model as described in [16]. For example, figure 3 shows the B operations for registering a service and the CSP for a use case that shows that after a service has registered under some service identifier, another entity may connect to that particular identifier. Effectively this is one possible and desired linear trace in the system.

The first two events in the CSP are internal to the used ProB tool and thus implementation details, but the `register_with_ResourceManager?Sname?Sid!` `REGISTRATION_OK` event is the first H_IN specific one and expresses that the particular event can occur successfully for some service id `Sid` and service name `Sname`. The register event is then followed eventually by a `connect` event that uses the same service id `Sid` to connect successfully. The RUN process that is executed concurrently states that all the events that are passed as its parameters do not constrain the CSP process (the list of events here is truncated). In effect we are saying that the events in the RUN process are ignored until the desired `connect` event with desired arguments is encountered. This same mechanism is also used to hide operations that are purely internal to the B machine, such as the `rm_register` above, so that only H_IN primitives are used to build the use cases. This way the use cases should be applicable to any system that has the same set of primitives.

```
                                              ss,err <-- rm_register(nn,aa) =
                                              PRE
                                                  nn : SERVICE_NAME &
                                                  aa <: ICNODE_ADDRESS
                                              THEN
                                                  CHOICE
                                                      err := REGISTRATION_ERROR ||
sid,err <-- register_with_ResourceManager(nn) =       ss :: SID
PRE                                               OR
        nn : SERVICE_NAME &                       ANY
        icnode_state = RUNNING                        newsid
THEN                                                  WHERE
    sid,err <-- rm_register(nn,icnode_address)            newsid : SID - rmsids
END ;                                                 THEN
                                                          ss := newsid ||
                                                          rmsids := rmsids \/ { newsid } ||
                                                          err := REGISTRATION_OK
                                                      END
                                              END
                                              END ;
```

```
CSP:   MAIN = initialise -> notify_resource_manager_location -> REGISTER;;
       REGISTER = register_with_ResourceManager?Sname?Sid!REGISTRATION_OK ->
                  (CONNECT ||| RUN[register_with_ResourceManager,send,nm_register,...]);;
       CONNECT(SS) = connect!SS!CONNECT_OK -> skip;;
```

Fig. 3. The "register" primitive expressed in B and a CSP property

By the time we completed this model, the specifications had already changed somewhat and the initial test model had turned out to be complex enough that we felt that constructing the chain of refinements between these models would be too time consuming, especially as this process might have to be repeated. Thus we decided to develop the models separately but nevertheless make an effort to make sure that the CSP use cases would be compatible with both.

We felt that it was still worth the while to continue, as a correct - in terms of verification - system does not guarantee that the system would have behaved in accordance with the customer's wishes - hence the need for testing at all levels of development.

It can be argued that correct development of the system would have been achieved if we had followed the refinement rules and constructed a concrete specification in the B0 subset of B which is translatable to a 'normal' imperative language (B0 sequentializes actions and removes non-determinism). There do exist code generators from B0 to C, C++ (and also Java and Ada). We faced three problems here, firstly the time spent developing and refining the specification would have been considerable and secondly we would have to have developed a code generator to LISP and ensured that it preserved the semantics of the model. Finally we felt that there would certainly be changes that would result from connecting the LISP system to the tester and all of these changes could not be done at the LISP level but higher up in the refinements chain. These facts together outweighed the potential benefits - of course if B (and it has now been superseded) would have been taken into more general use then this route might pay off in the future. Additionally, strict refinement based approaches do not cope with change in the specification well resulting in techniques such as retrenchment to preserve the mathematical link between now differing specifications.

The LISP model was constructed essentially as an event loop: the system reads in messages from incoming ports and these events are then processed. These events match the H_IN primitives. Notably, initially all H_IN events were accepted by the event loop. Since this model is used for testing this meant that given this model to a model based tester, it would generate any of these primitives to be sent to the implementation. The effects of this are described further below in section 4.2.

Figure 4 shows the LISP code that corresponds to the "register" primitive in figure 3. This function is called from the event loop after an event `Hactivate Service` with one parameter `sid` of integer type has been received.

If the parameter `sid` has been registered before (1), the event `HactivateService_ret` is sent back with parameter value zero signaling failure (3) and the event loop is re-entered. The model has been augmented with a tool specific checkpoint mechanism (2), which allows tagging parts of the model so that the tags are reported in test traces, but they can also be used as coverage guides. If the `sid` is new, it is then associated with the port from which the original message came from (4) and a new internal interface is created (5) and also associated with the `sid` (6) for later use. Then a random value non-zero is generated (7) and after some more bookkeeping and checkpoints (8), the return message is sent back (9) containing the random value. Again, since this is a model of the system, emitting a message with a random value means that the tester expects the implementation to produce that message with any value in place of the random value.

Each H_IN primitive has a similar function and furthermore there are functions for bookkeeping and internal data structures as well as functions whose purpose is to enforce the typing of the primitive parameters.

```
(define HandleActivateService
  (lambda (msg env in_port ret_port)
    (let*
        ((msg_name (ref msg 0)) (sid (ref msg 1))
1        (known_sid (r_known_sid? (env_rmap_port env) sid)))
      (if (known_sid)
          (begin
2          (checkpoint (tuple 'unable-to-register sid)) ; a named checkpoint
3          (output ret_port (tuple 'HactivateService_ret sid 0)) ; sending a message
           (h_in_router env))
          (begin
4          (r_dict_add (env_sid_to_outport env) sid ret_port)
5          (let ((sid_port (make-interface)))
             (begin
6              (r_dict_add (env_rmap_port env) sid sid_port)
7              (let ((pid (+ (random 254) 1))) ; make up a value
                 (begin
                   (r_dict_add (env_s2p_port env) sid pid)
                   (env_incr_service_ctr env)
8                  (checkpoint (tuple 'service-activated (print-pname in_port) sid))
9                  (output ret_port (tuple 'HactivateService_ret sid pid))))
                 (h_in_router env)))))))))
```

Fig. 4. The LISP code corresponding to the "register" primitive

```
(define send-hactivate                    (define send-connect
  (lambda ()                                (lambda (sid)
    (let* ((oport (any-oport)) (sid (random 254)))    (let ((oport (any-oport)))
      (begin                                    (begin
        (output oport (tuple 'HactivateService sid))   (output oport (tuple 'Hconnect_req sid))
        (tuple oport sid)))))                    #t )))) ;; we have reached our goal


                                          (define main
(define receive-hactivate_ret               (lambda ()
  (lambda (send-port sid)                     (if |usecase:use case 2|
    (let ((receive-port (oport-to-iport send-port) ))    (let ((a (send-hactivate)))
    (let ((inmsg (handshake receive-port #f)))       (let* ((s_oport (ref a 0))
      (let* ( (iport (gp inmsg)) (msg (gm inmsg)))        (sid (ref a 1)) )
        (begin                                    (begin
          ;; We make sure that message is what we want      (receive-hactivate_ret s_oport sid)
          (require (equal? (ref msg 0) 'HactivateService_ret) #f)   (send-connect sid))))))
          (require (equal? (ref msg 1) sid) #f)
          (require (> (ref msg 2) 0) #f)))))))
```

Fig. 5. The LISP use case corresponding to the B use case

The use cases are also written in LISP and they are similar to the CSP ones in that they are essentially sequences of events. Figure 5 shows the same use case as earlier in figure 3: the **main** function calls three functions, that perform the communication. This use case is expected to be running as an observer in parallel with the system model. The way this use case either influences the testing or adapts to it is elaborated below.

4.2   Use Cases, Data and Control of the System

As mentioned earlier, the model of the system contains no information about the content of the data it is handling, the only properties the model deals with is the size of the data. This is consistent with the B model where the structure of the data was abstracted out using a generic DATA type.

For instance, a camera service has registered itself to the system under the name CAM. In order to use that service, an application must know that name and furthermore be able to send across correct commands, potentially in a certain order. For H_IN the name of the service is any sequence of characters with a maximum length that is associated with some port. If someone asks for that particular string then H_IN can freely choose a number within some bounds to represent a connection to that service and then transmit data, which from H_IN point of view is a sequence of items in a buffer.

For theorem proving and model checking purposes this generality is not a problem and also if the H_IN can be tested in a test bench where no real world entities are running it is possible to use essentially random data. However, if there is a service implementation in the network, it becomes necessary to know its identity and also how to maintain communication with it by sending compatible messages in right order. So there should be a mechanism of dealing with the data.

Also, the system is built with a certain purpose in mind: H_IN should connect entities and transport data between them. The information how to do this is contained in the model, but the sequence of primitives that performs the data transfer is just as probable as any other sequence. For example a valid sequence of primitives always starts with a "register" primitive and is then followed by a "connect" primitive, but the model may always enable sending a "send" message with bogus arguments first. Of course this is due to the way the model has been constructed: all primitives should always be accepted. This generation of unexpected behavior is partly the reason why model based testing is powerful, but it also has the risk that the intended behavior is never completely exercised as there is always a high possibility of choosing a bogus message.

Both the data issue and sequencing of primitives can be resolved by modifying the model. The expected data can be hard-coded in the model so that whenever a message is to be sent or received from a given address, it has the desired format. The major downside of this is that this is completely bound to the particular application. The sequencing can also be enforced in the model: a boolean flag in the event loop can ensure that a "send" can only occur after a successful "connect". This is more general than the data hard-coding but it seems clear that if more complex sequences than two messages are considered, the system quickly becomes complex.

Both of these clearly limit the state space of the system so it might be possible to consider these models a refinement of the "generic" model, but the loss of generality is unappealing.

To solve this we put the application specific information to use cases and then count on the testing tool to be able to utilize the information in them. The use cases are used in two ways: to influence the test execution and to observe the test execution.

Figure 6 gives an example of both cases. This is part of the use case shown in figure 5 and on the left hand side of figure 6 the use case adapts and observes the

```
(define send-hactivate
  (lambda ()
    (let* ((oport (any-oport)) (sid (random 254)))
      (begin
        (output oport (tuple 'HactivateService sid))
        (tuple oport sid)))))
```

```
(define send-hactivate
  (lambda ()
    (let* ((oport port1) (sid 12))
      (begin
        (output oport (tuple 'HactivateService sid))
        (tuple oport sid)))))
```

Fig. 6. A use case that observes and another that influences

test run, here the **HactivateService** primitive that is used to register the sender of this primitive under the id given as parameter. There are parameter values that are both chosen randomly, the port where the message occurs and an integer variable corresponding to the service id. As we explained earlier, the random value indicates that any value produced by the test tool can occur here. So this states that we want to be able to observe primitive **HactivateService** for any port and service id and that we want to remember both for later use. The version the right hand side of figure 6 is similar, but both the port and service id have fixed values; we expect this use case to influence the tester tool so that it is able to generate the values given in the use case. For payload data, this is the mechanism we force the service commands to comply with possible existing systems.

For control flow, we can explicitly guide the implementation event by event, but we'd prefer to encode the same use case as in figure 3 and let the tester tool find the path to such a state. This allows us to guide the implementation to a state of high concurrency or otherwise unlikely situations and then let the tester tool proceed (semi-)randomly. Unfortunately the tester tool did not support this kind of use cases at the time so we could not try this approach. We expect this feature to appear in future. The potential downside of this feature is that it might require heavy computations, which are problematic when performing on-the-fly testing.

While not obvious from the use case listings, we expect that tool support will remove the work needed to write the use cases as code. The primitives for communicating with the implementation can be derived from interface specification, resulting in a user interface that allows specifying the use cases in an MSC-like format. The user only needs to fill in the sequence of actions and constraints on the data values.

4.3   Test Configuration

There are multiple parallel entities that operate over the H_IN layer, but the H_IN layer itself may cover several subsystems as described in section 3, so there needs to be a way of connecting the tester tool to all of them. The model considers H_IN as a single system: there may be several points of communication, but they all reside on the same H_IN layer and are independent of each other.

In practice each entity on the H_IN layer must reside on some subsystem. Also, every subsystem must implement part of the H_IN network that offers primitives to the users and is able to communicate to other similar subsystems within the H_IN network. There exists a C-language interface for these primitives and when model communicates with the outside world via port, this communication in terms of the tool run time system must be mapped to these C data

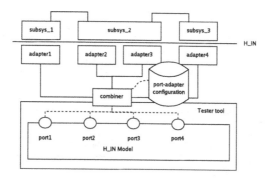

Fig. 7. Test configuration for a H_IN system consisting of three subsystems

types and function calls and back. Our aim was to construct the model used for testing at detailed enough level that we could claim that no information loss or information generation would be needed in these adapters. The only part where we are not entirely convinced about meeting this goal is the handling of asynchronous messages, which required its own bookkeeping mechanism.

Qtronic has the notion of an "adapter process" which communicates with the tester tool using a specific protocol, in this case over TCP/IP, and contains the user produced adaptation between the implementation and the Qtronic run time system. We have decided to make each adapter process correspond to one model port, rather than have one adapter process for each subsystem. This gives more flexibility in at the price of more overhead, which may be a problem for subsystems with low processing capabilities.

The traffic between different adapter processes is routed via a special "combiner adapter" that is configured with the address information of the port specific adapters. This configuration has to match the real world subsystem configuration. Figure 7 shows an example configuration, where the model has four ports and the system under test consists of three subsystems. Two of the ports have been mapped on the same subsystem, while other two are mapped to two different susbsytems.

We assume that we are able to execute the adapter processes on the subsystems, which is a valid assumption for in-house testing. However, if we want to test a subsystem implementation we have two possibilities: either demand the producer of the subsystem that there is an access to H_IN or then we must be able to exercise the service that exists on the subsystem using its own primitives. Use cases that contain the service specific data is one possibility, another one is to model the service in addition to the H_IN model.

5   Results and Conclusions

The abstract B model has 800 +/- 100 lines of code, the LISP model has 2100 +/- 200 lines of code and the C implementation of the H_IN has 20000 +/- 1000 lines of code. As usual, the modeling phase already uncovered some errors

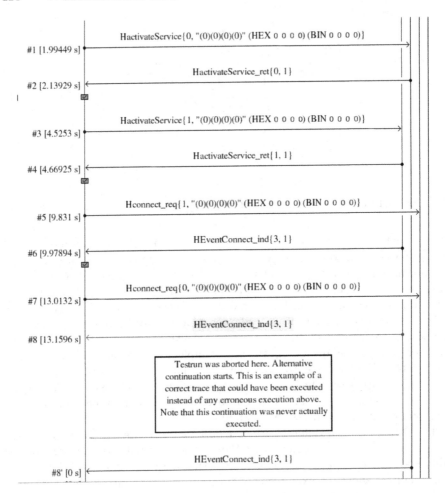

Fig. 8. An error trace involving three entities

and assumptions in the implementation. The modelling process (including requirements elicitation and revision, plus various versions of the model) took 4 man months to produce the first major release of the specification and the first feature complete tester also took about 4 man months.

During testing, one bug was found that essentially was due to the implementation not expecting out-of-band messages, for example a "send" before a "connect" had been made. These could have been found by writing a tester that would have produced messages randomly.

We found four bugs that were of a concurrent nature, the earliest one is shown in figure 8, where two services on same subsystem register themselves and then an application tries to connect to both producing a connect request only for the other one. This trace is the shortest one to that error.

Another error of a similar kind, occurred with the same configuration where the indication of arrived data was given to the wrong entity. The third error was related to closing down of connections when two entities on different subsystems closed down the connection at same time and the fourth error occurred when two connections were sending data in both directions requiring four entities where connections were across subsystems. Furthermore we found errors that had no concurrent cause and most likely would have been caught by any kind reasonable testing. Typically the faults were such that they manifested themselves as multiple reported errors and identifying the root cause took some human work. Also, at times it was necessary to modify the model so that these errors could be circumvented and testing could be continued.

The implementation under test had simple test programs that set up a service and then a client would connect to that service to transfer simple data and this could be repeated. There were three scenarios: client connects and sends data, client connects and receives data and client connects and both client and service send and receive data.

Our assumption was that model-based testing would be able to exercise the system better and that our approach would be more efficient when compared to taking the "traditional" approach of writing linear test cases or test programs, especially for errors that arise from concurrency.

Bugs were found and the mentioned parallel errors were such that save for the first one there were no explicit requirements that would have led to test cases uncovering the errors. In the first case, the requirement existed, but there were multiple potential configurations of subsystems which were not explicitly noted down. We feel that it is unlikely that there would have been hand written sequential test cases that would have caught these errors. Furthermore, we note that they would not have been repicated by repeated executions of the existing test programs.

However measuring the parallel goodness of the testing is not straightforward as complete coverage is most likely not going to be achieved. Given this, would it be possible to identify the part that was not tested and guide later testing to cover that?

We used `gprof` utility to obtain call information for C functions for the individual H_IN components on subsystems, but it seemed that while the results of the longer on-the-fly produced more coverage data, this could always be matched simply by running the existing simple tests more times. It may be that a coverage analysis with smaller granularity than function level is needed to note the changes. However, the C implementation relies on threads and callbacks, which means that the branching is not detectable on the C-code level. It seems that measuring the parallel quality of the testing based on this kind of metrics is not good enough.

Another possibility we considered is using the traces produced by testing to deduce how much of the potential state space has been exercised. We implemented a prototype tool that takes in a description of a use-case and then attempts to show how many of the potential parallel executions of those were seen in the testing. Other approach would have been feeding the traces back to the B model checker and obtaining relevant information there. However, both of these approaches required tool development for which we didn't have resources.

Yet another way is to add the desired information to the model or the use cases. The Qtronic test tool has its own coverage criteria which aims to cover the model as well as possible and as the model has been constructed so that observable events are part of the coverage, the tool is able to produce meaningful testing and report checkpoints that may have metadata associated with them. It is possible to modify the model with auxiliary constructs that keep track of its own parallel state and produce that as output. The downside here is that another model is embedded in the model of the actual system.

Our preferred approach would have been to drive the model to a desired state with high concurrency using a use case and then let the tester tool proceed with a random walk. Unfortunately the tool support for this feature was not available at the time.

Our approach managed to uncover errors that would otherwise most likely not been found, but at the price of creating the system essentially twice. Constructing the models is a different skill when compared to writing test cases and this may be the greatest obstacle in adoption of this kind of testing. Nevertheless the possibility of using use cases to drive the model may be useful in demonstrating the value in terms that are understandable to traditional testers. The tool support is nearly there to allow the use cases to act as a loose template for test execution which would allow the test engineer to write testcase-like constructs to exercise the implementation.

We did not set out to do comparisons with other existing approaches and tools, rather we were looking for experiences in combining components in a toolchain. There are alternative approaches for both the specification and the tester tool side, especially ToRX [17] and the Spec Explorer [18], but we did not evaluate these.

It is inevitable that errors, especially of a concurrent nature, are introduced during development through decisions (primarily architectural) made while implementing. In addition we see errors introduced through requirements change which can not be adequately modelled and verified at the more abstract levels of modelling. Even though we have had to take a pragmatic approach which has compromised some "formal methods ideals" we have seen our approach uncover errors earlier and provide more detail about those errors.

Acknowledgments

This work has been made in cooperation with the EU Rodin Project (IST-511599). We would like to thank Kimmo Nupponen and Antti Huima from Conformiq for valuable help.

References

1. Erl, T.: Service-Oriented Architecture. Prentice-Hall, Englewood Cliffs (2005) 0-13-185858-0.
2. Oliver, I.: Experiences in using B and UML in industrial development. In: Julliand, J., Kouchnarenko, O. (eds.) B 2007. LNCS, vol. 4355, pp. 248–251. Springer, Heidelberg (2006)

3. Abrial, J.R.: The B-book: assigning programs to meanings. Cambridge University Press, New York, USA (1996)
4. Vries, R.d., Tretmans, J.: On-the-Fly Conformance Testing using Spin. In: Holzmann, G., Najm, E., Serhrouchni, A. (eds.) Fourth Workshop on Automata Theoretic Verification with the Spin Model Checker. ENST 98 S 002, Paris, France, Ecole Nationale Supérieure des Télécommunications, pp. 115–128 (1998)
5. Conformiq Software Ltd.: Conformiq Qtronic, a model driven testing tool (2006–2007) http://www.conformiq.com/qtronic.php
6. Suoranta, R.: New directions in mobile device architectures. In: Ninth Euromicro Conference on Digital System Design: Architectures, Methods and Tools (DSD 2006) (30 August - 1 September 2006), Dubrovnik, Croatia, pp. 17–26. IEEE Computer Society, Los Alamitos (2006)
7. Suoranta, R.: Modular service-oriented platform architecture - a key enabler to soc design quality. In: 7th International Symposium on Quality of Electronic Design (ISQED 2006) (March 27-29, 2006), San Jose, CA, USA, pp. 11–13. IEEE Computer Society, Los Alamitos (2006)
8. Savolainen, J., Oliver, I., Mannion, M., Zuo, H.: Transitioning from product line requirements to product line architecture. compsac 01, 186–195 (2005)
9. Kruger, I.H., Mathew, R.: Systematic development and exploration of service-oriented software architectures. In: Proceedings of Fourth Working IEEE/IFIP Conference on Software Architecture WICSA 2004, pp. 177–187 (2004)
10. Kronlof, K., Kontinen, S., Oliver, I., Eriksson, T.: A method for mobile terminal platform architecture development. In: Proceedings of Forum on Design Languages 2006. Darmstadt, Germany (2006)
11. Leuschel, M., Butler, M.: ProB: A model checker for B. In: Araki, K., Gnesi, S., Mandrioli, D. (eds.) FME 2003. LNCS, vol. 2805, pp. 855–874. Springer, Heidelberg (2003)
12. Lee, D., Yannakakis, M.: Principles and methods of testing finite state machines - A survey. In: Proceedings of the IEEE. vol. 84, pp. 1090–1126 (1996)
13. Gnesi, S., Latella, D., Massink, M.: Formal test-case generation for uml statecharts. iceccs 00, 75–84 (2004)
14. Feijs, L., Goga, N.S.M., Tretmans, J.: Test Selection, Trace Distance and Heuristics. In: Schieferdecker, I., König, H., Wolisz, A. (eds.) Testing of Communicating Systems XIV, pp. 267–282. Kluwer Academic Publishers, Dordrecht (2002)
15. Pyhälä, T., Heljanko, K.: Specification coverage aided test selection. In: Lilius, J., Balarin, F., Machado, R.J. (eds.) Proceeding of the 3rd International Conference onApplication of Concurrency to System Design (ACSD'2003), Guimaraes, Portugal, IEEE Computer Society, Guimaraes, Portugal, pp. 187–195. IEEE Computer Society, Los Alamitos (2003)
16. Leuschel, M., Butler, M.: Combining CSP and B for Specification and Property Verification. In: Fitzgerald, J.A., Hayes, I.J., Tarlecki, A. (eds.) FM 2005. LNCS, vol. 3582, pp. 221–236. Springer, Heidelberg (2005)
17. Tretmans, G.J., Brinksma, H.: Torx: Automated model-based testing. In: Hartman, A., Dussa-Ziegler, K. (eds.) First European Conference on Model-Driven Software Engineering, Nuremberg, Germany, pp. 31–43 (2003)
18. Veanes, M., Campbell, C., Schulte, W., Tillmann, N.: Online testing with model programs. In: ESEC/FSE-13: Proceedings of the 10th European software engineering conference held jointly with 13th ACM SIGSOFT international symposium on Foundations of software engineering, pp. 273–282. ACM Press, New York, USA (2005)

Utilising Code Smells to Detect Quality Problems in TTCN-3 Test Suites

Helmut Neukirchen[1] and Martin Bisanz[2]

[1] Software Engineering for Distributed Systems Group,
Institute for Informatics, University of Göttingen,
Lotzestr. 16–18, 37083 Göttingen, Germany
neukirchen@cs.uni-goettingen.de
[2] PRODYNA GmbH, Eschborner Landstr. 42–50, 60489 Frankfurt, Germany
martin.bisanz@prodyna.de

Abstract. Today, test suites of several ten thousand lines of code are specified using the *Testing and Test Control Notation* (TTCN-3). Experience shows that the resulting test suites suffer from quality problems with respect to internal quality aspects like usability, maintainability, or reusability. Therefore, a quality assessment of TTCN-3 test suites is desirable. A powerful approach to detect quality problems in source code is the identification of *code smells*. Code smells are patterns of inappropriate language usage that is error-prone or may lead to quality problems. This paper presents a quality assessment approach for TTCN-3 test suites which is based on TTCN-3 code smells: To this aim, various TTCN-3 code smells have been identified and collected in a catalogue; the detection of instances of TTCN-3 code smells in test suites has been automated by a tool. The applicability of this approach is demonstrated by providing results from the quality assessment of several standardised TTCN-3 test suites.

1 Introduction

Current test suites from industry and standardisation that are specified using the *Testing and Test Control Notation* (TTCN-3) [1,2] reach sizes of around 40–60 thousand lines of code [3,4,5]. These test suites are either generated or respectively migrated automatically [6] or they are created manually [4,5]. In both cases, the resulting test suites need to be maintained afterwards. The maintenance of test suites is an important issue for industry [6] and standardisation [7,8]. A burden is put on the maintainers if the test suites have a low internal quality resulting from badly generated code or from inexperienced developers [6]. Hence, it is desirable to assess the quality of TTCN-3 test specifications.

According to the ISO/IEC standard 9126 [9], a software product can be evaluated with respect to three different types of quality: *internal quality* is assessed using static analysis of source code. *External quality* refers to properties of software interacting with its environment. In contrast, *quality in use* refers to the quality perceived by an end user who executes a software product in a specific context. In the remainder, we will focus on internal quality problems.

A. Petrenko et al. (Eds.): TestCom/FATES 2007, LNCS 4581, pp. 228–243, 2007.

A simple approach for the quality assessment of source code are metrics [10]. In earlier work, we have experienced that metrics are suitable to assess either very local [3] or very global [11] internal quality aspects of TTCN-3 test suites. However, properties of language constructs which are, for example, related but distributed all over the source code are hard to assess using simple metrics. Instead, a more powerful pattern-based approach is required to detect patterns of inappropriate language usage that is error-prone or may lead to quality problems. These patterns in source code are described by so called *code smells*.

This paper introduces TTCN-3 code smells and utilises them to detect internal quality problems in TTCN-3 test suites. The located quality problems can be used as input for the plain quality assessment of test suites and as well as a starting point for the quality improvement of test suites.

The structure of this paper is as follows: subsequent to this introduction, foundations and work related to smells in software are presented in Section 2. A survey of work concerning smells in tests is given in Section 3. As the main contribution, a catalogue of TTCN-3 code smells is introduced in Section 4. Then, in Section 5, a tool is described which is able to automatically detect instances of TTCN-3 code smells in test suites. Section 6 provides results from applying this tool to several huge standardised test suites. Finally, this paper concludes with a summary and an outlook.

2 Foundations

The metaphor of *"bad smells in code"* has been coined by Beck and Fowler in the context of refactoring [12]. Refactoring is a technique to improve the internal quality of software by restructuring it without changing its observable behaviour. As an aid to decide where the application of a refactoring is worthwhile, Beck and Fowler introduce the notion of smell: they define smells in source code as *"certain structures in the code that suggest (sometimes they scream for) the possibility of refactoring"* [12]. According to this definition, defects with respect to program logic, syntax, or static semantics are not smells, because these defects cannot be removed by a behaviour-preserving refactoring. This means, smells are indicators of bad internal quality with respect to (re-)usability, maintainability, efficiency, and portability.

Smells provide only hints: whether the occurrence of an instance of a certain smell in a source code is considered as a sign of low quality may be a matter that depends on preferences and experiences. For the same reason, a list of code structures which are considered as smell is never complete, but may vary from project to project and from domain to domain [13].

Beck and Fowler provide a list of 22 smells which may occur in Java source code. They describe their smells using unstructured English text. The most prominent smell is *Duplicated Code*. Code duplication deteriorates in particular the changeability of a source code: if code that is duplicated needs to be modified, it usually needs to be changed in all duplicated locations as well. Another example from Beck's and Fowler's list of smells is *Long Method* which relates to the fact that short methods are easier to understand and to reuse, because they do exactly one thing. A further example is the smell called *Data Class* which characterises classes that only have attributes and accessor methods, but the actual algorithms working on these data are wrongly located in methods of other classes.

Most of the smells from Beck and Fowler relate to pathological structures in the source code. Thus, to detect such structures, a pattern-based approach is required: for example, to identify duplicated code, pattern-matching is required; to detect data classes, it has to be identified whether the methods of a class are only simple get and set methods and whether methods in other classes do excessively manipulate data from that particular class. Such patterns cannot be detected by metrics — however, the notion of metrics and smells is not disjoint: each smell can be turned into a metric by counting the occurrences of a smell, and sometimes, a metric can be used to detect and locate an instance of a smell. The latter is, for example, the case for the *Long Method* smell which can be expressed by a metric which counts the lines of code of a method.[1]

Bad smells are also related to *anti-patterns* [14]. Anti-patterns describe solutions to recurring problems that are wrong and bad practice and shall thus be avoided. A well-known anti-pattern is the one called *Spaghetti Code*, i.e. software with little structure. Even though this and other anti-patterns relate to source code, anti-patterns do not refer to low-level code details as code smells do. In fact, the majority of the anti-patterns do not relate to source code at all, but to common mistakes in project management.

The awareness of problematic source code structures is older than the notion of smells in source code. For example, patterns of data flow anomalies which can be detected by static analysis have been known for a long time [15]. However, these older works mainly relate to erroneous, inconsistent, inefficient, and wasteful code constructs. The added value of smells is to consider also more abstract source code quality problems, for example those which lead to maintenance problems.

3   Smells in Tests

As stated in the previous section, the perception of what is considered as a smell may vary from domain to domain. Hence, for the testing domain, a separate investigation of smells is required. Van Deursen et al. and Meszaros studied smells in the context of tests that are based on the Java unit test framework *JUnit* [16].

Van Deursen et al. [17] introduce the term *test smell* for smells that are specific to the usage of *JUnit* as well as for more general JUnit-independent issues in test

---

[1] It has to be noted that Beck and Fowler state that for detecting instances of a smell *"no set of metrics rivals informed human intuition"*. This is obviously true for those smells where no corresponding metric exists. However, in the cases, where such a metric exists, this statement does in our opinion rather relate to the fact that reasonable boundary values for such a metric may vary from case to case and thus it is hard to provide a universally valid boundary value for that metric.

behaviour that can be removed by a refactoring. An example for a JUnit-specific test smell is *General Fixture* which refers to test cases that share unnecessarily the same fixture (i.e. test preamble), just because the test cases are collected in the same JUnit testcase class. A more general test smell is, for example, *Test Run War* which relates to the fact that test cases may behave non-deterministic due to shared test resources when several test campaigns run in parallel.

Meszaros [18] refines the notion of test smell by distinguishing between three kinds of smells that concern tests: *code smells* relate to test issues that can be detected when looking at source code, *behaviour smells* affect the outcome of tests as they execute, and *project smells* are indicators of the overall health of a project which do not involve looking at code or executing tests. Within this classification, smells of different kinds may affect each other; for example, the root cause of a behaviour smell may be a problem in the code. We regard this classification of test smells as reasonable and adopt this terminology as well.

Those test smells from Van Deursen et al. that are JUnit-specific (e.g. *General Fixture*) can be considered as code smells while others are more general (e.g. *Test Run War*) and can thus be regarded as behaviour smells. Meszaros does not only refine the notion of test smells, but also extends the list of test smells from Van Deursen et al. by further smells. An example for an additional code smell is *Conditional Test Logic* which refers to tests which are error-prone because they use complex algorithms to calculate test data and to steer test behaviour. A behaviour smell identified by Meszaros is, for example, *Fragile Tests*, which are tests that fail after non-relevant changes of the *System Under Test* (SUT). An example of a project smell is *Developers Not Writing Tests*.

4   A TTCN-3 Code Smell Catalogue

While code smells have been identified for tests written using the JUnit framework, smells have not yet been investigated in the context of TTCN-3. The project smells identified by Meszaros [18] are independent from any test language and can thus be used as well in projects that involve TTCN-3. Most of Meszaros' behaviour smells apply to TTCN-3 tests without change, however those behaviour smells whose root cause is a JUnit related code smell are only applicable after a reinterpretation. Only a subset of the JUnit related code smells can be reinterpreted in a way that they are applicable to TTCN-3. Hence, code smells related to TTCN-3 need further investigation.

We have started to identify TTCN-3 code smells which we use to assess the internal quality of TTCN-3 test specifications. When investigating possible smell candidates we have relaxed Beck's and Fowler's definition of smells in source code: We include not only internal quality problems in TTCN-3 source code that can be improved by a behaviour preserving refactoring, but we consider as well quality problems which obviously require a change of the behaviour. One example is a test case which never sets a test verdict. In this case, a statement that sets a verdict needs to be added. This cannot be achieved by applying a refactoring, since this is a change that would not be behaviour-preserving.

Though, we still adhere to the definition of code smell, in that we do not consider errors in TTCN-3 source code with respect to syntax or static semantics as a smell.

As a starting point for our investigations, we examined those code smells that were already known for implementation and testing languages. Even though the smells listed by Beck and Fowler [12] are intended for Java code, some of them proved to be suitable for TTCN-3 code. A further source was the TTCN-3 refactoring catalogue [3,19,20] which was in turn inspired by the JUnit refactorings and JUnit code smells published by Van Deursen et al. [17]. The refactorings collected in the TTCN-3 refactoring catalogue already refer briefly to code smell-like quality issues as a motivation for each refactoring.

In contrast to the plain listing of unstructured smell descriptions that is used by Beck and Fowler or by Van Deursen et al., we have catalogued our TTCN-3 code smells in a structured way. This structured presentation allows a more systematic and faster access to the smell descriptions. The entries in our TTCN-3 code smell catalogue are listed in the following format: each smell has a *name*; those smells which are derived from other sources have a *derived from* section which lists the corresponding references; a *description* provides a prose summary of the issue described by the smell; the *motivation* part explains why the described code structure is considered to have low quality; if several variants of a smell are possible (e.g. by relaxing or tightening certain requirements on a code structure), this is mentioned in an *options* section; one or more actions (typically refactorings) which are applicable to remove a smell are listed in the *related actions* section; finally, a TTCN-3 source code snippet is provided for each smell in the *example* section.

In our smell catalogue, the names of TTCN-3 code smells are emphasised using *slanted* type and TTCN-3 keywords are printed using **bold** type. The following overview on our TTCN-3 code smell catalogue gives an impression of the so far identified 38 TTCN-3 code smells. The overview provides the name and the summary of each smell and uses the same division into 10 sections as our TTCN-3 code smell catalogue:

Duplicated Code

- *Duplicate Statements:* A duplicate sequence of statements occurs in the statement block of one or multiple behavioural entities (functions, test cases, and altsteps).
- *Duplicate Alt Branches:* Different **alt** constructs contain duplicate branches.
- *Duplicated Code in Conditional:* Duplicated code is found in the branches of a series of conditionals.
- *Duplicate In-Line Templates:* Two or more in-line templates are very similar or identical.
- *Duplicate Template Fields:* The fields of two or more templates are identical or very similar.
- *Duplicate Component Definition:* Two or more test components declare identical variables, constants, timers, or ports.

– *Duplicate Local Variable/Constant/Timer:* The same local variable, constant, or timer is defined in two or more functions, test cases, or altsteps running on the same test component.

References

– *Singular Template Reference:* A template definition is referenced only once.
– *Singular Component Variable/Constant/Timer Reference:* A component variable, constant, or timer is referenced by one single function, test case, or altstep only, although other behaviour runs on the component as well.
– *Unused Definition:* A definition is never referenced.
– *Unused Imports:* An import from another module is never used.
– *Unrestricted Imports:* A module imports more than needed.

Parameters

– *Unused Parameter:* A parameter is never used within the declaring unit: **in**-parameters are never read, **out**-parameters are never assigned, **inout**-parameters are never accessed at all.
– *Constant Actual Parameter Value:* The actual parameter values for a formal parameter are the same for all references.
– *Fully-Parametrised Template:* All fields of a template are defined by formal parameters.

Complexity

– *Long Statement Block:* A function, test case, or altstep has a long statement block.
– *Long Parameter List:* The number of formal parameters is high.
– *Complex Conditional:* A conditional expression is composed of many Boolean conjunctions.
– *Nested Conditional:* A conditional expression is unnecessarily nested.
– *Short Template:* A template definition is very short.

Default Anomalies

– *Activation Asymmetry:* A default activation has no matching subsequent deactivation in the same statement block, or a deactivation has no matching previous activation.
– *Unreachable Default:* An **alt** statement contains an **else** branch while a default is active.

Test Behaviour

– *Missing Verdict:* A test case does not set a verdict.
– *Missing Log:* **setverdict** sets the verdict **inconc** or **fail** without calling **log**.
– *Stop in Function:* A function contains a **stop** statement.

Test Configuration

- *Idle PTC:* A *Parallel Test Component* (PTC) is created, but never started.
- *Isolated PTC:* A PTC is created and started, but its ports are not connected to other ports.

Coding Standards

- *Magic Values:* A literal is not defined as a TTCN-3 constant.
- *Bad Naming:* An identifier does not conform to a given naming convention.
- *Disorder:* The sequence of elements within a module does not conform to a given order.
- *Insufficient Grouping:* A module or group contains too many elements.
- *Bad Comment Rate:* The comment rate is too high or too low.
- *Bad Documentation Comment:* A documentation comment does not conform to a given format, e.g. T3Doc [21].

Data Flow Anomalies

- *Missing Variable Definition:* A variable or **out** parameter is read before a value has been assigned.
- *Unused Variable Definition:* An assigned variable or **in**-parameter is not read before it becomes undefined.
- *Wasted Variable Definition:* A variable is assigned and assigned again before it is read.

Miscellaneous

- *Over-specific Runs On:* A behavioural entity runs on a component but uses only elements of the super-component or no component elements at all.
- *Goto:* A **goto** statement is used.

To give an impression of how the entries in our TTCN-3 code smell catalogue look like, the smells *Duplicate Alt Branches* and *Activation Asymmetry* are subsequently presented in detail. In addition to the already mentioned style of typesetting TTCN-3 keywords and names of smells, references to refactorings from the TTCN-3 refactoring catalogue [3,19,20] are printed in *slanted* type as well.[2] Please refer to our complete TTCN-3 code smell catalogue [22] for a detailed description of all so far identified TTCN-3 code smells.

4.1   TTCN-3 Code Smell: *Duplicate Alt Branches*

Derived from: TTCN-3 refactoring catalogue [3,19,20].
Description: Different **alt** constructs contain duplicate branches.

---

[2] References to refactorings and to smells can still be distinguished, because the names of refactorings usually start with a verb followed by a noun, whereas the names of smells usually consist of an adjective and a noun.

Motivation: Code duplication in branches of **alt** constructs should be avoided just as well as any other duplicated code, because duplication deteriorates changeability. In particular, common branches for error handling can often be handled by default altsteps if extracted into an own altstep beforehand.

Options: Since analysability is increased if the path leading to a **pass** verdict is explicitly visible in a test case, **alt** branches leading to **pass** can be excluded optionally.

Related action(s): Use *Extract Altstep* refactoring to separate the duplicate branches into an own altstep. Consider refactoring *Split Altstep* if the extracted altstep contains branches which are not closely related to each other and refactoring *Replace Altstep with Default* if the duplicate branches are invariably used at the end of the **alt** construct.

Example: In Listing 1.1, both test cases contain an **alt** construct where the last branch (lines 6–10 and lines 19–23) can be found as well in the other **alt** construct.

```
1  testcase myTestcase1() runs on myComponent {
2      alt {
3          [ ] pt.receive(messageOne) {
4              pt.send(messageTwo);
5          }
6          [ ] any port.receive {
7              log("unexpected message");
8              setverdict(inconc);
9              stop;
10         }
11     }
12 }
13
14 testcase myTestcase2() runs on myComponent {
15     alt {
16         [ ] pt.receive(messageThree) {
17             pt.send(messageFour);
18         }
19         [ ] any port.receive {
20             log("unexpected message");
21             setverdict(inconc);
22             stop;
23         }
24     }
25 }
```

Listing 1.1. Duplicate Alt Branches

4.2 TTCN-3 Code Smell: *Activation Asymmetry*

Description: A default activation has no matching subsequent deactivation in the same statement block, or a deactivation has no matching previous activation.

Motivation: The analysability with respect to active defaults is improved if default activation and deactivation is done on the same "level", usually at the very beginning and end of the same statement block. Furthermore, this enables a static analysis of matching activation and deactivation.

Options: Because defaults are implicitly deactivated at the end of a test case run, statement blocks in test cases can be excluded optionally.

Related action(s): Default activation or deactivation should be added if missing, and matching default activation and deactivation should be moved to the same statement block.

Example: In Listing 1.2, the altstep "myAltstep" (lines 1–6) is used as default. Function "myFunction" (lines 8–10) activates this altstep as default, but no **deactivate** statement is contained in the statement block of this function. Even though it might be reasonable in some situations to move activation and deactivation of defaults into separate functions, this has to be considered as an asymmetric default activation. A further asymmetry can be found in the test case "myTestcase": the statement block of the **deactivate** statement in Line 20 consists of lines 13–15 and Line 20. This statement block contains no **activate** statement, since the activation of the default is performed within the statement block of the function "myFunction" that is called in Line 14.

```
1   altstep myAltstep() runs on myComponent {
2       [ ] any port.receive {
3           log("unexpected message");
4           setverdict(inconc);
5       }
6   }
7
8   function myFunction() return default {
9       return activate(myAltstep());
10  }
11
12  testcase myTestcase() runs on myComponent {
13      var default myDefaultVar := null;
14      myDefaultVar := myFunction();
15      alt {
16          [ ] pt.receive(messageOne) {
17              pt.send(messageTwo);
18          }
19      }
20      deactivate(myDefaultVar);
21  }
```

Listing 1.2. Activation Asymmetry

5   A Tool for Detecting TTCN-3 Code Smell Instances

Our TTCN-3 code smell catalogue can be utilised for the quality assessment of TTCN-3 test suites. One possibility is to use it as part of a checklist in a manual inspection of TTCN-3 code. However, the efficiency of such a code inspection can be significantly improved if the detection of instances of TTCN-3 code smells is automated by a tool.[3] This allows the code reviewers to focus

---

[3] All of our TTCN-3 code smells are intended to be detected by static analysis; however, those analyses required for smells related to test behaviour and data flow anomalies are —in the general case— undecidable and can thus only solved by static analysis heuristics (in the simplest case by neglecting any branching and assuming a linear control flow instead).

on high-level logical errors in the test suite, since instances of low-level code smells have already been detected automatically. However, in our experience an everyday usage of an automated issue detection outside of a formal inspection is even more beneficial: the push-button detection of smell instances allows test engineers to easily obtain feedback on the internal quality of the TTCN-3 test suites that they are currently developing.

We have implemented the automated detection of instances of TTCN-3 code smells into our open-source TTCN-3 Refactoring and Metrics tool *TRex* [20]. The initial version of TRex [23] has been developed in collaboration with the Motorola Labs, UK, to provide an *Integrated Development Environment* (IDE) for the quality assessment and improvement of TTCN-3 test suites. In that version, the quality assessment was based on metrics; for the quality improvement, refactoring is used [3]. Since then, we have extended the quality assessment capabilities of TRex by an additional automated detection of TTCN-3 code smell instances. So far, TRex provides rules to detect by static analysis instances of the following 11 TTCN-3 code smells:

- *Activation Asymmetry*,
- *Constant Actual Parameter Value* smells for templates,
- *Duplicate Alt Branches*,
- *Fully-Parametrised Template*,
- *Magic Values* of numeric or string types with configurable tolerable magic numbers,
- *Short Template* smells with configurable character lengths,
- *Singular Component Variable/Constant/Timer Reference*,
- *Singular Template Reference*,
- *Duplicate Template Fields*,
- instances of any local *Unused Definition*,
- an *Unused Definition* of a global template instance.

As stated in Section 2, whether a certain code structure is considered as a smell or not, may vary from project to project. Therefore, TRex supports enabling and disabling individual TTCN-3 code smell detection rules and to store these preferences as customised analysis configurations (Figure 1). Furthermore, it is possible to parametrise some smell detection rules. For example, for detecting instances of the *Magic Values* smell, a *Magic Number* detection rule and a *Magic String* detection rule are available; the *Magic Number* detection rule can be parametrised to exclude user defined values (e.g. 0 and 1 which are usually considered to be tolerable magic numbers) from the smell instance detection.

The results of the smell analysis are displayed as a tree in the *Analysis Results* view (Figure 2). The results are collected in a history, which allows to compare analysis results. Clicking on an entry of the analysis result jumps to the corresponding location in the TTCN-3 source code to allow a further manual inspection. Some rules, for example *Unused Definitions*, offer the possibility of invoking so called *Quick Fixes*. Quick Fixes automatically suggest the invocation of TTCN-3 refactoring to remove a detected instance of a smell. Since a couple of refactorings are implemented in TRex [23], this does not only allow an

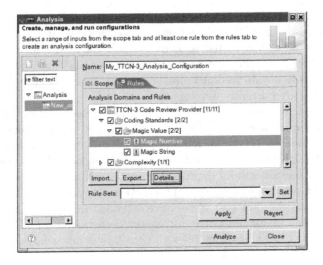

Fig. 1. TRex Smell Analysis Configuration

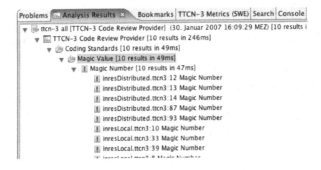

Fig. 2. TRex Smell Analysis Results View

automated quality assessment, but as well an automated quality improvement of TTCN-3 test suites.

5.1   Implementation

The implementation of the TRex tool is based on the Eclipse platform [24] as shown in Figure 3. Eclipse provides generic user interface and text editor components as well as a language toolkit for behaviour preserving source code transformation. As an infrastructure for the automated quality assessment and quality improvement functionality of TRex (blocks (2) and (3) of Figure 3), TRex creates a syntax tree and a symbol table of the currently opened test suites (Block (1) of Figure 3). For lexing and parsing the TTCN-3 core notation, *'ANother Tool for Language Recognition'* (ANTLR) [25] is used. A further description of the implementation of the quality improvement based on refactor-

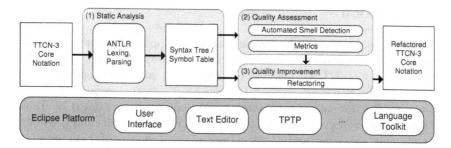

Fig. 3. The TRex Toolchain

ings and of the implementation of the quality assessment based on metrics can be found in earlier papers [3,23].

The smell analysis configuration dialogue and the smell analysis results view are provided by the *static analysis framework* which is part of the *Eclipse Test & Performance Tools Platform* (TPTP) [26]. In the context of the TPTP static analysis framework, each smell detection capability is represented by a rule. TPTP provides the underlying programming interface to add and implement rules and to call and apply the rules to files according to the user-defined analysis configuration. The actual smell instance detection is based on syntax tree traversals and symbol table lookups. For example, to detect *Duplicate Alt Branches*, the sub-syntaxtrees of all branches of **alt** and **altstep** constructs of a TTCN-3 module are compared. Currently, only exact sub-tree matches are detected; however, since the syntax tree does not contain tokens which do not have any semantical meaning, the detection of duplicates is tolerant with respect to formatting and comments. The *Unused Definition* rules make intensively use of the symbol table to check for every symbol whether it is referenced at least once or not. To ease the implementation of future smell detection rules, we have extracted frequently used helper algorithms into methods of a smell detection library.

5.2 Related Work

Approaches for the automatic detection of source code issues that are detectable by static analysis and go beyond the application of metrics have been known for a long time. The most prominent example is probably the *Lint* tool [27]. Even though Lint is older than the notion of smells, it detects issues which are nowadays considered as code smell. Current research shows that automatic detection of instances of a smell is still relevant [13,28]. In addition to this research, mature tools for detecting instances of smells in Java programs do already exist. Examples are tools like *FindBugs* [29] or *PMD* [30]. All the mentioned work deals with the detection of instances of smells in source code written in implementation languages like C or Java. Hence, this work does neither consider TTCN-3 related smells nor more general test specific smells at all. The only known work on the automated detection of instances of test smells is restricted to the detection of JUnit code smells [31].

6   Application

To evaluate the practicability of our approach, we applied TRex to several huge test suites that have been standardised by the *European Telecommunications Standards Institute* (ETSI). The first considered test suite is Version 3.2.1 of the test suite for the *Session Initiation Protocol* (SIP) [4], the second is a preliminary version of a test suite for the *Internet Protocol Version 6* (IPv6) [5]. Table 1 shows the number of detected instances of TTCN-3 code smells and provides as well some simple size metrics to give an impression of the size of these test suites.

Both test suites are comparable in size and in both, the same types of smells can be found. Magic numbers can be found quite often in both test suites. An excerpt from the SIP test suite is shown in Listing 1.3: the magic number "65.0" used in Line 10 occurs several times throughout the test suite. If that number must be changed during maintenance, it must probably changed at all other places as well which is very tedious.

The number of detected instances of the *Activation Asymmetry* smell is as well very high in both test suites. However, the number drops, if test cases are excluded from the detection. Even though the SIP test suite has less *Activation Asymmetry* smell instances, they still deteriorate the analysability of this test suite as shown in Listing 1.3: the altstep "defaultCCPRPTC" is activated in Line 6 and remains activated after leaving this function. Hence, calling this function leads to side effects that are difficult to analyse.

Finally, Listing 1.3 can be used to demonstrate occurrences of the *Unused Definition* smell in the SIP test suite: the local variable "v_BYE_Request" defined in Line 3 is never used in the function and thus just bloats the code, making it harder to analyse.

The instances of *Singular Component Variable/Constant/Timer Reference* smells can be neglected in both test suites. However, the high number of *Duplicate Alt Branches* in both test suites indicates that the introduction of further altsteps is worthwhile. For example, the branch in lines 9–11 of Listing 1.4 can be found as duplicate in several **alt** statements of the IPv6 tests suite.

Table 1. Instances of TTCN-3 Code Smells Found in ETSI Test Suites

Metric/TTCN-3 Code Smell	SIP	IPv6
Lines of code	42397	46163
Number of functions	785	643
Number of test cases	528	295
Number of altsteps	10	11
Number of components	2	10
Instances of *Magic Values* (Magic numbers only, 0 and 1 excluded)	543	368
Instances of *Activation Asymmetry* (Test cases included)	602	801
Instances of *Activation Asymmetry* (Test cases excluded)	73	317
Instances of *Duplicate Alt Branches* (Inside the same module only)	938	224
Instances of *Singular Component Variable/Constant/Timer Reference*	2	15
Instances of *Unused Definition* (Local definitions only)	50	156

This and our further analysis [22] of the detected smell instances give evidence that these instances are correctly considered as issues and can thus be used for quality assessment and as starting point to improve the internal quality of the respective test suites.

```
1   function ptc_CC_PR_TR_CL_TI_015(CSeq loc_CSeq_s ) runs on SipComponent
2   {
3       var Request v_BYE_Request;
4
5       initPTC(loc_CSeq_s);
6       v_Default := activate(defaultCCPRPTC());
7
8       tryingPTCBYE();
9
10      waitForTimeout(65.0*PX_T1);
11
12      notRepeatBYE(PX_TACK);
13
14  } //end ptc_CC_PR_TR_CL_TI_015
```

Listing 1.3. *Magic Values, Activation Asymmetry, Unused Definition* (SIP)

```
1   tc_ac.start;
2   alt {
3       [ ] ipPort.receive ( mw_nbrAdv_noExtHdr (
4                           p_paramsIut.lla,
5                           p_paramsRt01.lla ) ) {
6           tc_ac.stop;
7           v_ret := e_success;
8       }
9       [ ] tc_ac.timeout{
10          v_ret := e_timeout;
11      }
12  } // end alt
```

Listing 1.4. *Duplicate Alt Branches* (IPv6)

7   Conclusion

We presented a catalogue of 38 TTCN-3 code smells that can be utilised to detect code-level problems in TTCN-3 test suites with respect to internal quality characteristics like usability, maintainability, or reusability. Each of our entries in the TTCN-3 code smell catalogue provides a description of the considered code issue, a motivation why it is considered to have low quality, an action to remove the smell (typically using a TTCN-3 refactoring [3]), and an example. In this paper, we gave an overview of our TTCN-3 code smell catalogue and presented excerpts from the full version [22]. We have implemented the automated detection of instances of TTCN-3 code smells in our TRex tool and demonstrated the applicability of our approach by assessing the internal quality of standardised test suites.

In future, we intend to extend our TTCN-3 smell catalogue by further code smells and also by more sophisticated high-level smells (e.g. smells related to issues in a test architecture). In parallel, we will implement further smell detection rules in TRex and evaluate their validity. The current smell detection rules are implemented in an imperative style in Java. To ease the implementation of further smell detection rules it is desirable to specify the code pattern that is described by a smell in a declarative way like the PMD tool [30] supports for Java-specific smells. Finally, we believe that it is worthwhile to investigate smells for other test specification languages, for example the *UML 2.0 Testing Profile* (U2TP) [32].

Acknowledgements. The authors like to thank Jens Grabowski and the anonymous reviewers for valuable comments on improving this paper.

References

1. ETSI: ETSI Standard (ES) 201 873-1 V3.2.1 (2007-02): The Testing and Test Control Notation version 3; Part 1: TTCN-3 Core Language. European Telecommunications Standards Institute (ETSI), Sophia-Antipolis, France, also published as ITU-T Recommendation Z.140 (February 2007)
2. Grabowski, J., Hogrefe, D., Réthy, G., Schieferdecker, I., Wiles, A., Willcock, C.: An introduction to the testing and test control notation (TTCN-3). Computer Networks 42(3), 375–403 (2003)
3. Zeiss, B., Neukirchen, H., Grabowski, J., Evans, D., Baker, P.: Refactoring and Metrics for TTCN-3 Test Suites. In: Gotzhein, R., Reed, R. (eds.) SAM 2006. LNCS, vol. 4320, pp. 148–165. Springer, Heidelberg (2006)
4. ETSI: Technical Specification (TS) 102 027-3 V3.2.1 (2005-07): SIP ATS & PIXIT; Part 3: Abstract Test Suite (ATS) and partial Protocol Implementation eXtra Information for Testing (PIXIT). European Telecommunications Standards Institute (ETSI), Sophia-Antipolis, France (July 2005)
5. ETSI: Technical Specification (TS) 102 516 V1.1 (2006-04): IPv6 Core Protocol; Conformance Abstract Test Suite (ATS) and partial Protocol Implementation eXtra Information for Testing (PIXIT). European Telecommunications Standards Institute (ETSI), Sophia-Antipolis, France (April 2006)
6. Baker, P., Loh, S., Weil, F.: Model-Driven Engineering in a Large Industrial Context – Motorola Case Study. In: Briand, L.C., Williams, C. (eds.) MoDELS 2005. LNCS, vol. 3713, pp. 476–491. Springer, Heidelberg (2005)
7. ETSI: Specialist Task Force 296: Maintenance of SIP Test Specifications. European Telecommunications Standards Institute (ETSI), Sophia-Antipolis, France (2007)
8. ETSI: Specialist Task Force 320: Upgrading and maintenance of IPv6 test specifications. European Telecommunications Standards Institute (ETSI), Sophia-Antipolis, France (2007)
9. ISO/IEC: ISO/IEC Standard No. 9126: Software engineering – Product quality; Parts 1–4. International Organization for Standardization (ISO) / International Electrotechnical Commission (IEC), Geneva, Switzerland (2001-2004)
10. Fenton, N.E., Pfleeger, S.L.: Software Metrics. PWS Publishing, Boston (1997)
11. Zeiss, B., Vega, D., Schieferdecker, I., Neukirchen, H., Grabowski, J.: Applying the ISO 9126 Quality Model to Test Specifications – Exemplified for TTCN-3 Test Specifications. In: Bleek, W.G., Raasch, J., Züllighoven, H. (eds.) Proceedings of Software Engineering 2007 (SE 2007), Bonn, Gesellschaft für Informatik. Lecture Notes in Informatics, vol. 105, pp. 231–242. Köllen Verlag (2007)

12. Fowler, M.: Refactoring – Improving the Design of Existing Code. Addison-Wesley, Boston (1999)
13. van Emden, E., Moonen, L.: Java Quality Assurance by Detecting Code Smells. In: Proceedings of the 9th Working Conference on Reverse Engineering, pp. 97–106. IEEE Computer Society Press, Los Alamitos (2002)
14. Brown, W.J., Malveau, R.C., McCormick, H.W., Mowbray, T.J.: Anti-Patterns. Wiley, New York (1998)
15. Fosdick, L.D., Osterweil, L.J.: Data Flow Analysis in Software Reliability. ACM Computing Surveys 8(3), 305–330 (1976)
16. Gamma, E., Beck, K.: JUnit (February 2007) http://junit.sourceforge.net
17. van Deursen, A., Moonen, L., van den Bergh, A., Kok, G.: Refactoring Test Code. In: Extreme Programming Perspectives, pp. 141–152. Addison-Wesley, Boston (2002)
18. Meszaros, G.: XUnit Test Patterns. Addison-Wesley, Boston (to appear, 2007)
19. Zeiss, B.: A Refactoring Tool for TTCN-3. Master's thesis, Institute for Informatics, University of Göttingen, Germany (March 2006) ZFI-BM-2006-05
20. TRex Team: TRex Website (February 2007) http://www.trex.informatik.uni-goettingen.de
21. ETSI: ETSI Standard (ES) 201 873-10 V3.2.1: TTCN-3 Documentation Comment Specification. European Telecommunications Standards Institute (ETSI), Sophia-Antipolis, France (to appear, 2007)
22. Bisanz, M.: Pattern-based Smell Detection in TTCN-3 Test Suites. Master's thesis, Institute for Informatics, University of Göttingen, Germany (December 2006) ZFI-BM-2006-44
23. Baker, P., Evans, D., Grabowski, J., Neukirchen, H., Zeiss, B.: TRex – The Refactoring and Metrics Tool for TTCN-3 Test Specifications. In: Proceedings of TAIC PART 2006 (Testing: Academic & Industrial Conference – Practice And Research Techniques), Cumberland Lodge, Windsor Great Park, UK, pp. 90–94. IEEE Computer Society, Los Alamitos (29th–31st August 2006)
24. Eclipse Foundation: Eclipse (February 2007) http://www.eclipse.org
25. Parr, T.: ANTLR parser generator (February 2007) http://www.antlr.org
26. Eclipse Foundation: Eclipse Test & Performance Tools Platform Project (TPTP) (February 2007) http://www.eclipse.org/tptp
27. Johnson, S.: Lint, a C Program Checker. Unix Programmer's Manual, AT&T Bell Laboratories (1978)
28. Moha, N., Gueheneuc, Y.G.: On the Automatic Detection and Correction of Design Defects. In: Demeyer, S., Mens, K., Wuyts, R., Ducasse, S. (eds.) Proceedings of the 6th ECOOP Workshop on Object-Oriented Reengineering. LNCS, Springer, Heidelberg (to appear)
29. Pugh, B.: FindBugs (February 2007) http://findbugs.sourceforge.net
30. Dixon-Peugh, D.: PMD (February 2007) http://pmd.sourceforge.net
31. van Rompaey, B., du Bois, B., Demeyer, S.: Characterizing the Relative Significance of a Test Smell. In: Proceedings of the 22nd IEEE International Conference on Software Maintenance (ICSM 2006), Philadelphia, Pennsylvania, pp. 391–400. IEEE Computer Society, Los Alamitos (September 25–27, 2006)
32. OMG: UML Testing Profile (Version 1.0 formal/05-07-07). Object Management Group (OMG) (July 2005)

A Bounded Incremental Test Generation Algorithm for Finite State Machines

Zoltán Pap[1], Mahadevan Subramaniam[2], Gábor Kovács[3],
and Gábor Árpád Németh[3]

[1] Ericsson Telecomm. Hungary, H-1117 Budapest, Irinyi J. u. 4-20, Hungary
zoltan.pap@ericsson.com
[2] Computer Science Department, University of Nebraska at Omaha
Omaha, NE 68182, USA
msubramaniam@mail.unomaha.edu
[3] Department of Telecommunications and Media Informatics – ETIK,
Budapest University of Technology and Economics,
Magyar tudósok körútja 2, H-1117, Budapest, Hungary
kovacsg@tmit.bme.hu, rubrika@gmail.com

Abstract. We propose a bounded incremental algorithm to generate test cases for deterministic finite state machine models. Our approach, in contrast to the traditional view, is based on the observation that system specifications are in most cases modified incrementally in practice as requirements evolve. We utilize an existing test set available for a previous version of the system to efficiently generate tests for the current – modified – system.

We use a widely accepted framework to evaluate the complexity of the proposed incremental algorithm, and show that it is a function of the size of the change in the specification rather than the size of the specification itself. Thus, the method is very efficient in the case of small changes, and never performs worse than the relevant traditional algorithm – the HIS-method. We also demonstrate our algorithm through an example.

Keywords: conformance testing, finite state machine, test generation algorithms, incremental algorithms.

1 Introduction

Large, complex systems continuously evolve to incorporate new features and new requirements. In each evolution step – in addition to changing the specification of the system and producing a corresponding implementation – it may also be necessary to modify the testing infrastructure. Manual modification is an ad hoc and error prone process that should be avoided, and automatic specification-based test generation methods should be applied.

Although testing theory is especially well developed for finite state machine (FSM)-based system specifications, existing algorithms handle changing specifications quite inefficiently. Most research has been focusing on the analysis of rigid, unchanging descriptions. Virtually all proposed methods rely solely on a

A. Petrenko et al. (Eds.): TestCom/FATES 2007, LNCS 4581, pp. 244–259, 2007.

given specification machine to generate tests. These approaches are therefore incapable of utilizing any auxiliary information, such as existing tests created for the previous version of the given system. All test sequences have to be created from scratch in each evolution step, no matter how small the change has been.

In this paper we develop a novel, bounded incremental algorithm to automatically re-generate tests in response to changes to a system specification. In its essence the algorithm maintains two sets incrementally; a prefix-closed state cover set responsible for reaching all states of the finite state machine, and a separating family of sequences applied to verify the next state of transitions. The complexity of the algorithm is evaluated based on the bounded incremental model of computation of Ramalingam and Reps [1]. It is shown that the time complexity of the proposed algorithm depends on the size of the change to the specification rather than the size of the specification itself. Furthermore, it is never worse than the complexity of the most traditional algorithm – the HIS-method [2] [3] [4].

This research builds on our earlier work in [5] where we have developed a framework to analyze the effects of changes on tests based on the notion of consistency between tests and protocol descriptions. In the current paper, we have extended our focus to the test generation problem, which is a major step both in terms of complexity and practical importance.

The rest of the paper is organized as follows. A brief overview of our assumptions and notations is given in Section 2. In Section 3, we describe some relevant FSM test generation algorithms and the HIS-Method in particular. Section 4 describes the model of incremental computation. In Section 5 we introduce the incremental algorithm for maintaining a checking sequence across changes, provide a thorough analysis of its complexity and demonstrate it through an example. Sections 6 and 7 describe related work and conclusions, respectively.

2   Finite State Machines

Finite state machines have been widely used for decades to model systems in various areas. These include sequential circuits [6], some types of programs [7] (in lexical analysis, pattern matching etc.), and communication protocols [8]. Several specification languages, such as SDL [9] and ESTELLE [10], are extensions of the FSM formalism.

A finite state machine M is a quadruple $M = (I, O, S, T)$ where I is the finite set of input symbols, O is the finite set of output symbols, S is the finite set of states, and $T \subseteq S \times I \times O \times S$ is the finite set of (state) transitions. Each transition $t \in T$ is a 4-tuple $t = (s_j, i, o, s_k)$ consisting of start state $s_j \in S$, input symbol $i \in I$, output symbol $o \in O$ and next state $s_k \in S$.

An FSM can be represented by a state transition graph, a directed edge-labeled graph whose vertices are labeled as the states of the machine and whose edges correspond to the state transitions. Each edge is labeled with the input and output associated with the transition.

FSM M is said to be deterministic if for each start state – input pair (s, i) there is at most one transition in T. In the case of deterministic FSMs both the output and the next state of a transition may be given as a function of the start state and the input of the transition. These functions are referred to as the next state function δ: $S \times I \to S$ and the output function λ: $S \times I \to O$. Thus a transition of a deterministic machine may be given as $t = (s_j, i, \lambda(s_j, i), \delta(s_j, i))$.

For a given set of symbols A, A^* is used to denote the set of all finite sequences (words) over A. Let $K \subseteq A^*$ be a set of sequences over A. The prefix closure of K, written **Pref**(K), includes all the prefixes of all sequences in K. The set K is prefix-closed if **Pref**$(K) = K$.

We extend the next state function δ and output function λ from input symbols to finite input sequences I^* as follows: For a state s_1, an input sequence $x = i_1, ..., i_k$ takes the machine successively to states $s_{j+1} = \delta(s_j, i_j), j = 1, ..., k$ with the final state $\delta(s_1, x) = s_{k+1}$, and produces an output sequence $\lambda(s_1, x) = o_1, ..., o_k$, where $o_j = \lambda(s_j, i_j), j = 1, ..., k$. The input/output sequence $i_1 o_1 i_2 o_2 ... i_k o_k$ is then called a trace of M.

FSM M is said to be strongly connected if, for each pair of states (s_j, s_l), there exists an input sequence which takes M from s_j to s_l. If there is at least one transition $t \in T$ for all start state – input pairs, the FSM is said to be completely specified (or completely defined); otherwise, M is a said to be partially specified or simply a partial FSM.

We say that machine M has a reset capability if there is an initial state $s_0 \in S$ and an input symbol $r \in I$ that takes the machine from any state back to s_0. That is, $\exists r \in I : \forall s_j \in S : \delta(s_j, r) = s_0$. The reset is reliable if it is guaranteed to work properly in any implementation machine M^I, i.e., $\delta^I(s_j^I, r) = s_0^I$ for all states $s_j^I \in S^I$, and s_0^I is the initial state of M^I; otherwise it is unreliable.

Finite state machines may contain redundant states. State minimization is a transformation into an equivalent state machine to remove redundant states. Two states are equivalent written $s_j \cong s_l$ iff for all input sequences $x \in I^*$, $\lambda(s_j, x) = \lambda(s_l, x)$. Two states, s_j and s_l are distinguishable (inequivalent), iff $\exists x \in I^*$, $\lambda(s_j, x) \neq \lambda(s_l, x)$. Such an input sequence x is called a separating sequence of the two inequivalent states. A FSM M is reduced (minimized), if no two states are equivalent, that is, each pair of states (s_j, s_l) are distinguishable.

For the rest of the paper, we focus on strongly connected, completely specified and reduced deterministic machines with reliable reset capability. We will denote the number of states and inputs by $n = |S|$ and $p = |I|$, respectively.[1]

2.1   Representing Changes to FSMs

Although the FSM modeling technique has been used extensively in various fields, impact of changes on FSM models and their effects on test sets have only been studied recently following the observation that system specifications are in most cases modified incrementally in practice as requirements evolve. (See some of our earlier papers [11] [12] [5]).

A consistent approach for representing changes to FSM systems has been proposed in [12]. Atomic changes to a finite state machine M are represented by

---

[1] Therefore, $|T| = p * n$.

the means of edit operators $\omega^M : T \to T$.[2] An edit operator turns FSM $M = (I, O, S, T)$ into FSM $M' = (I, O, S', T')$ with the same input and output sets. We use the term "same states" written $s_j = s'_j$ for states that are labeled alike in different machines. Obviously, these states are not necessarily equivalent, written $s_j \cong s'_j$.

For deterministic finite state machines two types of edit operators have been proposed based on widely accepted fault models. A next state change operator is $\omega_n(s_j, i, o_x, s_k) = (s'_j, i, o_x, s'_l)$, where $\delta(s_j, i) = s_k \neq s'_l = \delta'(s'_j, i)$. An output change operator is $\omega_o(s_j, i, o_x, s_k) = (s'_j, i, o_y, s'_k)$, where $\lambda(s_j, i) = o_x \neq o_y = \lambda'(s'_j, i)$. It has been shown in [12], that with some assumptions the set of deterministic finite state machines with a given number of states is closed under the edit operations defined above. Furthermore, for any two deterministic FSMs M_1 and M_2 there is always a sequence of edit operations changing M_1 to M_2, i.e., to a machine isomorphic to M_2.

3   FSM Test Generation and the HIS-Method

Given a completely specified deterministic FSM M with n states, an input sequence x that distinguishes M from all other machines with n states is called a checking sequence of M. Any implementation machine $Impl$ with at most n states not equivalent to M produces an output different from M on checking sequence x.

Several algorithms have been proposed to generate checking sequences for machines with reliable reset capability [13] [14], including the W-method [15], the Wp-method [16] and the HIS-method [2] [3] [4]. They all share the same fundamental structure consisting of two stages: Tests derived for the first – state identification – stage check that each state presented in the specification also exists in the implementation. Tests for the second – transition testing – stage check all remaining transitions of the implementation for correct output and ending state as defined by the specification. The methods, however, use different approaches to identify a state during the first stage, and to check the ending state of the transitions in the second stage. In the following we concentrate on the HIS-method as it is the most general approach of the three.

The HIS-method derives a *family of harmonized identifiers* [4], also referred to as a *separating family* of sequences [3]. A separating family of sequences of FSM M is a collection of n sets $Z_i, i = 1, ..., n$ of sequences (one set for each state) satisfying the following two conditions: For every pair of states s_i, s_j: (I) there is an input sequence x that separates them, i.e., $\exists x \in I^*, \lambda(s_i, x) \neq \lambda(s_j, x)$; (II) x is a prefix of some sequence in Z_i and some sequence in Z_j. Z_i is called the separating set of state s_i. The HIS-method uses appropriate members of the separating family in both stages of the algorithm to check states of the implementation.

3.1   The HIS-Method

Consider FSM M with $|S| = n$ states, and implementation $Impl$ with at most n states. Let $Z = \{Z_1, ..., Z_n\}$ be a separating family of sequences of FSM

---

[2] $\omega(t)$ is used instead of $\omega^M(t)$ if M can be omitted without causing confusion.

M. Such family may be constructed for a reduced FSM the following way: For any pair of states s_i, s_j we generate a sequence z_{ij} that separates them using for example a minimization method [17]. Then define the separating sets as $Z_i = \{z_{ij}\}, j = 1...n$.

The state identification stage of the HIS-method requires a prefix-closed state cover set $Q = \{q_1, ..., q_n\}$ of FSM M, and generates test sequences $r \cdot q_i \cdot Z_i$, $i = 1...n$ based on it, where r is the reliable reset symbol and "$\cdot$" is the string concatenation operator. A Q set may be created by constructing a spanning tree[3] of the state transition graph of the specification machine M from the initial state s_0. Such a spanning tree is presented on Figure 1(a) in Section 5.1. A prefix-closed state cover set Q is the concatenation of the input symbols on all partial paths of the spanning tree, i.e., sequences of input symbols on all consecutive branches from the root of the tree to a state.

If *Impl* passes the first stage of the algorithm for all states, then we know that *Impl* is similar to M, furthermore this portion of the test also verifies all the transitions of the spanning tree. The second, transition testing stage is used to check non-tree transitions. That is, for each transition (s_j, i, o, s_k) not in the spanning tree the following test sequences are generated: $r \cdot q_j \cdot i \cdot Z_k$.

The resulting sequence is a checking sequence, starting at the initial state (first a reset input is applied) and consisting of no more than pn^2 test sequences of length less than $2n$ interposed with reset [3]. Thus the total complexity of the algorithm is $O(pn^3)$, where $p = |I|$ and $n = |S|$.

4   Incremental Computation Model

A *batch* algorithm for a given problem is an algorithm capable of computing the solution of the problem $f(x')$ – the output – on some input x'. Virtually all traditional FSM-based conformance test generation algorithms [13] [14] are such *batch* algorithms. Their input is the specification of a system in form of an FSM model and the output is a checking sequence that is (under some assumptions) capable of determining if an implementation conforms to the specification FSM.

An incremental algorithm intends to solve a given problem by computing an output $f(x')$ just as a batch algorithm. Incremental computation, however, assumes that the same problem has been solved previously on a slightly different input x providing output $f(x)$, and that the input has undergone some changes since, resulting in the current input $x + dx = x'$. An incremental algorithm takes the input x and the output $f(x)$ of the previous computation, along with the change in the input dx. From that it computes the new output $f(x + dx)$, where $x + dx$ denotes the modified input. A batch algorithm can be used as an incremental algorithm, furthermore, in case of a fundamental change (take

---

[3] A spanning tree of FSM M rooted from the initial state is an acyclic subgraph (a partial FSM) of its state transition graph composed of all the reachable vertices (states) and some of the edges (transitions) of M such that there is exactly one path from the initial state s_0 to any other state.

$x = null$ input for example) the batch algorithm will be the most efficient incremental algorithm.

4.1 Evaluating the Complexity of an Incremental Algorithm

The complexity of an algorithm is commonly evaluated using asymptotic worst-case analysis; by expressing the maximum cost of the computation as a function of the size of the input. While this approach is adequate for most batch algorithms, worst-case analysis is often not very informative for incremental algorithms. Thus, alternative ways have been proposed in the literature to express the complexity of incremental algorithms. The most widely accepted approach has been proposed in [1]. Instead of analyzing the complexity of incremental algorithms in terms of the size of the entire current input, the authors suggest the use of an adaptive parameter capturing the extent of the changes in the input and output. The parameter Δ or "$CHANGED$" represents the size of the "$MODIFIED$" part of the input and the size of the "$AFFECTED$" part of the previous output. Thus Δ represents the minimal amount of work necessary to calculate the new output. The complexity of incremental algorithm is analyzed in terms of Δ, which is not known a priori, but calculated during the update process. This approach will be used in this paper to evaluate the complexity of the presented algorithm and to compare it to existing batch and incremental methods.

5 Incremental Test Generation Method

This section presents a novel incremental test generation algorithm. The algorithm – in contrast to traditional (batch) test generation methods – is capable of maintaining a checking sequence across changes in the specification, thus avoiding the need of regenerating a checking sequence from scratch at each stage of an incremental development.

We focus on the following problem: Consider a system specification given as a reduced, completely specified and deterministic FSM M. There exists a complete checking sequence for M capable of detecting any fault in an implementation $Impl$, which has the same input I and output O alphabet as M and has no more states than M. The specification is modified to M' by a unit change, i.e., by applying a single – output or a next state – change operator. The problem is to create a complete checking sequence for the new specification M' if such exists.

We concentrate on systems with reliable reset capability, and we assume the HIS-Method as a reference point in creating an incremental algorithm and evaluating its performance. The HIS-method is essentially the superposition of two completely independent algorithms. One is used to build a set of input sequences responsible for reaching all states of the finite state machine (a prefix-closed state cover set). The other is applied to create a set of input sequences to verify the next state of the transition (a separating family of sequences).

Our incremental test generation method likewise involves two completely autonomous incremental algorithms. Note that these algorithms may also be applied independently for various purposes. They could be used to detect undesirable effects of a planned modification during development, such as subsets of states becoming equivalent or unreachable.

It has to be emphasized that a given change to the specification FSM may affect the two algorithms differently. Therefore two separate Δ parameters (see Section 4.1) have to be used to capture the extent in which the changes affect the two algorithms.

5.1   Incremental Algorithm for Maintaining a Prefix-Closed State Cover Set

Given a specification FSM M, a prefix-closed state cover set Q of M and a change $\omega(s_m, i, o, s_j)$ to FSM M turning it to M' our purpose is to create a new valid prefix-closed state cover set Q' for M'.

The problem can be reduced to maintaining a spanning tree of the state transition graph of the specification machine rooted from the initial state s_0 (see Section 3.1). Assuming the spanning tree ST of FSM M representing the Q set – i.e., input sequences on all partial paths of ST are in Q – we intend to produce a new valid spanning tree ST' of FSM M'.

Let us call a transition an ST-transition iff it is in ST. A subtree of ST rooted from a state $s_i \neq s_0$ is a proper subtree of the spanning tree ST and will be referred to as ST_{s_i}.

Given the change $\omega(s_m, i, o, s_j)$ we will refer to state s'_m of FSM M' as *modified* state, since a transition originating from state s_m of FSM M is modified by the change. In this paper we focus on unit changes; at each incremental step there is a single *modified* state, i.e., the cardinality of the set of *modified* states $MODIFIED$ abbreviated as MOD is one: $|MOD| = 1$.

A state s'_i of FSM M' is affected by the change with respect to the Q set iff for input sequence $q_i \in Q$ corresponding to state s_i: $\delta(s_0, q_i) \neq \delta'(s'_0, q_i)$. Such a state is said to be a *q-affected* state.[4] The algorithm identifies the set of *q-affected* states $AFFECTED_Q$ abbreviated as AFF_Q, where $0 \leq |AFF_Q| \leq n$. If AFF_Q is not an empty set – $|AFF_Q| > 0$ – then ST must be adapted to M'.

We define the set $CHANGED_Q \subseteq S'$ to be $MOD \cup AFF_Q$ and denote $|CHANGED_Q|$ as Δ_Q. The set $CHANGED_Q$ will be used as a measure of the size of the change to the specification, and the complexity of the incremental algorithm will be expressed as a function of parameter Δ_Q, where $1 \leq \Delta_Q \leq n$.

The input of the algorithm is the original machine M, the change operator and the spanning tree ST of M. It provides M', the new spanning tree ST' of

---

[4] Other definitions of *q-affected* state could be used depending on the assumed testing algorithm. A more relaxed definition could be for example the following: A state s'_i of FSM M' is affected by the change with respect to the Q set iff there exists no path from s'_0 to s'_i in ST' after the change. This definition should be assumed in case the same set (for example a distinguishing sequence or W-set) is used to check each ending state.

M' and the set of unreachable states as output. The algorithm consists of two phases and handles output and next state changes separately in the first phase. The first phase marks all q-affected states of FSM M' then collects them in the set AFF_Q. If $|AFF_Q| = 0$ then ST is a valid spanning tree of M' and the algorithm terminates, otherwise the second phase completes the spanning tree for all *q-affected* states.

Phase 1 – Output Change. Take output change $\omega_o(s_m, i, o_x, s_j)=(s'_m, i, o_y, s'_j)$, $o_x \neq o_y$. Create FSM M' by applying the change operator. Initialize AFF_Q as an empty set, and the spanning tree ST' of M' as $ST' := ST$.

As an output change operator is applied to FSM M, it only changes an edge label of the state transition graph of FSM M, but does not affect its structure. That is, $\delta(s_m, i) = \delta'(s'_m, i)$, and the change does not affect any states with respect to the Q set. AFF_Q is not extended.

Phase 1 – Next State Change. Take next state change $\omega_n(s_m, i, o_x, s_j) = (s'_m, i, o_x, s'_k)$, $s_j \neq s'_k$. Create FSM M' by applying the change operator. Initialize AFF_Q as an empty set, and the spanning tree ST' of M' as $ST' := ST$.

- $(s_m, i, o, s_j) \notin ST$: If the transition upon input i at state s_m is not an ST-transition then any change to it can not affect the spanning tree of FSM M. AFF_Q is not extended.
- $(s_m, i, o, s_j) \in ST$: If the transition upon input i at state s_m is an ST-transition then the change affects the spanning tree. The *q-affected* states are identified walking the $ST'_{s'_j}$ subtree. All states of $ST'_{s'_j}$ (including s'_j) are marked as *q-affected* states. The AFF_Q set can be determined using a simple breadth-first search of the $ST'_{s'_j}$ subtree with a worst case complexity of $|AFF_Q|$.

Phase 2: Determining a spanning tree of M'. Phase 2 of the algorithm takes the set AFF_Q from Phase 1 and completes the spanning tree for each member of AFF_Q to create a spanning tree ST' of M'.

If $|AFF_Q| = 0$ (there are no *q-affected* states) then ST is a spanning tree of M'. Return ST' and the algorithm terminates.

If $|AFF_Q| > 0$ then we apply the following method: All transitions of ST' leading to *q-affected* states are removed from ST' along with the modified transition (s'_m, i, o, s'_k). Then we extend ST' as follows.

For all s'_x in AFF_Q we start checking the transitions leading to s'_x in M' until either a transition originating from an unaffected state is found or there are no more inbound transitions left. If a transition (s'_i, i, o, s'_x) such that $s'_i \notin AFF'_Q$ is found then: (I) $ST' := ST' \cup (s'_i, i, o, s'_x)$, (II) $AFF_Q := AFF_Q \setminus \{s'_x\}$, (III) if there is transition (s'_x, i, o, s'_y) where $s'_y \in AFF_Q$ then repeat Steps I-III on s'_y.

The algorithm stops after all s'_x in AFF_Q has been checked, then return ST' and AFF_Q; the algorithm terminates. At the end of the last turn ST' will be a spanning tree of M', and any s'_z remaining in AFF_Q is unreachable from s'_0 in M'.

Q-set example. Take FSM M on Figure 1(a) where bold edges represent the spanning tree and the double circle denotes the initial state.

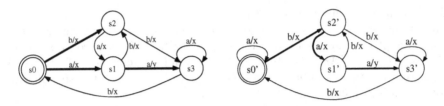

(a) FSM M with its spanning tree ST (b) Modified FSM M' with the updated spanning tree ST'

Fig. 1. Example for maintaining the preamble

Initially let $ST' = ST$ and $AFF_Q = \emptyset$. The modification $\omega_n(s_0, a, x, s_1) = (s'_0, a, x, s'_0)$ is a next state change. As transition (s_0, a, x, s_1) is in ST, we need to determine the set of q-affected states by walking the $ST'_{s'_1}$ subtree. We get $AFF_Q = \{s'_1, s'_3\}$. In Phase 2 transitions leading to q-affected states – (s'_0, a, x, s'_1) and (s'_1, a, y, s'_3) – are removed from ST'. Then one of the states – say s'_1 – is selected from AFF_Q. Transition (s'_2, a, x, s'_1) is identified, which is a link originating from a not affected state s'_2. We add it to ST' and remove s'_1 from AFF_Q. We then check transitions originating from s'_1 and find (s'_1, a, y, s'_3) that leads to a q-affected state. We add (s'_1, a, y, s'_3) to ST' and remove s'_3 from AFF_Q. Now, $AFF_Q = \emptyset$, so the algorithm terminates and returns ST', see Figure 1(b).

Theorem 1. *The incremental algorithm for maintaining a spanning tree described above has a time complexity of $O(p * \Delta_Q)$, where $1 \leq \Delta_Q \leq n$.*

Proof. Phase 1 of the algorithm has worst case complexity of $O(|AFF_Q|)$.

Phase 2 of the algorithm first searches a path from the unaffected states of M' to the q-affected states. There are exactly $p * |AFF_Q|$ transitions originating from the q-affected states. Therefore there can be at most $p * |AFF_Q|$ steps that do not provide a path from unaffected states of M' to the q-affected states summarized over all backward check turns of Phase 2. Thus there are no more than $(p + 1) * |AFF_Q|$ backward check turns.

If a link is found from an unaffected state to an affected state s'_x then the algorithm adds all states of AFF_Q reachable from s'_x via affected states. Again, there can be at most $p * |AFF_Q|$ such steps summarized over all forward check turns of Phase 2.

As any of the $p * |AFF_Q|$ transitions are processed at most twice by the algorithm, less than $2 * (p + 1) * |AFF_Q| \approx O(p * |AFF_Q|)$ steps are necessary to complete Phase 2. The total complexity of the algorithm is $O(p * |AFF_Q|) \leq O(p * \Delta_Q)$ □

The new set Q' of M' contains $|AFF_Q|$ modified sequences: Input sequences of ST' leading from s'_0 to s'_i for all s'_i in AFF_Q.

5.2 Incremental Algorithm for Maintaining a Separating Family of Sequences

We are again given the specification FSM M, and the change $\omega(s_m, i, o, s_j)$ turning M to M'. We also have a separating family of sequences of FSM M (a separating set for each state): $Z = \{Z_1, ..., Z_n\}|Z_i = \{z_{ij}\}, j = 1...n$, where z_{ij} is a separating sequence of states s_i, s_j of FSM M. Our objective is to create a new separating family of sequences Z' for M'. Note that we consider a somewhat structured separating family of sequences as discussed later. This, however, does not restrict the generality of the approach as each incremental step generates a separating family according the assumed structure.

Informally speaking, to maintain a separating family of sequences we have to identify all separating sequences affected by the change. Then for all such state pairs a new separating sequence has to be generated. Notice that this is a problem over state pairs rather than states. Therefore we introduce an auxiliary directed graph A^M with $n(n+1)/2$ nodes, one for each unordered pair (s_j, s_k) of states of M including identical state pairs (s_j, s_j). There is a directed edge from (s_j, s_k) to (s_l, s_m) labeled with input symbol i iff $\delta(s_j, i) = s_l$ and $\delta(s_k, i) = s_m$ in M. The auxiliary directed graph A^M is used to represent and maintain separating sequences of FSM M. The graph is updated by our algorithm at each incremental step.

We define a *separating state pair* as an unordered pair of states (s_x, s_y) such that $\lambda(s_x, i) \neq \lambda(s_y, i)$ for some $i \in I$. A machine M is minimal iff there is a path from each non-identical state pair $(s_j, s_k), j \neq k$ to a *separating state pair* in its auxiliary directed graph A^M. The input labels along the route concatenated by the input distinguishing the *separating state pair* form a separating sequence of states s_j and s_k.

We make the following assumptions on the separating sequences of FSM M: (I) Each *separating state pair* (s_x, s_y) has a single separating input $i|\lambda(s_x, i) \neq \lambda(s_y, i)$ associated to it. If a given pair has multiple such inputs, then the input to be associated is chosen randomly. (II)The set of separating sequences of FSM M is prefix-closed.

Then separating sequences of FSM M form an acyclic subgraph of the auxiliary directed graph A^M, such that there is exactly one path from each state pair $(s_x, s_y), x \neq y$ to a *separating state pair*. That is, separating sequences form a forest over the non-identical state pairs of A^M, such that each tree has a *separating state pair* as root and all edges of the given tree are directed toward the root – see Figure 2(a) below for example. Let us refer to this forest (a subgraph of A^M) as SF. We call an edge of A^M an SF-edge iff it is in SF. A subtree of SF having state pair (s_i, s_j) as root is a proper subtree of the forest SF and will be referred to as SF_{s_i, s_j}. Note that by walking such a tree (or its subtree) we always assume that it is explored opposing edge directions from the root (or an inner node) toward leaves.

Thus the problem of deriving the separating family of sequences for FSM M can be reduced to maintaining separating state pairs, their associated separating input and a forest SF over non-identical state pairs of A^M across changes.

Given the change $\omega(s_m, i, o, s_j)$ turning M to M' all state pairs that include state s_m are modified to construct the auxiliary directed graph $A^{M'}$ of FSM M'. Accordingly all unordered state pairs of $A^{M'}$ involving s'_m are referred to as *z-modified* state pairs. As a result of the unit change assumption the cardinality of the set of *z-modified* state pairs $MODIFIED_Z$ abbreviated as MOD_Z is n: $|MOD_Z| = n$.[5]

The algorithm derives the set of state pairs affected by the change. Such state pairs are said to be *z-affected*. The set of *z-affected* state pairs is referred to as $AFFECTED_Z$ abbreviated as AFF_Z, where $0 \leq |AFF_Z| \leq n(n-1)/2$. We define the set $CHANGED_Z \subseteq S' \times S'$ as $MOD_Z \cup AFF_Z$. The complexity of the incremental algorithm will be expressed as a function of parameter $|CHANGED_Z|$ referred to as Δ_Z, where $n \leq \Delta_Z \leq n(n-1)/2$.

The input of the algorithm is the auxiliary directed graph A^M of FSM M, the change operator and the forest SF of A^M representing separating sequences of M. The output is $A^{M'}$, the new forest SF' of $A^{M'}$ and a set containing pairs of equivalent states.

The algorithm consists of two phases and handles output and next state changes separately in the first phase.

Phase 1 – Output Change. Take output change $\omega_o(s_m, i, o_x, s_k) = (s'_m, i, o_y, s'_k), o_x \neq o_y$. Initialize AFF_Z as an empty set, $A^{M'} := A^M$ and $SF' := SF$.

For state pairs $\forall s'_i \in S' : (s'_m, s'_i)$ apply the change to $A^{M'}$ and:

- If state pair (s'_m, s'_i) is a new *separating state pair* then mark it and associate i as separating input.
- If i has been the separating input of *separating state pair* (s_m, s_i) in A^M but $\lambda'(s'_m, i) = \lambda'(s'_i, i) = o_y$ then all state pairs of the tree with (s'_m, s'_i) root – including (s'_m, s'_i) – are added to AFF_Z (marked as *z-affected*). These states can be identified by walking the given tree from the root.
  - If there is another input $i_1 | \lambda'(s'_m, i_1) \neq \lambda'(s'_i, i_1)$ then (s'_m, s'_i) remains a *separating state pair* with i_1 associated as separating input. State pair (s'_m, s'_i) is removed from AFF_Z.
  - If $\forall i \in I : \lambda'(s'_m, i) = \lambda'(s'_i, i)$ then (s'_m, s'_i) is no longer a *separating state pair*, thus the *separating state pair* marking is removed from (s'_m, s'_i).
- Do nothing otherwise.[6]

Phase 1 – Next State Change. Take next state change $\omega_n(s_m, i, o_x, s_k) = (s'_m, i, o_x, s'_l), s_k \neq s'_l$. Initialize AFF_Z as an empty set, $A^{M'} := A^M$ and $SF' := SF$.

For state pairs $\forall s'_i \in S : (s'_m, s'_i)$ apply the change to $A^{M'}$ and:

- If the edge of A^M marked by input i at state pair (s_m, s_i) is an SF-edge then the modification affects the given tree of the spanning forest. All state pairs of the $SF'_{s'_m, s'_i}$ subtree are *z-affected* states (including (s'_m, s'_i)). Thus

---

[5] Pairs s'_i, s'_i of identical states are also modified here.

[6] One could assume different definitions for affected state pairs as a design choice.

the $SF'_{s'_m, s'_i}$ subtree is explored using a simple breadth-first search, all state pairs are added to AFF_Z (marked as *z-affected*).
- Do nothing otherwise.

Phase 2. Phase 2 of the algorithm takes the set AFF_Z from Phase 1 and updates the forest SF' for each member of AFF_Z.

If $|AFF_Z| = 0$ then SF is a valid forest over $A^{M'}$ representing a separating sequence for each non-identical state pair of M'. Return SF' and the algorithm terminates.

If $|AFF_Z| > 0$ then the following method is applied: All edges of SF' originating from *z-affected* state pairs are removed from SF'. Then we extend SF' as follows. We examine all edges of $A^{M'}$ originating from a *z-affected* state pair and construct a subgraph of $A^{M'}$ denoted as $A^{M'}_{AFF}$ the following way: (I) For each *z-affected* state pair there is a corresponding node in $A^{M'}_{AFF}$. (II) For each edge between *z-affected* state pairs there is an edge in $A^{M'}_{AFF}$. (III) If there is an edge originating from a *z-affected* state pair leading to a not affected state pair then we mark the given *z-affected* state pair at the head of the edge.

Next we explore $A^{M'}_{AFF}$ opposing edge directions from marked state pairs using breadth-first search to create a spanning forest over $A^{M'}_{AFF}$ with marked state as root nodes. All state pairs covered by the spanning forest are removed from AFF_Z. Finally SF' is expanded simply appending the spanning forest of $A^{M'}_{AFF}$. Each tree of the forest of $A^{M'}_{AFF}$ is linked to SF' by an edge leading to a not affected state pair from its marked root node. Return SF' and AFF_Z; the algorithm terminates.

At the end of the algorithm AFF_Z contains any pairs of equivalent states for which no separating sequence exists. Each partial path of SF' represents a separating sequence of M': Given a path from node (s'_i, s'_j) to *separating state pair* (s'_x, s'_y) the input labels along the route concatenated by the separating input of s'_x, s'_y form a separating sequence z'_{ij} of states s'_i and s'_j. The separating family of sequences of FSM M' is given as $Z' = \{Z'_1, ..., Z'_n\}|Z'_i = \{z'_{ij}\}, j = 1...n$.

Z set example. The auxiliary graph A^M of M is presented on Figure 2(a). Bold edges represent the forest SF of M, separating state pairs are shown in bold ellipses, separating inputs are represented by bigger sized edge labels, while the dotted edges between identical state pairs are maintained but have no importance for the algorithm.

Initially $AFF_Z = \emptyset$ and let $SF' = SF$. Edges labeled with input a originating from state pairs $(s'_0, s'_0), (s'_0, s'_1), (s'_0, s'_2), (s'_0, s'_3)$ are modified to create $A^{M'}$. (s'_0, s'_1) is a *separating state pair* and is therefore not affected. The a-labeled edge originating from state pair (s_0, s_2) is not in SF thus (s'_0, s'_2) is not affected either. (s'_0, s'_0) is irrelevant. Therefore only state pair (s'_0, s'_3) is *z-affected*: $AFF_Z = \{(s'_0, s'_3)\}$. In Phase 2 the a-labeled edge originating from (s'_0, s'_3) is removed from SF'. Then edges originating from (s'_0, s'_3) are checked and an edge $\langle(s'_0, s'_3), (s'_0, s'_2)\rangle$ leading to a non-affected state pair is found. The given edge is added to SF' and (s'_0, s'_3) is removed from AFF_Z. Now, $AFF_Z = \emptyset$, thus the algorithm terminates and returns SF', see Figure 2(b). All separating sequences

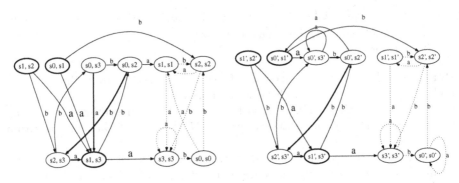

(a) The auxiliary graph A^M of FSM M (b) The updated auxiliary graph $A^{M'}$ of FSM M'

Fig. 2. Auxiliary graphs

are unchanged except the one of states s'_0 s'_3, which is changed from $a \cdot a$ to $b \cdot b \cdot a \cdot a$.

Theorem 2. *The incremental algorithm for maintaining a separating family of sequences described above has a time complexity of $O(p * \Delta_Z)$, where $n \leq \Delta_Z \leq n^2$.*

Proof. Regardless of change operator type Phase 1 of the algorithm involves $|MOD_Z|$ modification steps and $O(|AFF_Z|)$ steps to identify affected state pairs. Phase 2 of the algorithm first creates a subgraph in $p * |AFF_Z|$ steps and then creates a spanning forest over it with $O(p * |AFF_Z|)$ complexity. Thus the total complexity of the algorithm is $O(p * \Delta_Z)$.   □

5.3   Total Complexity of the Incremental Testing

Our algorithm – just as the original HIS-method – derives actual test sequences in two stages by concatenating sequences from the sets Q and Z_i. Each test sequence – a part of the checking sequence – is either a sequence $r \cdot q_i \cdot z_{ij}$ (stage 1) or a sequence $r \cdot q_i \cdot i \cdot z_{xy}$ (stage 2). A test sequence must be regenerated after a change if either the q-sequence, or the z-sequence of the given test sequence is modified by our incremental algorithms above. That is, a test sequence $r \cdot q_i \cdot \ldots \cdot z_{ij}$ of M is modified iff $s'_i \in AFF_Q$ or $(s'_i, s'_j) \in AFF_Z$. Such sequences are identified using links between the sets Q', $Z'_i, i = 1 \ldots n$ and test sequences. The number of modified test sequences is less or equal than $p * n^2$, i.e., in worst case the number of test cases to be generated is equivalent to those generated by a batch algorithm. The resulting test set is a complete test set of M' that is no different than one generated using the *batch* HIS-method. It consists of the same set of test sequences generated using valid Q' and Z' sets of M'.

Note that the concatenation operation is – from the complexity point of view – a quite expensive step. Concatenation, however, only has to be performed as a part of the testing procedure itself. If no actual testing is needed after a change to the specification then we should only – very efficiently – maintain the sets Q and Z according to each modification and do the concatenation just as necessary.

6   Related Work

Nearly all test generation approaches in the FSM test generation literature propose batch algorithms, i.e., they focus on building test sets for a system from scratch without utilizing any information from tests created for previous versions of the given system. One of the few exceptions – the most relevant research – has been the work of El-Fakih at al. [18]. Similarly to our approach, the authors assume a specification in form of a complete deterministic FSM M, which is modified to a new specification M'. The problem in both cases is to generate a checking sequence to test if an implementation $Impl$ conforms to M'. Our approach, however, involves some fundamental improvements on El-Fakih's method. The most important are:

1. El-Fakih's algorithm does not intend to create a complete test set for the modified specification M', instead it may be used to "generate tests that would only test the parts of the new implementation that correspond to the modified parts of the specification" [18]. This necessitates the following quite restrictive assumption: "the parts of the system implementation that correspond to the unmodified parts of the specification have not been changed" [18]. Thus, it is presumed that no accidental or intentional (malicious) changes are introduced to supposedly unmodified parts of the implementation. Such faults could remain undetected as the test set generated by the algorithm is not complete. Note that the assumption above is unavoidable as not even the union of the existing test set of M and the incremental test set generated by the algorithm provide a complete test set for M'.

   Our algorithm, on the other hand, maintains a complete test set across the changes to the specification. The algorithm modifies the existing test set of M to create a complete test set for M' – if such exists – capable of detecting any fault in $Impl$.

2. El-Fakih's algorithm is not a bounded incremental algorithm in the sense that it uses traditional batch algorithms to create a state cover set and a separating family of sequences for a given FSM upon each modification. Therefore its complexity is the function of the size of the input FSM, not the extent of the change.

   Our method in turn is a bounded incremental algorithm, its complexity is dependent on the extent of the change. It identifies how the modification affects the existing test set of the original specification machine M. New tests are only generated for the affected part and the calculation is independent of the unaffected part. The complexity of our algorithm is no worse than the complexity of the corresponding batch algorithm.

7 Conclusion

We have presented a bounded incremental algorithm to generate test cases for deterministic finite state machine models. Our approach assumes a changing specification and utilizes an existing test set of the previous version to efficiently maintain a complete test set across the changes to the specification. For each update of the system a complete test set is generated with the same fault detection capability as that of a traditional *batch* algorithm. The complexity of the algorithm is evaluated based on the bounded incremental model of computation of Ramalingam and Reps [1]. The time complexity of the proposed algorithm is shown to be bounded; it is a function of the size of the change to the specification rather than the size of the specification itself. It is never worse than the complexity of the relevant traditional algorithm – the HIS-method. Furthermore, the two autonomous incremental algorithms building up the incremental test generation method may also be applied independently for various purposes during development.

In the future we plan to further experiment with the presented algorithm to gain sufficient performance data for practical analysis. Our current focus has been on time complexity but the approach leaves space for fine-tuning and optimizations in several aspects that will have to be studied. The research reported here is regarded as a first step in developing efficient incremental testing algorithms. We plan to investigate if this approach can be extended to different problems and models.

References

1. Ramalingam, G., Reps, T.: On the computational complexity of dynamic graph problems. Theoretical Computer Science 158(1-2), 233–277 (1996)
2. Sabnani, K., Dahbura, A.: A protocol test generation procedure. Computer Networks and ISDN Systems 15(4), 285–297 (1988)
3. Yannakakis, M., Lee, D.: Testing finite state machines: fault detection. In: Selected papers of the 23rd annual ACM symposium on Theory of computing, pp. 209–227 (1995)
4. Petrenko, A., Yevtushenko, N., Lebedev, A., Das, A.: Nondeterministic state machines in protocol conformance testing. In: Proceedings of the IFIP TC6/WG6.1 Sixth International Workshop on Protocol Test systems, vol. VI, pp. 363–378 (1994)
5. Subramaniam, M., Pap, Z.: Analyzing the impact of protocol changes on tests. In: Proceedings of the IFIP International Conference on Testing Communicating Systems, TestCom, pp. 197–212 (2006)
6. Friedman, A.D., Menon, P.R.: Fault Detection in Digital Circuits. Prentice-Hall, Englewood Cliffs (1971)
7. Aho, A.V., Sethi, R., Ullman, J.D.: Compilers: Principles, Techniques, and Tools. Addison-Wesley, London (1986)
8. Holzmann, G.J.: Design and Validation of Protocols. Prentice-Hall, Englewood Cliffs (1990)
9. ITU-T: Recommendation Z.100: Specification and description language (2000)
10. TC97/SC21, I.: Estelle – a formal description technique based on an extended state transition model. international standard 9074 (1988)

11. Subramaniam, M., Chundi, P.: An approach to preserve protocol consistency and executability across updates. In: Davies, J., Schulte, W., Barnett, M. (eds.) ICFEM 2004. LNCS, vol. 3308, pp. 341–356. Springer, Heidelberg (2004)
12. Pap, Z., Csopaki, G., Dibuz, S.: On the theory of patching. In: Proceedings of the 3rd IEEE International Conference on Software Engineering and Formal Methods, SEFM, pp. 263–271 (2005)
13. Lee, D., Yiannakakis, M.: Principles and methods of testing finite state machines – a survey. In: Proceedings of the IEEE, vol. 84(8), pp. 1090–1123 (1996)
14. Bochmann, G.V., Petrenko, A.: Protocol testing: review of methods and relevance for software testing. In: ISSTA '94: Proceedings of the 1994 ACM SIGSOFT international symposium on Software testing and analysis, pp. 109–124. ACM Press, New York, USA (1994)
15. Chow, T.: Testing software design modelled by finite-state machines. IEEE Transactions on Software Engineering 4(3), 178–187 (1978)
16. Fujiwara, S., Bochmann, G.v., Khendec, F., Amalou, M., Ghedamsi, A.: Test selection based on finite state model. IEEE Transactions on Software Engenieering 17, 591–603 (1991)
17. Kohavi, Z.: Switching and Finite Automata Theory. McGraw-Hill, New York (1978)
18. El-Fakih, K., Yevtushenko, N., von Bochmann, G.: FSM-based incremental conformance testing methods. IEEE Transactions on Software Engineering 30(7), 425–436 (2004)

Experimental Testing of TCP/IP/Ethernet Communication for Automatic Control

Przemyslaw Plesowicz and Mieczyslaw Metzger

Faculty of Automatic Control, Electronics and Computer Science,
Silesian University of Technology,
ul. Akademicka 16, 44-100 Gliwice, Poland
{przemyslaw.plesowicz, mieczyslaw.metzger}@polsl.pl

Abstract. The TCP/IP/Ethernet protocol is considered not suitable for use in real-time control systems. It deals with a lack of time determinism, which characterizes fieldbuses. Nevertheless several corporations propose networking based on the TCP/IP/Ethernet even for control purposes with some modifications of the standard however. This paper examines possibility of application of the TCP/IP/Ethernet communication without modifications (introducing also Internet as one of tested cases) for feedback control purposes. Experimental investigations have been performed in four stages. In the beginning tests of network properties, including tests of transmission time and packet loss measurements have been performed. Three following stages show experimental testing of feedback control, when TCP/IP transmission occurs between PI controller and control plant. Three representative platforms have been chosen for testing: LabVIEW, RSLogix and Simatic. The main and original contribution presented in this paper is design and construction of three test stands as well as methodology of testing experiments. Standard control over analog channel has been also presented as comparison. The results of testing show acceptable performance of control via TCP/IP/Ethernet networking.

Keywords: TCP/IP communication testing, Ethernet TCP/IP, networks, network-based feedback control.

1 Introduction

Nowadays automation systems designed for industrial plants became complex and usually consist of many components such as instrumentation, software and networking. A growing need for advanced industrial networking techniques for complex applications in engineering and research results in more and more sophisticated technologies such as Profibus, Modbus, ControlNet, DeviceNet, CAN, FIP and many others. Mentioned standards have an important advantage over the widely used Ethernet standard — they are time-deterministic. Unfortunately, the application of fieldbuses has been limited due to very high cost of appropriate hardware and software and due to incompatibility of multivendor products. This situation has pushed engineers toward attempts to apply worldwide-used and in consequence inexpensive Ethernet standard. This is the reason, why appropriate

A. Petrenko et al. (Eds.): TestCom/FATES 2007, LNCS 4581, pp. 260–275, 2007.

testing techniques are crucial for such based on instrumentation, software and networks control systems.

Fundamentals of basic testing techniques for communication systems are well defined in recent publications over the last decade (see for example [1], [2], [3]). Although such methods are very convenient for complex communication systems in a general case, the specialised experimental tests dealing with development of laboratory stands seems to be more adequate for testing communication in automatic control. This paper presents such instrumentation and tests — designed, developed and carried out for testing TCP/IP/Ethernet communication in feedback control.

Remote data-acquisition and monitoring of non-critical plants can be achieved without deterministic characteristics, hence the TCP/IP/Ethernet can be used for remote monitoring without problems, with some requirements and modifications of the standard however — see for example [4], [5], [6]. The control theory proposes some analytical discussions of the problem of delays introduced by networking in control systems — see e.g. [7], [8], [9], [10].

This paper examines possibility of application of the TCP/IP/Ethernet communication without modifications (introducing also Internet as one of tested cases) for feedback control purposes. Experimental investigations have been performed in four stages. In the beginning, tests of network properties have been performed (including measurements of transmission time and packet loss). Three following stages show experimental testing of feedback control, in which TCP/IP transmission occurs between PI controller and control plant. Three representative platforms have been chosen for testing: LabVIEW (*National Instruments*), RSLogix (*Rockwell-Allen-Bradley*) and Simatic (*Siemens*). The hardware (PLC and distributed I/O) used during experiments was especially chosen, to represent solutions most popular in industry (*Rockwell-Allen-Bradley, Siemens*) and most popular in scientific research and education (*National Instruments*). The main and original contribution presented in this paper is design and construction of three test stands as well as methodology of testing experiments. As comparison, standard control over analog channel has been also presented. The results of testing show acceptable performance of control via TCP/IP/Ethernet networks.

2 Motivation

Using TCP/IP protocols, it is possible to build modular automatic control systems, where controllers are connected with plant using TCP/IP through SCADA (supervisory control and data acquisition) mediating software [11] (Fig. 1). Modular design of plant-controller system allows easy testing of various automatic control algorithms. Using more sophisticated mediating software, it is possible to connect more clients to the plant: remote automatic controllers, own SCADA systems, historical modules, databases and other applications using control/measurement data (Fig. 3). Usually however, such testing activities are preceded by simulation. Using simulator (Fig. 2) in place of real plant assures similar conditions during experiments, but also additionally *repeatability of testing conditions*. Using TCP/IP also in this case allows taking advantage of modular system design. Additionally, plant simulation for educational purposes provides protection against physical damage due to improper control of real (physical) plant. During tests presented in this paper, use of real (physical) plants was possible, taking however above presented statements in consideration, *presented tests have been performed using virtual (simulated) plants.*

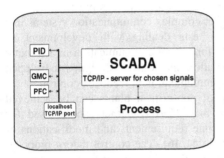

Fig. 1. Plant with SCADA software and .con trollers connected [11]

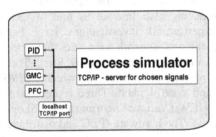

Fig. 2. Process (plant) simulator

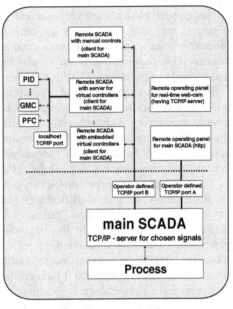

Fig. 3. Advanced architecture, allowing creation of individual SCADA systems [11]

3 Tests of TCP/IP/Ethernet network Properties

3.1 Tests of Transmission Time

One of the most important parameters describing the operation of computer network as a part of automatic control system is transmission delay. The purpose of these experiments was to measure the influence of software (operating system) and hardware on transport delay of TCP/IP/Ethernet-based local area network.

Materials and methods: The test environment consisted of two PCs, equipped with FastEthernet network cards, Ethernet switch, and crossed cable. All connections have been configured to 100Mb/s, Full Duplex. During the experiments operation of *Microsoft Windows XP, QNX v.6.2* and *Linux* (kernel 2.4) have been tested.

Experiments: To minimize measurement errors, tests have been conducted with minimal load of tested systems (absolute minimal number of process running). *Round Trip Time* has been measured (using *ping* tool) for the following parameters:

- packet size: 64[B],
- packet sending frequency: $1, 10, 100, 1000[s^{-1}]$,
- connection via: Ethernet switch (*"switch"*), crossed cable (*"cross"*),
- operating system: Windows XP – *"win"*, Linux – *"lin"*, QNX 6.2 – *"qnx"*.

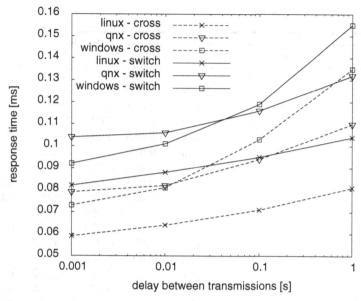

Fig. 4. Comparison of *mean* response time of *Linux, QNX, Windows XP*, operating systems

Results: Table 1 shows comparison of measured *RTTs* for different operating systems and different connections. Considering the results it is possible to conclude that:

- transmission time changes in obvious way for different operation system,
- opposed to anticipations, QNX operating system advertised as *real-time OS* required the longest time to respond. Analyzing time deviations, it is possible to state that QNX does not present higher determinism than other operating systems (in area of network transmission),
- for higher packet sending rate, transmission time is even 40% ÷ 50% *shorter* compared to slower sending rates. This phenomenon occurs in both cases (*"cross"* and *"switch"*), and is caused probably by operating system allocating more resources for more resource demanding processes (or not removing the process code from cache memory),
- the delay of 15 ÷ 25μs — introduced by Ethernet switch is clearly visible,
- in case of unmodified version of *Linux*, removing unnecessary processes and using closed network enables shortening bidirectional transmission time to <0.15[*ms*], where deviation is lower than 0.05[*ms*],

3.2 Packet Loss Measurement

Second important factor influencing operation of transmission channel is the number of errors occurring. Knowledge of error ratio and transmission time (without acknowledgments) allows calculation of effective transmission time when retransmission occurs.

Table 1. Comparison of Round Trip Time — various operating systems

connection	sending speed [packet/s]	operating system	time (min/mean/max/dev)			
cross	1	lin	0.073	0.081	0.109	0.013
		qnx	0.100	0.110	0.159	0.016
		win	0.098	0.135	0.147	0.008
	10	lin	0.062	0.071	0.093	0.011
		qnx	0.086	0.094	0.156	0.011
		win	0.089	0.103	0.119	0.008
	100	lin	0.060	0.064	0.081	0.010
		qnx	0.075	0.082	0.102	0.009
		win	0.071	0.081	0.106	0.010
	1000	lin	0.057	0.059	0.093	0.009
		qnx	0.075	0.079	0.131	0.008
		win	0.070	0.073	0.164	0.013
switch	1	lin	0.092	0.104	0.115	0.014
		qnx	0.120	0.132	0.167	0.015
		win	0.114	0.155	0.166	0.009
	10	lin	0.089	0.095	0.110	0.009
		qnx	0.108	0.116	0.132	0.010
		win	0.107	0.119	0.137	0.015
	100	lin	0.083	0.088	0.107	0.012
		qnx	0.099	0.106	0.128	0.005
		win	0.091	0.101	0.122	0.010
	1000	lin	0.079	0.082	0.122	0.010
		qnx	0.099	0.104	0.167	0.006
		win	0.088	0.092	0.166	0.009

Table 2. Comparison of transmission errors

connection	sending speed [packet/s]	packet number (sent/lost)		time (min/mean/max) [ms]		
switch	10^3	10^8	388	0.072	0.076	2.088
	10^2	10^7	39	0.072	0.084	0.864
	10^1	10^6	4	0.090	0.106	0.243
	10^0	10^5	1	0.126	0.137	0.261
cross	10^3	10^8	1	0.058	0.062	1.890
	10^2	10^7	0	0.058	0.070	0.857
	10^1	10^6	0	0.075	0.091	0.207
	10^0	10^5	0	0.111	0.122	0.216

Materials and methods. To minimize measurement errors, tests have been conducted with minimal load of tested systems (absolute minimal number of process running). Number of lost packets (and *Round Trip Time*) has been measured (using *ping* tool) for the following parameters:

- packet size: 36[B],
- packet number: 100,
- packet sending frequency: $1, 10, 100, 1000[s^{-1}]$,
- connection via: Ethernet switch (*"switch"*), crossed cable (*"cross"*),

Results and conclusions. Because of the test formula, it is necessary to remember, that single measurement consists of bidirectional packet transmission, so the real error ratio is equal approximately 0.5 of measured value. Analyzing the results, several phenomena are observable:

- Ethernet switch is main cause of packet loss — the ratio of packet lost when using switch compared to transmission using crossed cable is ≈ 400 to 1,
- during switched packet transmission, probability of error (packet loss) is ≈$2*10^{-6}$ (2 ppm), but probability of error in transmission using crossed should be considered as value under measurement accuracy.
- there is no observable influence of transmission speed on amount of errors in tested range (1 ÷ 1000 packets/s).

Additionally, result similar to shown in subsection (3.1) has been observed:
- using switch, the packet transmission time is approximately 15÷25μs longer, than in case of crossed cable,

Final conclusion is, that in real switch-based local area network expected value of packet loss is approximately 1 ÷ 10ppm on every switch. For higher packet sending rate, transmission time is even 40% ÷ 50% *shorter* compared to slower sending rates. This phenomenon occurs in both cases (*"cross"* and *"switch"*), and is caused probably by operating system allocating more resources for more resource demanding processes (or not removing the process code from cache memory). Occurrence of longer transmission time for lower packet sending frequency shows clearly, that it is not only TCP/IP/Ethernet what contributes to longer transmission delays. Longer and unpredictable delays are caused mainly by resource management strategy and timesharing (multitasking) procedures of operating system.

4 Influence of Transmission Time on Control Quality (LabVIEW Platform)

To evaluate applicability of TCP/IP/Ethernet based networks for automatic control purposes, several tests have been performed. During tests, influence of transmission delay introduced by computer network on control quality in automatic control system (*Plant-Controller*) has been measured.

The first-order plus dead-time (FOPDT) dynamics has been chosen as control plant. Most of process control plants have or can be approximated by this form of the model. Also PI controller has been chosen because its simplicity and popularity in industry.

4.1 Materials and Methods

The test environment consisted of two PCs, equipped with FastEthernet network cards and AD/DA interface cards (*National Instruments*). The test applications used during experiments (virtual plant, virtual controller and TCP/IP networks) have been built using *LabVIEW* environment (*National Instruments*). Standard, unmodified TCP/IP/Ethernet has been used for signal transmission.

LabVIEW has been chosen, during the test software design process, because its position as *de facto* standard in area of automatic control test software.

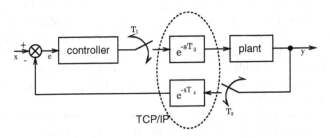

Fig. 5. PI controller connected via TCP/IP to FOPDT plant

When the controller and the plant are realized as real-time applications (in LabVIEW or LabVIEW-RT environments) the control system under consideration can be treated as a virtual control system. Such a way of simulations is very helpful for testing of control algorithms in conditions closest to the industrial reality. Additionally the virtual parts of the system can be replaced by corresponding commercial real-world components. In standard applications investigated controller and plant should be connected by industry standard signals (4-20 mA). Such a way of testing requires very expensive I/O hardware. The connection based on the TCP/IP can be low-cost alternative of automatic control systems testing. When values transmitted via digital channel are equal to values of the industry standard, the tested automation component can be easily used for connection with real-world component by industry standard.

The process model and controller have been realized as real-time simulator in the LabVIEW environment. A more detailed description of how to realize real-time simulation of the system (presented in Fig. 5) can be found in [12].

The term "virtual controller" seems to be well defined. The virtual controller must include at least all professional controller features, such as anti-reset windup action as well as bumpless switching between manual and automatic control. Without possibility of I/O connection, such a controller should be treated as only simulated controller (for simulation of control systems). For use as the virtual controller, this controller must include the connections with an I/O-PC-board or I/O modular system. For investigations presented here, the controller must include the

TCP/IP socket as well. The PI controller has been built using the NI-LabVIEW environment.

4.2 Experimental Tests

Several experimental tests have been made, to discover possible differences between automatic control qualities. For a comparison, additional test without TCP/IP has been performed (communication using local variables).
Following connection channels have been tested:

- locally connected (in one computer, without TCP/IP) –"no-net"
- locally TCP/IP-connected (in one computer) – "localhost"
- TCP/IP-connected in intranet (local based network) – "local-net"
- TCP/IP-connected in Internet – "internet"
- TCP/IP-connected via network simulator –"netsim"

Following connection channels have been tested:

- time constant of the plant: $T_1 = 1[s]$; $20[s]$
- plant dead-time: $T_0 = 0$; $0.25T_1$; $0.5T_1$
- process variable sampling time: $RT_i = 0.05[s]$; $0.2[s]$
- selected pairs of controller gain and integration time: $(kR;T_i)$

The data for charts has been collected with rate of 40 samples per second. Round Trip Time in tested intranet was measured: $\approx 0.3ms$, and Internet: $\approx 35ms$.

The Wide Area Network connection (Gliwice-Warsaw, Poland; 350 km distance) used in these experiments was low quality link — with high packet loss, and long maximal transmission time (up to $\approx 1500ms$).

4.3 Observations and Conclusions

The results presented in Fig. 6,7 show the system response after step change of set point from 40% to 50%.

Presented results show the satisfactory control quality in all cases except the Internet, however by making additional assumptions it is also possible to obtain satisfactory control quality using Internet. It is noticeable, that there are almost no differences present between the first three characteristics ("no-net", "localhost", "local-net"). Thus, it is possible, to make the following conclusion: for not too fast plants (time constant higher than 1s), TCP/IP based feedback loop in intranet offers control performance and quality very similar to the best performance available for discrete systems. This can be derived from the fact, that in most cases no difference between control over LAN and control in one computer (even without use of TCP protocol stack) is observed.

Problems with satisfactory control quality occurred, only in case of Internet connection (very low link quality) of controller and fast plant (time constant of 1s, Fig. 6). Having tested plant with longer time constant (20s and higher), almost no difference in control quality had been observed compared to local area network — even when low quality WAN link had been used (see Fig.7).

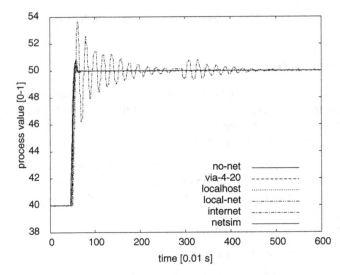

Fig. 6. Typical system response. Parameters: $kR = 12.0$; $T_i = 0.6[s]$; $RT_i = 0.05[s]$; $T_1 = 1.0[s]$; $T_0 = 0$.

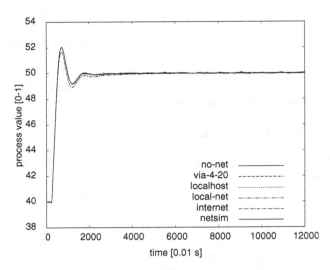

Fig. 7. Selected worst-case system response. Parameters: $kR = 3.0$; $T_i = 24[s]$; $RT_i = 0.05[s]$; $T_1 = 20.0[s]$; $T_0 = 5.00[s]$.

5   Influence of Transmission Time on Control Quality (Logix Platform)

Use of Logix platform *(Rockwell Automation)* was caused by desire to test highest-class automation hardware and software widely used in industry, with communication channels based on *CIP — EtherNet/IP — TCP/IP* protocols and *Ethernet* devices. Results here presented, have also been initially signalized in [13].

5.1 Materials and Methods

A special laboratory stand has been developed for experimental testing. The test environment consisted of one PC, equipped with FastEthernet network card and AD/DA interface cards (*National Instruments*), programmable logic controller *FlexLogix 5434* (*Rockwell Automation*), with Ethernet/IP communication module and Ethernet/IP-equipped distributed I/O — *FlexIO*. As OPC server *RS-Linx* has been used.

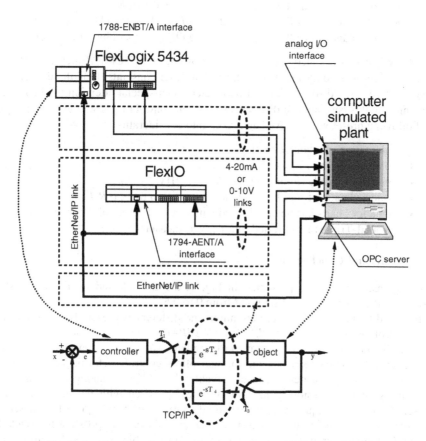

Fig. 8. Architecture of comparative test setup (Logix, EtherNet/IP) [13]

The applications used during testing experiments (plant simulator) have already been described in section (3.1), but additionally OPC communication routines have been added. For plant control, standard PID controller available in *FlexLogix 5434* PLC has been used.

Because of earlier made thesis, stating that speed instabilities of PLC's and computers has got significant influence on tests, special test environment has been prepared to eliminate this influence. This resulted in design presented in Fig. 8, in which signal transmission occurred via *0-10V*, or *EtherNet/IP+0-10V* (to allow differential measurements). This solution allows also obtaining independence from AD/DA converters delay and other delays not associated with transmission. Thus, the tests have been performed in following setups:

- no tests with local connection (in one computer) have been made — simulation of PLC's internal PID controller characteristics was not possible,
- plant and controller connected using *0-10V* channel,
- plant and controller connected using *EtherNet/IP* and *0-10V* channel,
- plant and controller connected using *OPC* protocol based channel,

Results of tests presented in section 4 showed, that TCP/IP/Ethernet influence becomes significant only in case of short plant time constants, thus the experiments have been performed for all defined parameters with special attention to "bad" plant dynamics. In this case biggest differences between control responses can be expected.

Following preset values have been taken into consideration.

- plant time constant: $T_1 = 1[s]$
- plant dead-time: $T_0 = 0; 0.25T_1; 0.5T_1$
- process variable sampling time: $RT_i = 0.01[s]$
- selected pairs of regulator gain and integration time: $(kR; T_i)$
- data for charts has been collected with rate of 100 samples per second,
- Round-Trip-Time in tested intranet was measured: $\approx 0.3ms$.

5.2 Results and Conclusions

The responses, which are presented in Fig. 9, are the selected, worst-case system responses after step change of set point from 0 to 0.5.

It should be noticed, that there are almost no differences present between the first two control responses (*"0–10V"*, *"EtherNet/IP+0–10V"*). Thus, it is possible, to make following conclusion: for not too fast plants (plant time constant higher than 1s), Ethernet/IP based feedback via intranet offers control performance and quality very similar to the best performance available (*"0–10V"* — analog connection). In case of signal transmission using OPC protocol — even with the best parameters chosen — the control quality degrades significantly. This fact is probably caused by data exchange desynchronization in OPC server. Hence, it is possible to conclude, that OPC protocol server (in case of *RS-linx*) is not suitable as feedback transmission channel of systems with short time constant. It should be noted however, that this statement should be verified with experiment — the control quality degradation could have been caused by OPC server version of particular vendor (*Rockwell Automation-Allen-Bradley*).

As shown in this paper, even though Ethernet-TCP/IP network (and consequently *Allen-Bradley's EtherNet/IP*) is considered to be time-nondeterministic — it is possible to build automatic control system using this type of intranet. In local area networks (intranets), with Ethernet switch used for collision domains separation *EtherNet/IP* seems to be good communication medium for automatic control.

Presented experimental results show, that considering typical local network time delays and plant time constants, automatic control using *EtherNet/IP* channel is similar to the quality of system using analog channel. The use of *OPC* protocol for control of plants having short time constants seems to be problematic however.

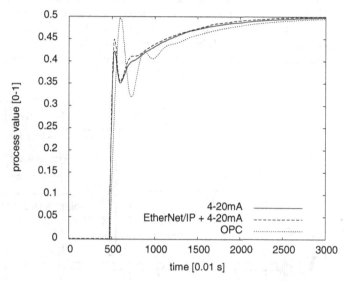

Fig. 9. Selected worst-case system response. Parameters: $kR = 1.4$; $T_i = 18[s]$; $RT_i = 0.01[s]$; $T_1 = 1.0[s]$; $T_0 = 0.5[s]$.

During the tests interesting phenomena have been observed — large execution time instabilities of main program loop. In case of *FlexLogix 5434*, execution time (and its variation) of the same program was much bigger ($0.2 - 2.3ms$) than transmission time in local area network ($0.07 - 0.15ms$). This leads to conclusion, that often it is not computer network that causes control quality deterioration. Another fact has also been observable — due to low computational power of PLC hardware, their response time was much longer than PC with the same communication interfaces.

6 Influence of Transmission Time on Control Quality in Simatic Platform

In this section, results of *Simatic* (*Siemens*) hardware and software have been presented. Hardware and software of this manufacturer was chosen because of its popularity in European industry.

6.1 Materials and Methods

A special laboratory stand has been developed for experimental testing. The test environment consisted of one PC, equipped with FastEthernet network cards and

AD/DA interface card (*National Instruments Lab-PC+*), *Simatic S7-300* programmable logic controller (*Siemens*).

The test applications used during experiments (plant simulator) have already been described in section (3.1). Standard, unmodified TCP/IP/Ethernet was used for signal transmission. For plant control, standard PID controller available in *Simatic S7-300* PLC had been used.

Because of earlier statement that speed instabilities of PLCs and computers has got significant influence on tests, special test environment has been prepared to eliminate this influence. This resulted in design presented in Fig. 10, in which signal transmission occurred via *0-10V*, *TCP/IP+0-10V*, and *TCP/IP* — connected via proxy in industrial computer — for analog (*0-10V*) digital conversion (*TCP/IP/Ethernet*).

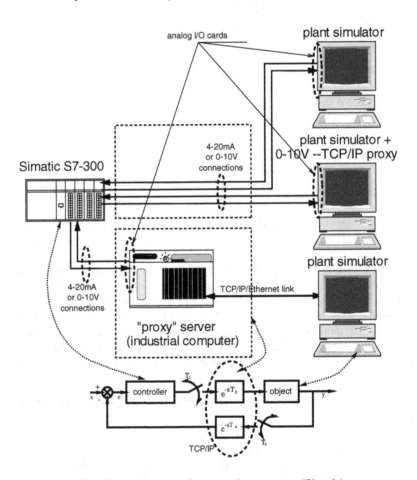

Fig. 10. Architecture of comparative test setup (Simatic)

This solution allows differential measurements and obtaining independence from AD/DA converters delay and other delays not associated with transmission.

The tests have been performed in following setups (Fig. 10):

- no tests with local connection (in one computer) have been made — simulation of PLC's internal PID controller characteristics was not possible,
- plant and controller connected using *0-10V* channel (*"localhost"* in figure),
- plant and controller connected using *0-10V* and *TCP/IP* channel, in one operating system, using proxy (*"proxy"* in figure),
- plant and controller connected using *0-10V* and *TCP/IP* channel, in local area network, using proxy (*"intranet"* in figure),

Because results of tests presented in section (3) showed, that TCP/IP/Ethernet influence becomes significant only in case of short plant time constants, the experiments have been performed for all defined parameters with special attention to "bad" plant dynamics (similar to those presented in section 3). In this case biggest differences between control responses can be expected.

Following preset values have been taken into consideration.

- plant time constant: $T_1 = 1[s]$
- plant dead-time: $T_0 = 0; 0.25T_1; 0.5T_1$
- process variable sampling time: $RT_i = 0.01[s]$
- selected pairs of regulator gain and integration time: $(kR;T_i)$
- data for charts has been collected with rate of 100 samples per second,

6.2 Results and Conclusions

The result presented in Fig. 11 is the selected, worst-case system response after step change of set point from 0 to 0.5.

It is noticeable, that control quality of system using *TCP/IP+0-10V* connection is worse than system using analog connection only (*0-10V*).

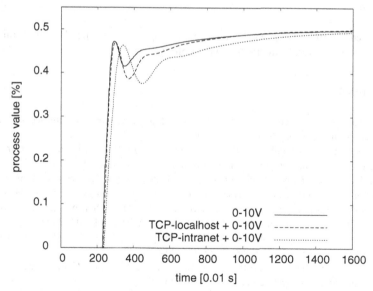

Fig. 11. Selected worst-case system response. Parameters: $kR = 1.4$; $T_i = 3s$]; $RT_i = 0.01[s]$; $T_1 = 1.0[s]$; $T_0 = 0.5[s]$.

Deterioration of control quality occurs not only because of transmission delays in computer network, but also due to delays and desynchronization of data exchange in proxying software.

Thus, the final conclusion is, that during designing of automatic control systems, unless inevitable, it is better to avoid excessive data passing.

7 Concluding Remarks

Currently the analog transmission is the most popular transmission in the industrial feedback loops. Although the networked communication is used in higher-level automatic control (monitoring, SCADA systems) such kind of transmission is too expensive for application in control loops. That is why an effort to apply the most popular Ethernet TCP/IP transmission is very promising. In the paper this kind of transmission has been tested experimentally. Opposed to majority of papers dealing with transmission in control and computer science this paper focuses rather on presenting method for testing of automatic control systems, and showing example test results. The main contribution presented in this paper is design and construction of three test stands as well as methodology of testing experiments. Each test stand consists of plant, communication channel and automatic controller.

Based on the experiments, it is possible, to make the following conclusion: for not too fast plants (time constant higher than 1s), TCP/IP based feedback loop in intranet offers control performance and quality very similar to the best performance available for discrete systems. It also is possible to control plants with longer time constant (20s and higher) via the Internet with almost no signs of control quality deterioration.

In general, presented results show promising perspectives not only for intranets, but also for Wide Area Networks (Internet) for plants with medium and long time constants.

Acknowledgments. This work has been supported by the Polish Ministry of Scientific Research and Information Technology.

References

1. Tretmans, J.: Test generation with inputs, outputs and repetitive quiescence. Software Concepts and Tools 17(3), 103–120 (1996)
2. Petrenko, A., Yevtushenko, N., Bochman, G.v., Dssouli, R.: Testing in context: framework and test derivation. Computer Communications 19, 1236–1249 (1996)
3. Grieskamp, W., Tillmann, N., Veanes, M.: Instrumenting scenarios in a model-driven development environment. Information and Software Technology 46, 1027–1036 (2004)
4. Flammini, A., Ferrari, P., Sisinni, E., Marioli, D., Taroni, A.: Sensor interfaces: from field-bus to Ethernet and Internet. Sensors and Actuators A 101, 194–202 (2002)
5. Maciel, C.D., Ritter, C.M.: TCP/IP Networking in Process Control Plants. Computer Industrial Engineering, vol. 35(3-4), pp. 611–614
6. Vitturi, S.: On the Use of Ethernet at Low Level of Factory Communication System. Computer Standards & Interfaces 23, 267–277 (2001)
7. Belle Isle, A.P.: Stability of Systems with Nonlinear Feedback Through Randomly Time-Varying Delays. IEEE Transactions On. Automatic Control AC-20(1), 67–75 (1975)

8. Krtolica, R., Özgüner, Ü., Chan, H., Göktaş, H., Winkelman, J., Liubakka, M.: Stability of Linear Feedback Systems with Random Communication Delays. International Journal of Control 59(4), 925–953 (1994)
9. Lee, K.C., Lee, S.: Performance evaluation of switched Ethernet for real-time industrial communications. Computer Standards & Interfaces 24, 411–423 (2002)
10. Decotignie, J-D.: Etherne-Based Real-Time and Industrial Communications. In: Proceedings of the IEEE, vol. 93(6), pp. 1102–1117 (2005)
11. Metzger, M.: Virtual controllers improve Internet-based experiments on semi-industrial pilot plants. In: Proceedings of the 16-th IFAC Triennal World Congress, CD, Elsevier, Amsterdam (2005)
12. Metzger, M.: Modelling and simulation of the sampled-data control of the nonlinear, continuous, distributed parameter plant. 15th IMACS World Congress, Wissenschaft und Technik Verlag, Berlin, Systems Engineering, vol. 5, pp. 161–166 (1997)
13. Plesowicz, P., Metzger, M.: Experimental Evaluation Of Ethernet/IP-Interconnected Control Systems. In: Proceedings of the IFAC Workshop on Programmable Devices and Systems (2006)

Towards Systematic Signature Testing

Sebastian Schmerl and Hartmut Koenig

Department of Computer Science
Brandenburg University of Technology Cottbus
PF 10 13 44, 03013 Cottbus, Germany
{sbs, koenig}@informatik.tu-cottbus.de

Abstract. The success and the acceptance of intrusion detection systems essentially depend on the accuracy of their analysis. Inaccurate signatures strongly trigger false alarms. In practice several thousand false alarms per month are reported which limit the successful deployment of intrusion detection systems. Most today deployed intrusion detection systems apply misuse detection as detection procedure. Misuse detection compares the recorded audit data with predefined patterns, the signatures. These are mostly empirically developed based on experience and knowledge of experts. Methods for a systematic development have been scarcely reported yet. A testing and correcting phase is required to improve the quality of the signatures. Signature testing is still a rather empirical process like signature development itself. There exists no test methodology so far. In this paper we present first approaches for a systematic test of signatures. We characterize the test objectives and present different test methods.

1 Motivation

The increasing dependence of human society on information technology (IT) systems requires appropriate measures to cope with their misuse. The enlarging technological complexity of IT systems increases the range of threats to endanger them. Besides preventive security measures reactive approaches are more and more applied to counter these threats. Reactive approaches allow responses and counter measures when security violations happened to prevent further damage. Complementary to preventive measures intrusion detection and prevention systems have proved as important means to protect IT resources. Meanwhile a wide range of commercial intrusion detection products is offered, especially for misuse detection. Nevertheless intrusion detection systems (IDSs) are not still deployed in a large scale. The reason is that the technology is considered not matured enough. Lacking reliability often resulting in high false alarm rates questions the practicability of intrusion detection systems [9].

The security function intrusion detection deals with the monitoring of IT systems to detect security violations. The decision which activities have to be considered as security violations in a given context is defined by the applied security policy. Two main complementary approaches are applied: anomaly and misuse detection. *Anomaly detection* aims at the exposure of abnormal user behavior. It requires a comprehensive

A. Petrenko et al. (Eds.): TestCom/FATES 2007, LNCS 4581, pp. 276–291, 2007.

set of data describing the normal user behavior. Although much research is done in this area it is difficult to achieve so that anomaly detection has currently still a limited practical importance. *Misuse detection* focuses on the (automated) detection of known attacks described by patterns, called *signatures*. These patterns are used to identify an attack in an audit data stream. This approach is applied by the majority of the systems used in practice. Their effectiveness, however, is also still limited. There are several reasons for this. On the one hand, many systems mainly confine themselves to detecting simply structured network based attacks, often still in a post-mortem mode. Multi-step or distributed attacks which are getting an increasing importance are not covered. On the other hand, the success and the acceptance of misuse detection systems essentially strongly depend on the conciseness and the topicality of the applied signatures. Imprecise signatures heavily confine the detection capability of the intrusion detection systems and lead to false alarms. The reasons of this detection in-accuracy can only in part imputed to qualitative restrictions of the audit functions of the monitored system or network. They must be rather sought in the signature derivation process itself. In particular, the derivation of signatures starting from given exploits often appears as weak point. An attack represents a sequence of actions that exploits a vulnerability in a program, operating system, or network. The derivation of a signature to detect the attack is mostly based on experience and expert knowledge. Methods for a systematic derivation have scarcely reported yet. Automated approaches to reusing design and modeling decisions of available signatures also do not exist. This results in relative long development times for signatures causing inappropriate vulnerability intervals [9].

In order to improve the accuracy of the derived signatures the signatures must be tested and corrected. The objective of a signature test is to prove, whether the derived signature is capable to exactly detect an attack in an audit trail. As the derivation process itself the testing of signatures is still rather empirical. There exist no test approaches and methods yet. This paper focuses on the testing of signatures. It present first approaches for a systematic test of signatures. The paper is structured as follows. In Section 2 we consider the signature derivation process and outline the reasons for the detection shakiness of current signatures. Section 3 backs up the need for a signature tests and outlines the two main issues signature tests have to cope with. In Section 4 we present four test strategies to testing signatures and describe their procedures. Section 5 sketches the application of one test strategy to a concrete signature. Some final remarks conclude the paper.

2 On the Derivation of Signatures

An Attack consists of a set of related security relevant actions or events in a system, e.g. a sequence of system calls or network packets. The task of an audit function is to capture information about the execution of each of these actions by generating audit data records that can be used for analysis. Misuse detection systems try to detect se-quences that correspond to known signatures. Thereby it is assumed that security violations do manifest themselves in distinct audit data records, i.e. they are observable, and that they can be detected on the basis of these audit data, i.e. they are detectable.

Fig. 1 depicts these relations. To run an attack the attacker uses *exploits* which are usually known shortly after their appearance. These are programs or pieces of codes to execute an attack which exploit vulnerabilities (e.g. coding faults, configuration errors etc) of the target system. Exploits have various appearances, e.g. program code, protocol packets or scripts. They contain of a sequence of operations or actions (at least one) which cause an abnormal behavior of the attacked host or network.

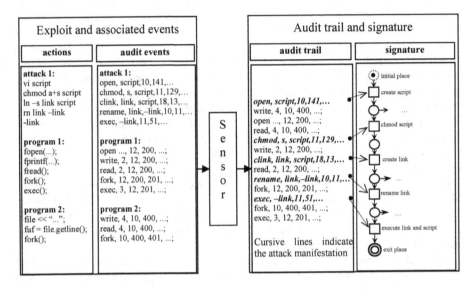

Fig. 1. Exploits, attack manifestations, and signatures

An attack represents a sequence of security relevant actions. They can be usually divided in three steps:

    (1) to transfer the attacked system in a vulnerable state,

    (2) to exploit the vulnerability to intrude the system, and

    (3) to access to the compromised system and/or to change its system data. (This is the proper objective of the attack.)

A signature can only detect step (1) and (2) of the attack. They are predictable and describable based on the knowledge about the vulnerability. The proper concern of the attacker cannot be described because it is not predictable. Thus signatures comprise only the first two steps of an attack.

The execution of attacks leaves traces which can be audited by IDS sensors. These traces are called *manifestations of the attacks*. Fig. 1 shows the traces for the example exploits. The traces are not stated separately. They are hidden in the audit trail. The latter consists of a sequence of records which contain the traces of all actions executed by the system. In order to separate the attack manifestations the audit trail is searched for attack patterns. These patterns are defined by signatures. A *signature* of an attack describes the criteria (patterns) required to identify the manifestation of an attack in an audit trail. It is possible that several attacks of the same type are executed simultaneously and proceed independently. Therefore it is necessary to be able to distinguish different instances of an attack. A *signature instance* identifies the manifestation of an attack instance in the audit data stream. Signatures are usually described by means of finite state automata, Petri nets, or special attack description languages [7]. Typically each intrusion detection system uses its own language which is customized to the applied analysis method. Fig. 2 shows an example of a signature modeling in a Petri net like languages described in [8].

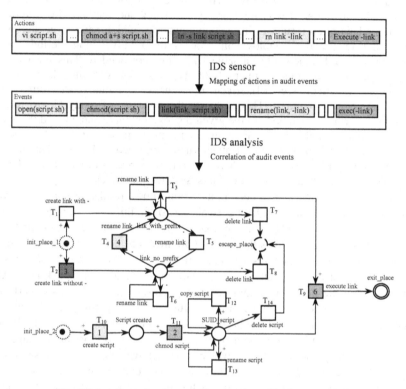

Fig. 2. Modeling of a signature in a Petri net like language [8]

The detection power of signature analysis depends on the accuracy of the signatures applied. To estimate the detection quality of intrusion detection systems usually two measures are applied: the number of security violations not detected (*false negatives*) and the frequency of false alarms (*false positives*). Not detected security violations are caused by over specification of the signatures, whilst false alarms are triggered by inaccurate specifications. The experience shows that not detected security violations have a more grave impact on the systems behaviour than false alarms. Nevertheless, a high false positives rate reveals as a severe problem for running intrusion detection systems in practice. Since misuse detection systems apply deterministic methods, the search for signature patterns, strictly speaking, excludes false positives per definition (assuming an effective audit function). The reality, however, is different, e.g. [3] reports about 10.000 false alarms per month for the use of commercial intrusion detection systems. Other evaluations [2], [5], confirm this experience. Small false positive rates are an important presumption for the acceptance of misuse detection systems in practice. Inaccurate signatures, therefore, strongly confine the detection power and acceptance of misuse detection systems.

The reasons for the detection shakiness of signatures are only in part caused by qualitative shortages of the used audit functions. They lie in the signature derivation process itself. The derivation of signatures from exploits is the actual weak point. Signatures are mostly empirically derived based on long-term experience of the security administrators. There are scarcely heuristics and methods for a systematic derivation. This often results in inaccurate signatures which have to be step-by-step refined during practical deployment. Therefore, relative long periods for the derivation of good, practically valuable signatures are needed. This means, on the other hand, long vulnerability intervals of the respective systems which cannot be accepted in practice (see [6]). Even if accurate signatures are found further adaptations may be required. This is due to the diversity of today IT environments which force further adaptations to the given deployment environment and the security policies applied. Additional adaptations and enhancements of the signatures are needed when new vulnerabilities or attack mutations become known. The derivation and the maintenance of signatures, therefore, represent one of the most complex tasks for the development and deployment of misuse detection systems.

Only a few approaches have been reported up to now on the systematic derivation of signatures from exploits. Cheung et al. try to simplify the signature design by applying attack models [1]. This approach corresponds to the design patterns of software engineering. It allows the reuse of architectural design decisions. The reuse of concrete modeled signatures or signature fragments is, however, not possible. Rubin et al. describe how mutants can be generated for a given attack [10]. Attack mutants exploit the same vulnerabilities as basis attack without, however, performing the same security relevant actions. If a signature for an attack mutant is supposed to be developed the signature of the basis attack could be reused, if available. Rubin et al. further describe in [11] a refinement of signatures based on formal languages. This approach can help the signature developer to remove triggers for false positives caused by imprecise signatures. The procedure, however, assumes an almost error-free reference signature. In [13] an approach is proposed to use diversity to modeling an implicit complete attack model. This has the advantage of an improved model, however multiple specifications are needed. Larson et al. [4] present a tool for extracting the significant events of an attack from the audit trail. It executes the attack and records the respective audit data. Then the differences between these audit data and an attack free audit trail are derived. The problem of deriving a signature from this difference, however, remains unsolved. In [12] the authors presented an approach to reusing patterns of existing signatures for the development of new signatures. It exploits the fact that similar attacks produces similar traces so that existing signatures may provide an informative basis for the development of new signatures. The approach is based on an iterative abstraction of signatures. Based on a weighted abstraction tree it selects those signatures or signature fragments, respectively, which possess similarities with the novel attack. The reuse of proved structures may not only reduce the efforts of the signature derivation process but it can also considerably shorten the costly test and correction phase.

3 On the Test of Signatures

Inaccurate signatures strongly limit the detection power of misuse detection systems as well as their economic profitability. As discussed the signature development process is complicated and tedious. Systematic derivation procedures are scarcely available. A certain inaccuracy is, therefore, inherent to the derived signatures. A testing and correcting phase is indispensable to improve the quality of the signatures. This phase is an essential part of the signature development process independently of the fact, whether the signatures are derived systematically or by experience.

The objective of a signature test is to prove the accuracy of the given signature by applying it to an audit trail which contains traces of the respective attack. If the signature does not completely detect all traces it must be corrected to approximate the *ideal signature*, i.e. the signature which describes all manifestations M_I of the attack. Normally the signature derivation process does not induce ideal signatures. The derived signatures are either under or over specified.

Under specified signatures describe beside action sequences which are required for a successful attack also actions which either correspond to legitimate behavior or which do not exploit the vulnerability. That means they describe a manifestation set M_U which represents a superset of the manifestations of the ideal signature M_I, i.e. $M_I \subset M_U$. Under specified signatures thus increase the false positives rate. A test strategy to detect under specified signatures has to investigate, whether the actions recorded in the audit trail really exploit the vulnerability. To derive test data actions have complementarily to be assigned to the audit events. These action sequences are then tested on a dedicated system concerning the exploitation of the vulnerability. If the vulnerability is not exploited a specification error exists and the signature must be corrected. The difficulties and limits of this approach lie beside the derivation of the action sequence in the assessment, whether the vulnerability is really exploited.

Over specified signatures do not detect all variants of the attack, i.e. there exist action sequences which successfully exploit the vulnerability but are not captured by the signature. The set of detected manifestations M_O is a subset of the manifestations of the ideal signature M_I, i.e. $M_O \subset M_I \subset M_U$. Over specified signatures induce not detected security violations, i.e. they increase the *false negatives* rate. The objective of a test and correction strategy for detecting over specified signatures is to enhance the signature to approximate the ideal signature. This can be achieved by extending the signature, by substituting actions, and by changing the order of the actions. The test strategy has to ensure that the detection of the attack remains guaranteed, if some actions are replaced by semantically equivalent actions, and that the vulnerability is further exploited.

4 Methods for the Test of Signatures

In this section we present four methods for testing signatures. The main approaches deal with the tests for under and over specified signatures. Furthermore, we present a preliminary test and a test of escape events.

4.1 Preliminary Test

The objective of preliminary tests is to ensure that the derived signatures do not contain grave errors.

Test method: The test consists of two steps. *Test step* (1): Assuming a newly derived signature S of attack A. This attack is first executed on a dedicated system. The resulting audit trail T is recorded and analyzed to determine the events representing traces of A. We call these events characterizing events $CE \subset T$ here. In test step (1) S is tested against CE, i.e. it is proved, whether a misuse detection systems containing signature S detects A. If the test fails a grave error in the signature specification can be assumed.

Test step (2): Now S is tested against the whole audit trail T, i.e. it is proved, whether the misuse detection system triggers an alarm when A is executed. If the test passes the test procedure can be continued. A negative test outcome usually indicates that the newly derived signature does not correctly correlate the characterizing events CE in T. Reasons for this are not exactly or too weakly specified signature conditions.

The test method can be mostly automated depending on the applied attack description language.

4.2 Tests for Under Specified Signatures

Under specified signatures contain specifications of action or event sequences which correspond to legitimate behavior or which do not exploit system vulnerabilities. They cause false positives.

The test methods presented in the following to detect under specified signatures is based on the mapping relation δ of the IDS sensor which maps the various security relevant actions into audit events. This relation is usually bijective realized in intrusion detection systems, i.e. there is a δ^{-1}. This means that the corresponding action sequences of the attack can be derived from the audit events demanded by the signature. The objective of the tests is to validate whether these action sequences corresponds to a successful attack. If not, the signature contains a specification error which triggers a false alarm.

Test method: The proposed test method comprises 3 steps. In *step* **(1)** appropriate test cases are derived. Since there is a wide range of conditions which have to be fulfilled between the correlating audit events it is not possible like in many other tests to exhaustively test all possible action sequences. Therefore an appropriate subset of test cases has to be selected depending of the coverage aimed at. Here either path or test coverage criteria can be applied as in software testing.

Step **(2)** derives for the selected test cases the corresponding action sequences from the signatures by means of δ^{-1}. Signatures specify the properties of the audit events, e.g. type, parameters etc during the attack. They also define conditions regarding the appearance and the context of the correlated audit events. Furthermore, the temporal order of the audit events can be demanded. If these conditions are taken into account the corresponding attack can be re-established.

In *step* **(3)** each derived action sequence is executed on a vulnerable system to prove whether they correspond to attacks. This first requires that correct attack conditions are established, especially temporal constraints have to be preserved if necessary. This is a decisive precondition because attacks are only successful, when certain conditions are fulfilled, e.g. a load situation. However, not all attacks depend on additional conditions. We distinguish *deterministic successful attacks, attacks with preconditions* and *brute force attacks*. Former attacks are independent of special system or application circumstances, therefore correct execution of a deterministic attack is always successful. Consequently this class of attacks is unrestricted testable with this test strategy. In the case of attacks with preconditions, the vulnerability is only exploitable if specific system parameters are fulfilled, e.g. special system load situations. Therefore the necessary preconditions must be synthesized before the test. The class of brute force attacks summarizes all attacks which do not exploiting a concrete vulnerability, e.g. brute force attacks of single authentication systems. Signatures for this kind of attacks are not testable with this test strategy.

The detection of a successful attack execution depends on the attack strategy and the exploited vulnerability, respectively. For disclosing this, four approaches may be generally applied. (a) *Instrumentation of the exploit*, i.e. the derived action sequence has to be changed so that a significant system change is observable. This can be done, for instance, by appending additional actions to the derived attack actions, e.g. setting up a root file or starting/terminating a privileged system process. (b) *Instrumentation of the vulnerability*, i.e. the vulnerable system or code is changed so that the exploitation of the vulnerability becomes observable. This is, for instance, useful, if the program code is available but the vulnerable system (e.g. a technical facility) cannot be patched. (c) *Instrumentation of the whole system*: This approach compares normal system behaviour recorded by an appropriate audit component with the system behaviour after executing an attack. Unlike the other approaches a detailed observation of the system is required. On the other hand, no interference of the attack and the system is needed. (d) *Using a test oracle* which passively examines the system behaviour. This approach though is not able to detect attacks which do not destroy systems functions, e.g. backdoors.

The test method can be automated if certain constraints are fulfilled which is often given. Table 1 contains these constraints. However, there are also shortages which limit the practicality of the method. One problem is that not every action generates an audit event so that δ^{-1} does not always re-establish the complete action sequence. It is not always required to correlate all events of an attack to detect the attack. The crucial issue, however, is to re-establish the attack preconditions. This requires a detailed knowledge of the system behaviour and the attack strategy by the test engineer.

Table 1. Automation degree of test steps

Step		Objective	Can be automated?
1		Test case selection	yes, using typical approaches of software testing
2		Derivation of action sequences	yes, if bijective IDS sensors are deployed
3	I	Establishing the correct attack conditions	yes, for deterministic attacks yes, if constraints can be automated re-established
	II	Execution of action sequences	yes
	III	Test of success	yes, if detection of a successful attack can be automated

4.3   Tests for over Specified Signatures

Over specified signatures do not capture all action sequences which successfully exploit a given vulnerability. Attackers often replace one or more actions of the attack by semantically equal actions. The aim of this transformation is to change the traces of the attack so significantly that the attack is not detected by the intrusion detection system. The proper attack strategy, however, remains preserved, i.e. the given vulnerability is further exploited for running the attack. If the signature does not recognize these attacks it produces false negatives. Test strategies for over specified signatures aim at detecting this detection weakness. Before describing the test strategy we first have to introduce the different types of transformations to change the attack.

There are three types of transforming an attack: No-Op insertion, permutation of actions, and action substitution.

No-Op insertion: In this transformation redundant actions are added which do not change the attack but its traces in the audit trail. This transformation tries to exploit deficiencies of the intrusion detection system to correlate audit events. The evasion of intrusion detection by means of No-Op insertion due to a signature error can be excluded as long as the signature does not strictly demand a direct timed sequence of certain audit events. This demand is only useful in very seldom, specific scenarios. Therefore a signature test can be waived for this transformation.

Permutation of actions: This transformation changes the order of certain actions of the attack and thus the sequence of their related audit events. Two actions can be changed if their execution and their influence on the attack do not depend on each other. These transformations allow bypassing signatures with over specified event sequences. The reason for this kind of over specification is mostly a not correct understanding of the semantics of the attack actions.

Action substitution: This transformation replaces single actions or action sequences by semantically equal action sequences. Thus the IDS sensor registers different actions and events, respectively, although the result of the actions remains the same. A simple example is the substitution of the file renaming operation *mv file1 file2* by a copy and an erase operation: *cp file1 file2*; *rm file1*.

Permutations and substitutions of attack actions produce isomorphic action sequences without enlarging the attack/exploit by redundant actions. Now we present a strategy to test a signature on the detection of isomorphic attack sequences. First we introduce some needed basic notions.

An action a of an attack changes, erases, creates, or uses a certain type of system resource, an *object*. The *object type O* characterizes the type of the system resource, e.g. a process, a socket, or a file. It is represented by the tuple (P, A, r, f) where P defines the set of *object properties*. The object type *file*, for instance, possesses among others the properties file name/path, creation date and access rights. A describes the set of actions of the object type O. Some of them can change the properties of O. The relation r: $a \rightarrow p$ with $a \in A$, $p \subseteq P$ describes for each action $a \in A$ the subset $p_a \subseteq P$ of the object properties which are changed by a. Finally f defines a relation f: $a, p \rightarrow A_a$, with $a \in A$, $p \subseteq P$, $A_a \subseteq A$ which defines for a given action a the set of semantically equivalent actions A_a preserving the properties of the object type unchanged. An object $o \in O$ is an instance of O which is characterized by the concrete property values. The object types together with the associated relations can be defined for a concrete system by an expert with average effort. These definitions have to be performed mostly once per system (e.g. with host based intrusion detection systems once per operating system or monitored application) and can be used for all signature tests concerning the system.

We now describe rules for transforming action sequences into semantically equal sequences using object types. Each action belongs to a certain object type. The information which action can be executed on an object is sufficient. This information can be easily derived from the signature by the test engineer. If the object types are given each action or action sequence, respectively, can be assigned an object type. Thus an action sequence $a_1a_2a_3....a_n$ with $a_i \in A$, can be mapped on objects $o_1o_2o_3...o_n$ (with $o_i=o_j$, if a_i and a_j relate on the same object). An action a_i can be replaced by a semantically equal action $á_i$ if (1) $á_i$ compared to a_i does not change additional properties of the respective object, or (2) if the additionally changed object properties by $á_i$ are not changed by former or later actions in the action sequence. Thus all substitutable action sequences $á_1á_2á_3...á_n$

with $á_i \in f\left(a_i, X / Y\right)$ and $X = \left(\bigcup_{l=1}^{i-1} r(a_l)\right), Y = \left(\bigcup_{k=i}^{n} r(a_k)\right)$ can be generated

from the original sequence $a_1a_2a_3....a_n$ by means of the respective object types and the relations r and f. The approach can be extended without loss of generality to replace single actions by action sequences and vice versa.

Permutations of actions in an action sequence require additional specifications by the test engineer to indicate dependencies between objects and between the actions of an object. In many cases though the following semantics preserving permutation can be performed: An action a_i is almost always exchangeable with action a_j ($i<j$), if a_i and a_j relate to the same object and a_i influences other properties than a_j ($r(a_i) \cap r(a_j)$ $= \emptyset$). Further there exists no action a_l with $i<l<j$ which uses the object associated with a_i and a_j. If such an action exists and a_i is exchanged with a_j then action a_l is executed under different conditions. The order of actions which create and erase objects remains unchanged due to the above mentioned rule that they change all properties. If the test engineer further specifies which objects are independent of each other so the associated actions can be exchanged as long as the before mentioned condition is fulfilled. This specification requires though certain knowledge about the system behaviour and the attack strategy. It is required once per signature.

The described substitution and permutation rules do not cover the whole range of possible action sequence transformations. There are certain types of attacks which can be transformed into action sequences which do correspond to a valid attack. These exceptional cases must be handled by the test engineer.

Test method: For the test, all action sequences are derived which distinguish concerning action sequence. This can be done analogously to step (1) and (2) of test method 2 whereby path coverage is applied in step (1). Next all possible combinations are generated for each action sequences according to the above given rules. Thereafter it is proved using an intrusion detection system, whether one of the derived action sequences is not detected by the signature. In this case the intrusion detection system does not trigger an alarm for this signature and action sequence. If the signature does not detect a transformed action sequence it has to be checked, whether this action sequence corresponds to a valid attack sequence. This test can be performed analogously to test method 2. If the test outcome is positive the signature has to be completed so that this sequence is also detected.

4.4 Test of Escape Events

Signatures only describe action sequences which represent successful attacks. When during analysis events are recognized which make the successful completion of an attack impossible the analysis has to be stopped to avoid false negatives and, of course, for performance reasons. Actions which prevent the attack to be completed are called *escape events*. They transfer the signature into the initial state. Many escape events are implicitly given by contrary events. For example, in Solaris OS the system calls *fork* and *exit* for creating and terminating processes are complementary events. If the creation of a new process is an indispensable condition for the success of an attack *fork* is a significant part of the signature. The corresponding escape event is *exit*. Escape events are, therefore, an indispensable part of the signature to stop or to re-initialize attack tracking. Consequently, their handling has to be tested.

For this, all events specified in the signature are again converted into actions according to step (2) test strategy 2 (comp. Section 4.2). Next the contrary events are assigned to each signature event using lists of actions with their corresponding contrary actions. The resulting sequence of contrary actions is then again converted into the corresponding events. This can be done using an IDS sensor. In the last step it is proved, whether the signature handles each contrary event. If this is not the case, the escape event is generally not modeled in the signature.

5 Example: Test for Under Specified Signatures

Signatures are specified using various description languages. Therefore the test scenarios has to be adapted to the given signature description language or semantic model, respectively. We now demonstrate this for the test method for under specified signatures with a concrete signature description language. We use EDL (*Event Description Language*) [8] which is based on a Petri-net like modeling approach. It supports the specification of complex multi-step attacks and possesses a high expressiveness and nevertheless allows for efficient analysis. Before describing the test procedure we first outline some essential features of EDL. More details can be found in [8].

5.1 Modeling Signatures in EDL

The descriptions of signatures in EDL consist of places and transitions which are connected by directed edges. Places represent states of the system which are traversed by the related attack. Transitions represent the state changes. They describe the specific events which cause the state change, e.g. security relevant actions. These events are contained in the audit data stream recorded during the attack. The signature execution is represented by tokens which flow from state to state. *Tokens* represent concrete signature instances. They can be labeled with values as in colored Petri-nets.

Places describe the relevant system states of an attack. They are characterized by a set of features and a place type. Features specify the properties of the tokens which are located in a place. The information contained in a token can change from place to place. EDL distinguishes four place types: *initial, interior, escape,* and *exit places. Initial places* are the starting places of a signature. They are marked with an initial token at the start of analysis. Each signature has exactly one *exit place* which describes the final place of signature. If a token reaches this place, then the signature has identified a manifestation of an attack in the audit data stream. *Escape places* indicate an analysis stop of an attack instance. They are reached if events occur which make the completion of the attack instance impossible. Tokens which reach these places are discarded. All other places are *interior places.* Fig. 3 shows a simple signature with places P_1 to P_4 for illustration.

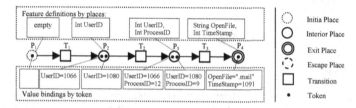

Fig. 3. Features and places

Transitions represent events which trigger state changes of signature instances. A transition is characterized by input places, output places, event type, conditions, feature mappings, consumption mode, and actions. *Input places* of transition t are places with an edge leading to the transition t. They describe the required state of the system before the transition can fire. *Output places* of transition t are places with an incoming edge from the transition t. They characterize the system state after the transition has fired. A change between system states requires a security relevant event. Therefore each transition is associated with an event type. Further, a system change can require additional conditions which specify that certain features of the event (e.g. user name) are assigned with particular values (e.g. root). Conditions can require distinct relationships between event and token features on input places (e.g. same values).

If a transition fires, then tokens are created on the transition's output places. These tokens describe the new system state. To bind values to the features of the new tokens the transitions contain *feature mappings.* These are bindings which can be parameterized with constants, references to event features, or references to input place features. The *consumption mode* (cf. [8]) of a transition controls whether tokens that activate the transition remain on the input places after the transition fired. This mode can be individually defined for each input place. The consumption mode can be considered as a property of a connecting edge between input place and transition. Only in the consuming case the tokens which activate the transition are deleted on the input places.

Fig. 4 illustrates the properties of a transition. The transition T_1 contains two conditions. The first condition requires that feature *Type* of event *E* contains the value *FileCreate*. The second condition compares feature *UserID* of input place P_1, referenced by *"P1.UserID"*, and feature *EUserID* of event type *E*, referenced by *"EUserID"*. This condition demands that the value of feature *UserID* of tokens on input place P_1 is equal to the value of event feature *EUserID*. Transition T_1 contains two feature mappings. The first one binds the feature *UserID* of the new token on the output place P_2 with the value of the homonymous feature of the transition activating token on place P_1. The second one maps the feature *Name* from the new token on place P_2 to event feature *EName* of the transition triggering event of type *E*.

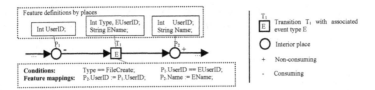

Fig. 4. Transition properties

5.2 Test Steps

We now explain the test for under specified signatures according Section 2.2 for a shell-link-attack which is described in EDL. A **shell-link-attack** exploits a special shell feature and the SUID (*Set-User-ID*) mechanism. If a link to a shell script is created and the link name starts with "-", then it is possible to create an interactive shell by calling the link. In old shell versions regular users could create an appropriate link which points to a SUID-shell-script and produce an interactive shell. This shell runs with the privileges of the shell-script owner (maybe root).

Fig. 5 depicts the respective EDL-signature.

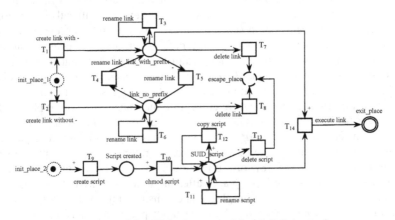

Fig. 5. Simplified EDL-signature of the shell-link-attack

Applying the test method of Section 4.2 to the test of shell-link-attack signature the following test steps have to be executed:

Step1: For test case selection, we use the path coverage criteria (C4). Thus every possible path from the initial to the exit place will be selected. In our example this results in 19 different paths (e.g. $T_1, T_5, T_4, T_9, T_{10}, T_{12}, T_{14}$), if each loop is passed maximum once.

Step2: Based on the selected paths action sequences are assigned to the events required by the transition by means of the inverse relation δ^{-1}. In this case timed independencies between the action sequences of transitions T_1 to T_8 and T_9 to T_{13} can be neglected. This restriction is possible, since the actions of the two transition paths from places *init_place_1* and *init_place_2*, respectively, to the *exit_place* are concurrent. Only transition T_{14} synchronizes the two concurrent action sequences. Since the shell-link-attack represents an attack which is executed within a shell, the inverse relation δ^{-1} assigns shell commands to the associated events. The transition conditions are used to implement the parameters of the shell commands.

We show this as example for transition T_3: T_3 fires, when a *rename_link-event* occurs and there is a token on place *link_with_prefix* and the event fulfils the two transition conditions (*link_with_prefix.link_name* == *new_link_name*) and (*new_link_name* == *RegExp*("-.*")). The associated *rename_link-event* is mapped by δ^{-1} onto the shell command *mv*. This command has two parameters: the old (*old_name*) and the future name (*new_name*) of the link or the file, respectively. The first transition condition determines the *old_name* parameter of *mv* command with the value of the *link_name* feature from the token of the place *link_with_pefix*. The second parameter (*new_name*) of the *mv* command is arbitrary, but due the second transition condition it is restricted to a name which begins with "–".

Step3a (*Establishing the correct attack conditions*): The step is dropped, since the shell-link-attack is a deterministic successful attack.

Step3b: Because the derived action sequence are shell commands they are simply executed by means of a scripts in a shell.

Step3c: The successful exploitation of the vulnerability by the action sequence can be proved by means of the shell command *id* which outputs the effective UserID. In case of a successful attack it should correspond to the UserID of the script owner. Therefore each action sequence has to be completed by appending an *id*-statement for comparing the UserID. Thus a successful attack execution can be determined automatically.

Test outcome: The execution of the 19 action sequences showed that all sequences which containing copy statements for triggering transition T_{12} don't leading to a successful attack. Accordingly the transition T_{12} must be incorrect. The analysis of this transition and the associated copy command *cp* revealed in a short time that *cp* during copying removes the SUID bit set before with *chmod* (T_{10}). Accordingly the signature must be corrected so that the outgoing edge of transition T_{12} leads to *script_created* and not as up to now to place *SUID_scrip*.

6 Final Remarks

The derivation of signatures from new exploits is still a tedious process which requires much experience. Systematic approaches are still rare. Newly derived signatures often possess therefore significant detection shakiness. Inaccurate signatures strongly limit the detection power of misuse detection systems as well as their acceptance in practice. A longer test and correction phase is needed until qualitative and accurate signatures can be applied which implicates an unacceptable vulnerability window. Systematic test methods can help to accelerate the signature development process and to reduce the vulnerability period of affected systems. In this paper we presented first approaches for a systematic signature test.

The detection shakiness of newly derived signatures is the result of heuristic derivation procedures. Even the rare systematic approaches scarcely induce ideal signatures due to the broad range of system details to be taken into account. Normally derived signatures are either under or over specified. We presented two test methods to detect these both kinds of variances as well as preliminary tests and a test on escape events. Test methods for signature tests require a strong involvement of the test engineer in details of the considered system. Unlike other tests signature tests do not require specific test architecture. All tests are executed on the vulnerable system and the monitoring intrusion detection system. A central issue of signature testing is to re-establish the conditions needed for successfully running an attack. This requires a lot of experience and limits the practicability of the tests. Beside this or in the case of deterministic successful attacks the test engineer needs only sparsely knowledge about the concrete attack and signature to accomplish the tests. We are currently investigating the proposed test methods with concrete signatures. Further we look for other test strategies.

References

1. Cheung, S., Lindqvist, U., Fong, M.: Modeling Multistep Cyber Attacks for Scenario Recognition. In: Proc. of the 3rd DARPA Information Survivability Conf. and Exposition, pp. 284–292. IEEE Computer Society Press, Washington (2003)
2. Debar, H., Morin, B.: Evaluation of the Diagnostic Capabilities of Commercial Intrusion Detection Systems. In: Wespi, A., Vigna, G., Deri, L. (eds.) RAID 2002. LNCS, vol. 2516, pp. 177–198. Springer, Heidelberg (2002)
3. Julisch, K.: Dealing with False Positives in Intrusion Detection. In: Debar, H., Mé, L., Wu, S.F. (eds.) RAID 2000. LNCS, vol. 1907, Springer, Heidelberg (2000)
4. Larson, U., Lundin Barse, E., Jonsson, E.: METAL - A Tool for Extracting Attack Manifestations. In: Julisch, K., Krügel, C. (eds.) DIMVA 2005. LNCS, vol. 3548, pp. 85–102. Springer, Heidelberg (2005)
5. Lippmann, R., Haines, J.W., Fried, D.J., Korba, J., Das, K.: The 1999 DARAP Off-Line Intrusion Detection System Evaluation. In: Debar, H., Mé, L., Wu, S.F. (eds.) RAID 2000. LNCS, vol. 1907, pp. 162–182. Springer, Heidelberg (2000)
6. Lippmann, R., Webster, S., Stetson, D.: The Effect of Identifying Vulnerabilities and Patching Software on the Utility of Network Intrusion Detection. In: Wespi, A., Vigna, G., Deri, L. (eds.) RAID 2002. LNCS, vol. 2516, pp. 307–326. Springer, Heidelberg (2002)

7. Meier, M., Bischof, N., Holz, T.: SHEDEL - A Simple Hierarchical Event Description Language for Specifying Attack Signatures. In: Proc. of the 17th IFIP International Conference on Information Security, pp. 559–571. Kluwer, Dordrecht (2002)
8. Meier, M., Schmerl, S., Koenig, H.: Improving the Efficiency of Misuse Detection. In: Julisch, K., Krügel, C. (eds.) DIMVA 2005. LNCS, vol. 3548, pp. 188–205. Springer, Heidelberg (2005)
9. Ranum, M.J.: Challenges for the Future of Intrusion Detection. In: Wespi, A., Vigna, G., Deri, L. (eds.) RAID 2002. LNCS, vol. 2516, Springer, Heidelberg (2002)
10. Rubin, S., Jha, S., Miller, B.: Automatic Generation and Analysis of NIDS Attacks. In: Proc. of. 20th Annual Computer Security Applications Conference, Tucson, AZ, USA, pp. 28–38. IEEE Computer Society Press, Los Alamitos (2004)
11. Rubin, S., Jha, S., Miller, P.B.: Language-based generation and evaluation of NIDS signatures. In: Proc. of the IEEE Symposium on Security and Privacy, Oakland, CA, USA, pp. 3–17. IEEE Computer Society Press, Los Alamitos (2005)
12. Schmerl, S., König, H., Flegel, U., Meier, M.: Simplifying Signature Engineering by Reuse. In: Müller, G. (ed.) ETRICS 2006. LNCS, vol. 3995, pp. 436–450. Springer, Heidelberg (2006)
13. Totel, E., Majorczyk, F., Mé, L.: COTS Diversity Based Intrusion Detection and Application to Web Servers. In: Valdes, A., Zamboni, D. (eds.) RAID 2005. LNCS, vol. 3858, pp. 43–62. Springer, Heidelberg (2006)

TPLan-A Notation for Expressing Test Purposes

Stephan Schulz[1], Anthony Wiles[1], and Steve Randall[2]

[1] European Telecommunications
Standards Institute (ETSI)
650 Route de Lucioles
F-06921 Sophia Antipolis Cedex
{Stephan.Schulz, Anthony.Wiles}@etsi.org
[2] PQM Consultants
4 The Myrtles
Tutshill, Chepstow
U.KSteve.Randall@pmqconsultants.com

Abstract. To this day test purposes are predominately defined in practice using natural language. This paper describes a more formal approach based on a notation which has been recently developed and standardized at the European Telecommunications Standards Institute (ETSI) called TPLan. We introduce here the motivation and main concepts behind this new notation, and share our experiences gathered from its application in the development of standardized test specifications. We also discuss how TPLan can help to make test development as a whole more efficient – especially in the context of suite based test specification.

1 Introduction

TPLan has been developed and standardized [1] by the ETSI Technical Committee Methods for Testing and Specification (TC-MTS). Members of this group include leading testing experts from industry and academia and it receives support from ETSI's own Protocol and Testing Competence Centre. For more than a decade MTS has been involved in the design of languages, methodologies, frameworks, and guidelines [2,3,4] to help rapporteurs to increase quality and effectiveness of their specifications. The Test Purpose notation, TPLan, was conceived when investigating approaches to improving efficiency in the development of test specifications based on patterns [5]. Here, it was realized that patterns could or should be identified much earlier than at the time of test case writing, i.e., when identifying test purposes.

Much of the current research on test specification development has focused on the development and use of either suite-based [2] or model-based [6] testing technologies. Test purposes have been anchored as a concept in conformance testing methodology [7] for a long time but, as such, have received little attention in the testing research community. Formal approaches to test purpose specification [9,10,11,12] have been proposed but these have yet to be deployed successfully in industry. Graphical approaches based on Message Sequence Charts (MSC) [13, 14] for specifying test purposes have had only limited success – arguably due to their limitation in

A. Petrenko et al. (Eds.): TestCom/FATES 2007, LNCS 4581, pp. 292–304, 2007.

expressing behaviour only in terms of interactions. In our experience natural language still dominates the specification of test purposes.

It was our intention with TPLan to make test purpose specification more structured but not completely formal. Evidence of this approach can be found in the notation where many of the base keywords have been selected from preferred writing styles used in ETSI's test purpose specifications. Another design criterion was to keep the core notation as independent as possible from any specific application area and testing technology while making it easily extensible. This opens TPLan to a wide range of applications from, for example, telecommunication to civil engineering. It also allows it to be used in conjunction with both suite-based and model-based testing.

After an introduction to test purposes and how they fit into test specification development we will discuss their specification with our notation in section 3 of this paper in more detail. TPLan has already been used within ETSI to specify more than a thousand test purposes in the context of test development for the Internet Protocol version 6 (IPv6) [15] and digital Public Mobile Radio (dPMR) [16]. Section 4 presents first experiences from these projects which have shown that TPLan can help to enforce uniformity of test purpose specifications and to identify inconsistencies in standard documents at an early stage before costly test case specification and validation. We believe, however, that this notation may have even more potential in the test specification process by reducing development times and increasing productivity. Section 5 proposes some ideas for more sophisticated tools that may achieve such additional gains.

2   About Test Purposes

As with any other development activity, better test specifications can be produced when a structured approach is followed. For more than 15 years ETSI has applied the methodology prescribed by [7,8] where the development of a complete test specification is broken down into five discrete steps as shown in Table 1. These steps can be understood as different levels of abstraction that bridge the large intellectual gap between a base specification and the final executable test suite. They are not only an essential framework to the test engineer but also enable a common understanding of the complete test suite between different target audiences.

Table 1. Steps in test specification development

Test Specification Step	Means of Specification	Question answered
Requirement (RQ) identification	Text, Tables	Which requirements are to be tested?
Test Purpose (TP) specification	Text, Tables, **TPLan**	What is to be tested?
Test Description (TD)	Text, Tables, MSCs, etc	How is it to be tested? (informally)
Test Case (TC) specification	TTCN-3, C, Java, Perl, Python, MSC, etc	How is it to be tested? (executable)
Test validation	-	Is test implemented correctly?

Test purposes are derived from the requirements stated in one or more base specifications that define the implementation. This direct relationship to the requirements makes it possible to make an early assessment of test coverage of the specification and to determine the inter-dependencies between different requirements. Each test purpose usually focuses on one specific requirement. Within ETSI, these base specifications are most often protocol standards.

Test purposes provide an essential abstraction of a test that specifies what is to be tested without going into the details of how a test is to be implemented. Test purposes are not test steps; they specify pass verdict criteria. Test purposes are written using the language and terminology of the base specification(s) and are independent of any particular programming language, test system or platform on which corresponding tests might eventually be executed. They need to be developed, discussed and stabilized prior to any test case specification.

Test purpose specification results in a rigid assessment of the requirements with which they are associated and can identify problems in base specifications long before any test is ever implemented or executed against an Implementation Under Test (IUT). Not all requirements will lead to test purposes due to the limitations imposed by the chosen type of testing, e.g., conformance or interoperability testing.

Test purposes serve an important role as a basic documentation tool. They do not only bridge the gap between original requirements and test case specification but also between technology experts (who are not necessarily test engineers), managers, and the test engineers. At ETSI this aspect is very important since test specifications are often reviewed and approved by standards working groups. These groups need to understand the requirements which are being tested without having to read detailed test case specifications. In addition, ETSI is an environment where test purposes and corresponding test case specifications are developed for a wide variety of technologies in a distributed (and multi-cultural) environment. Such an environment clearly has a need for a consistent and uniform approach to test purpose specification, i.e., a notation which provides a common and recognisable level of understanding.

3   Test Purpose Specification with TPLan

TPLan has been designed to make test purpose specification more formal without inhibiting the expressive power of prose. The intent was to enable a consistent and structured representation of test purposes across a wide range of application domains and cultural backgrounds while retaining the informal "look and feel" of a natural language. It is for this reason that the core TPLan syntax and semantics have been kept small and left open. Of course this flexibility or "freedom of expression" inevitably results in weaker semantics and limits the checks that a tool can perform purely on the basis of the TPLan definition itself.

3.1   Test Purpose Structure

A TPLan test purpose comprises two segments as shown in Figure 1: a header and a body. The header provides a unique identifier for the test purpose and, optionally, references to other useful information for the understanding of the test purpose. These can include the requirement(s) covered by the test purpose, the type of test purpose, dependencies with other test purposes, and the tested role of the IUT.

The body of a test purpose specifies the specific initial IUT condition required for the test purpose to be valid and critical verdict criteria for a test - in the form of a stimulus and response - to ensure that the requirement(s) are met. The structuring of the test purpose body into the "with", "when" and "then" clauses clearly shows the roots of this notation in black box testing. A test purpose body is usually written from the perspective of the IUT, i.e., pre-conditions refer to the required initial state of the IUT, etc.

```
-- test purpose header
TP id: <string>
< other test purpose headers (optional) >

-- test purpose body
with { <pre-conditions> }   -- optional clause

ensure that {

  when { <stimulus> }

  then { <response> }

}
```

Fig. 1. Basic structure of a TPLan test purpose specification

Table 2. Key TPLan concepts

Concept	Definition
Entity	A physical or logical actor which applies a stimulus or receives response, and vice versa.
Event	The measurable basis of a stimulus or response which may be parameterized with Values
Value	An abstract identifier representing either – a literal constant; – a numeric constant; – a field or other container
Unit	A concrete qualifier to a number which helps to indicate the relative size or quantity of the number.
Condition	An abstract expression of the status or state of the entity or entities under test.
Word	Any other natural language element useful for the specification of the test purpose body, for example, – an action – an article, preposition, adjective, adverb, etc

The keywords "when" and "then" should not be misunderstood to require a complete specification of accurate test sequence(s). Stimulus and response in a test purpose should focus and isolate only the directly relevant parts of information needed to assess if a requirement is indeed fulfilled by an IUT, for example, message types and critical information element values. Again, the level of information content and language used should be identical to the one of the requirement definition in the base specification.

3.2 Fundamental Building Blocks

The initial conditions, the stimulus and the response in a test purpose body are constructed using the concepts which are listed in Table 2.

Instances of these concepts can be created using quoted strings containing free form text. Some instances for entities and words such as "IUT", "sends" or "containing" have been pre-defined as keywords in the notation. An example of a basic TPLan test purpose is shown in Figure 2.

```
TP id    : CW_U01_002
Summary  : 'A busy user with information channel control
           but no B-Channel responds to an incoming
           SETUP'
RQ ref   : Section 9.5.1
IUT role : user
with { IUT in 'an information channel control state'
     and 'no B-Channel free'
     }

ensure that {

  when { the IUT receives 'a valid and compatible SETUP'
       from the TESTER
         containing 'a channel identification IE'
         indicating 'no B-channel available' }

   then { the IUT sends 'ALERTING' to the TESTER }

}
```

Fig. 2. A complete example of a test purpose

The drawback of quoted strings is, however, that it is impossible to associate much meaning with them. It is also not possible to check whether quoted strings specify instances of these concepts in the correct order; for example, that ALERTING in Figure 2 is really an event.

3.3 User Defined Extensions

TPLan allows users to extend or customize its vocabulary based on the concepts introduced in the previous section with keywords which are relevant to their own specific application domain. This concept makes TPLan much more powerful than other forms of test purpose specification. Although the notation does not support an explicit definition of the semantics associated with a word or phrase, such semantics can often be implied from application domain within which TPLan is being used.

As an example, assume that we define a new word "accepts". When we use this new word in a TPLan "when" clause, e.g., "when { the IUT accepts 'this message' }", then TPLan itself does not define or restrict what "accepts" actually means or how such acceptance is measured in an eventual test case specification. The word "accepts" could mean any of the following actions:

- the IUT displays a message to the user.
- no error is displayed to the user.
- the IUT will continue interacting normally.
- the IUT does nothing that is externally observable

However, the meaning of "accepts" is likely to be obvious to technology experts as well as test engineers familiar with the domain or technology. As a result, this word is a valid abstraction of either one or possibly more interactions with the user or internal or external entities.

---

xref CW_U { ETS_300_058_1 } -- ETSI standard reference

def condition information_channel_control_state

def event   SETUP { Channel_identification_IE }
def event   ALERTING

def value   no_B_channel_available

TP id   : CW_U01_002
Summary : 'A busy user with information channel
       control but no B-Channel responds to an
       incoming SETUP'
RQ ref : Section 9.5.1
IUT role : user

with { IUT **in an** information_channel_control_state
     and 'no B-Channel free'
  }

ensure that {

  when { the IUT **receives a valid and compatible** SETUP
      from the TESTER
        containing a Channel_identification_IE
        indicating no_B_channel_available }

  then { the IUT **sends** ALERTING **to the** TESTER }

}

Fig. 3. A complete example of a test purpose with user definitions

Users have to declare specific instances of the main concepts shown previously in Table 2 when they use them in the test purpose definition. Figure 3 illustrates an example test purpose from the telecommunication domain written for conformance testing. The user has included a cross-reference to identify the ETSI standard ETS 300 058-1 as the base specification and then defined one initial condition and two events representing the different message types.

The definition of the SETUP event shows one parameter. That does not mean that in practice the message that this event represents only has one parameter. It means that only this parameter is significant in determining whether the IUT fulfills the referenced requirement. In test purposes, events are abstract representations of exchanged or observed information; they are not complete message instances. Similarly the user defined value in the example is an abstract representation of a concrete value.

Note also that within TPLan, user defined conditions, events and values are expressed as identifiers, i.e., they must not contain spaces. In our example we have chosen underscores to preserve a feel of natural language to the identifiers but this is only our naming convention. Finally, notice that one initial condition has been specified for the sake of this example using a quoted string. Quoted strings can still be useful in cases where, for example, a pre-condition is very complex.

By means of a simple notation, the user is also able to restrict the syntactical context in which user-defined words can be used. Within a context definition statement, any word prefixed with a tilde character (~) may only be used in that context, any word surrounded by square brackets is considered optional and any unencumbered word can be used in any other syntactical context. As an example, the following definitions can be made:

def word requested

def context is [not] ~requested to

Here, the words "is", "not" and "to" are included in the predefined TPLan vocabulary. The "context" statement constrains the user-defined keyword "requested" so that it is only syntactically correct in the contexts "is requested to" and "is not requested to".

3.3   Arrangement of Test Purpose Definitions

In most cases and for a variety of reasons test purpose specifications need a logical structuring. To assist users in such structuring TPLan offers two complementary mechanisms which are grouping and inclusion.

Test purposes can be arranged into logical, hierarchical groups by using the "Group" and "End Group" statements as shown in Figure 4. These groups as well as individual test purposes can also be collected together into a single specification referred to as a Test Suite Structure (TSS) which also contains a header of its own.

In those cases where a number of test purpose writers are involved in a project, it will be necessary to maintain a single source of vocabulary extensions. For this purpose, the notation allows a TSS to include other TPLan files by means of a **#include** statement as shown in Figure 5. This mechanism uses a simple replacement method so that the content of the identified file is inserted into the file in place of the **#include** statement. Additionally, the inclusion mechanism can be used to construct a complete TSS from separate group files developed by multiple test purpose writers.

```
TSS     : COR_IOP --- identifier for all test purposes
Title   : 'RFC2460 IPv6 Core Specification'
Version : 1.0.1
Date    : 05.10.2006
Author  : 'Steve Randall (ETSI TC-MTS)'

Group 1 'Initialization functions'

Group 1.1 'System startup'

Group 1.1.1 'Memory check'
...
<test purpose definitions>
...
End Group 1.1.1

Group 1.1.2 'Media check'
...
<test purpose definitions>
...
End Group 1.1.2

End Group 1.1

End Group 1

...
```

Fig. 4. Example test purpose structuring using TSS header and grouping

```
TSS     : COR_IOP -- identifier for all test purposes
Title   : 'RFC2460 IPv6 Core Specification'
Version : 1.0.1
Date    : 05.10.2006
Author  : 'Steve Randall (ETSI TC-MTS)'

#include c:\include\SIUnitDefs.tplan
#include c:\include\IOPDefs.tplan
#include c:\include\IPv6Defs.tplan

#include c:\include\IPv6Group1.tplan -- Initialization

#include c:\include\IPv6Group2.tplan -- Outgoing call

#include c:\include\IPv6Group3.tplan -- Incoming call
```

Fig. 5. TPLan specification constructed from #include statements

4   First Experiences

TPLan has been used by ETSI for the specification of test purposes in its IPv6 and dPMR test development projects. In both cases test purposes have been specified for two types of testing, conformance and formalized interoperability testing [17].

Two examples, a dPMR conformance and an IPv6 interoperability test purpose, are shown in Figures 6 and 7. These examples illustrate how TPLan vocabulary can be customized for these specific application domains and adapted to different types of testing. Note that required TPLan user definitions have been omitted from the figures. Also, almost all test purpose header lines are optional. The ones chosen in these examples provide further information about a test purpose summary, the type of test purpose, a reference to the catalogued requirement that the test purpose pertains to, the role or type of equipment being the subject of test, as well as a reference to the test architecture or configuration in which the IUT or Equipment Under Test (EUT) is embedded.

```
TP id    : TP_PMR_0406_01
summary  : 'Header frame acknowledges connect request'
TP type  : conformance
RQ ref   : RQ_001_0406
IUT Role : CSF  -- Configured Service Function (CSF)
config ref: CF_dPMR_CSF_01 -- CSF Implementation Under
                  -- Test (IUT) and TESTER


with   { IUT in standby }


ensure that {
  when { IUT receives a Connection_Request }
  then { IUT sends an Acknowledgement_Frame }
}
```

Fig. 6. Example dPMR conformance test purpose

Our experiences with the first prototype version of TPLan (which allowed the use of a non extensible pre-defined set of keywords and quoted strings) were that writers felt limited in their ability to fully express test purposes. The language used by the standard document differed too much from the language that could be constructed from pre-defined TPLan keywords. Consequently, writers frequently requested new keywords to be added to the notation and made heavy use of quoted strings in test purpose specification. That, in turn, reduced the ability of project managers to ensure the quality and consistency of test purposes.

The introduction of user defined extensions to TPLan radically changed this situation. The ability to define a domain specific vocabulary not only gave writers more freedom in specifying test purposes but also made it easier to detect the misuse or misspelling of significant words. We noticed that, independent of the project type, the user defined vocabulary initially grows quite rapidly during the specification of first test purposes. After that, however, the need for new definitions levels off quickly. The test purpose writers also found it useful that user defined keywords could be explained or clarified with comments at one central place, i.e., their definition.

```
TP id     : TP_COR_8231_01
summary   : 'EUT uses at least two of the connected
             routers as its default routers '
TP type   : interoperability
RQ ref    : RQ_COR_8231
EUT role  : Host, Router -- = either Host or Router
config ref: CF_033_I -- 2 Routers and 1 Node as
                     -- Qualified Equipment (QE1/2/3) +
                     -- Equipment Under Test (EUT)
                     -- connected via 2 links
TD ref    : TD_COR_8231_01

with { QE1 having 1 unique unicast_address on each link

  and QE2 having 1 unique unicast_address on each link

  and EUT and QE3 able to communicate

  and QE1 'having disabled one of its interfaces' }


ensure that {

  when { (  QE1 disables 1 interface
       or QE2 disables 1 interface )
       and EUT is requested to send a packet to QE3 }

  then {   EUT sends the packet to QE3 }

}
```

Fig. 7. Example Ipv6 core interoperability test purpose

The extra effort spent in structured writing helped to reveal many problems or inconsistencies in the base specification prior to any test case specification or execution. This property of the notation became especially apparent during dPMR test purpose specification where writers were experts in the technology but novices in testing. Automated syntax checking with a simple parser [18] gave a first level of assurance on test purpose quality. A manual check of test purposes was nevertheless still required to assure their correctness. For a proficient English speaker it was easy to identify incorrect or badly written test purposes as these were not minor grammatical or spelling errors, the test purposes just obviously read incorrectly. It is not clear at this point if an improved syntax checker or further tool support could eliminate the need for this second grammar check.

5   Improving of Test Specification Efficiency

In this section we want to show how TPLan offers a foundation to build on. Remember that it has only recently been standardized and is still in its infancy. So far there is much interest in it but only limited tool support. Based on our early experience with TPLan we believe that additional tool support could help to further improve the speed and quality of test purpose as well as test specification development as a whole.

Sophisticated editor support is probably one of the more important issues. Context sensitive editors could assist, extend and manage TPLan user definitions and provide features such as syntax highlighting, keyword completion and other forms of vocabulary management. This kind of tool would help users to avoid writing incorrect TPLan test purposes to begin with. More advanced parsers which go beyond simple syntax checks are needed to help pinpoint incorrect test purposes early on in the specification process. It may also be possible to extend the analysis of test purposes by incorporating some of the English grammar checking technologies used by modern word processing software.

Test purposes that have been checked for correctness can be used as input to other forms of processing. One of these is the identification of recurring patterns in preconditions and the interactions between entities. Such patterns can be used in a variety of ways:

1. the identification of potentially reusable segments of test case specifications derived from the test purposes;
2. an assessment of test purpose variation;
3. an estimation of the possible complexity of the eventual test case specification;
4. an estimation of the effort likely to be required for implementing the eventual test specifications.

When used with suite-based test technologies TPLan test purposes can serve as the basis for test purpose publication or other presentation formats. TPLan test purposes seem especially attractive for generation of test specification stubs. They contain a considerable amount of information regarding initial conditions, verdict criteria and interaction of entities. Nevertheless there are many details which are not specified in test purposes but which are required for test case specification such as preamble implementation, complete message values, guarding against unexpected behaviour, postamble implementation, etc. To make test case generation as complete as possible we expect it to almost certainly be based on and driven by domain-specific TPLan vocabulary and semantics as well as other external sources of input. But once a clever approach is found it will be possible to develop code generators for many different testing languages since TPLan is independent of a specific test case specification language.

Another interesting application for future TPLan tools is the automatic validation of manually written test cases against TPLan test purposes to determine whether or not a test case implementation fulfils the criteria specified in the associated test purposes. This could be an interesting idea, e.g., for companies that define test purposes but subcontract test case specification.

When used in a model-based testing context, TPLan test purposes can be used in the definition of coverage criteria or testing directives for test generation from formal executable models. Here, the test purpose would define a path through model behaviour. Similarly as in the case of test case stub generation, the abstract nature of test purposes has to be again taken into account in model-based test generation. A stimuli or response of a test purpose specification may correspond to only one state transition but also to a path or even multiple paths in the model (see our discussion of "accepts" in Section 3.3). Secondly, the faithful use of data specified in a test purpose is non-trivial to handle in test case generation. Data is often not hard-coded but computed during the execution of models. Therefore, for example, it has to be ensured that the event parameter values specified in a test purpose are truly sent or expected by the generated test case.

6   Conclusions

In this paper we have introduced the new notation TPLan which has been developed and standardized by ETSI for expressing test purposes. TPLan attempts to formalize the specification of test purposes by requiring a certain structure and composition based on a set of well defined concepts, i.e., entities, event, value, units, conditions and words. It is independent of a specific testing technology or application domain. A key concept in TPLan is that the pre-defined vocabulary can be extended and customized by users for specific application domains. User definitions make it possible to add more meaning to test purposes and to customize the notation for a specific application domain.

This notation has already been used extensively by ETSI in its IPv6 and dPMR test specification developments. Experiences have been positive in that the quality of test purpose specifications was easier to monitor and affect. Further study is however still required to investigate other impacts of TPLan use such as its effect on overall test specification process. We expect that TPLan will have an even bigger impact on test specification development once more sophisticated tools for handling test purposes become available. Most interest seems to be in the generation of test case specification stubs from test purposes as well as the use of test purposes as a driver for model-based test generation. Some tools, such as a free simple parser and syntax highlighter, are already publicly available at [18].

In the future we see that TPLan standardization is likely to be extended. Currently the creation of TPLan profiles for specific application areas is under discussion, for example, for communicating systems. Such profiles will essentially just extend the pre-defined vocabulary and define semantics for the later. In addition, we are planning to study further existing work on requirements definition languages which are closely related to definition of test purposes.

Acknowledgments

We would like to thank all experts who took part in TPLan test purpose specification in the ETSI IPv6 and dPMR projects for their constructive comments and feedback on the notation itself. In addition, we would like to recognize Dr. Thomas Deiß for his extensive reviews of TPLan standard drafts.

References

[1] ETSI ES 202 553: Methods for Testing and Specification (MTS); TPLan: A Notation for expressing Test Purposes, European Telecommunications Standards Institute, Sophia Antipolis (2007)

[2] Willcock, C., et al.: An Introduction to TTCN-3. Wiley & Sons, Chichester (2005)

[3] Moseley, S., Randall, S., Wiles, A.: Experience within ETSI of the combined roles of conformance testing and interoperability testing. In: Proceedings of 3rd Conference on Standardization and Innovation in Information Technology (SIIT), Delft, The Netherlands, October, pp. 177–89 (2003)

[4] Randall, S.: Descriptive SDL., Telektronikk,Telenor AS, (4), pp. 107–12 (2000)

[5] Neukirchen, H., Dai, Z.R., Grabowski, J.: Communication Patterns for Expressing Real-Time Requirements Using MSC and their Application to Testing. In: Groz, R., Hierons, R.M. (eds.) TestCom 2004. LNCS, vol. 2978, pp. 144–159. Springer, Heidelberg (2004)

[6] Hartma, A., Nagin, K.: The AGEDIS tools for model based testing. In: Proceedings of the 2004 ACM SIGSOFT international Symposium on Software Testing and Analysis (ISSTA '04), Boston, MA, pp. 129–132. ACM Press, New York (2004)

[7] Information Technology - Open Systems Interconnection - Conformance Testing Methodology and Framework - Part 1: General concepts, Geneva (1994) ISO/IEC 9646-1

[8] Information Technology - Open Systems Interconnection - Conformance Testing Methodology and Framework - Part 2: Abstract Test Suite Specification, Geneva (1994) ISO/IEC 9646-2

[9] Jard, C., Jeron, T.: TGV: theory, principles and algorithms. In: Proceedings of 6th World Conference on Integrated Design and Process Technology (IDPT 2000), Pasadena, California, USA (June 2002)

[10] Desmoulin, A., Viho, C.: Formalizing Interoperability for Test Case Generation Purpose. In: Proceedings of IEEE Nasa ISoLA Workshop on Leveraging Applications of Formal Methods, Verification, and Validation, Columbia, MD, USA (September 2005)

[11] Tretmans, J.: A Formal Approach to Conformance Testing, Ph.D. Thesis, University of Twente, The Netherlands (1992)

[12] Deussen, P., Tobies, S.: Formal Test Purposes and The Validity of Test Cases. In: Peled, D.A., Vardi, M.Y. (eds.) FORTE 2002. LNCS, vol. 2529, Springer, Heidelberg (2002)

[13] Grabowski, J., Hogrefe, D., Nahm, R.: Test Case Generation with Test Purpose Specification by MSCs. In: Faergemand, O., Sarma, A. (eds.) SDL'93 - Using Objects North-Holland (October 1993)

[14] Object Management Group: UML 2.0 Testing Profile Specification (2003)

[15] Deering, S., Hinden, R.: Internet Protocol, Version 6 (IPv6) Specification. IETF RFC 2460 (December 1998)

[16] ETSI TS 102 490 (V1.3.1): Electromagnetic compatibility and Radio spectrum Matters (ERM); Peer-to-Peer Digital Private Mobile Radio using FDMA with a channel spacing of 6,25 kHz with e.r.p of up to 500 mW, European Telecommunications Standards Institute, Sophia Antipolis (2006)

[17] ETSI TS 102 237-1 (V4.1.1): Telecommunications and Internet Protocol Harmonization Over Networks (TIPHON) Release 4; Interoperability test methods and approaches; Part 1: Generic approach to interoperability testing, European Telecommunications Standards Institute, Sophia Antipolis (2003)

[18] http://www.tplan.info

Testing Nondeterministic Finite State Machines with Respect to the Separability Relation

Natalia Shabaldina[1], Khaled El-Fakih[2], and Nina Yevtushenko[1]

[1] Tomsk State University, 36 Lenin Str., Tomsk, 634050, Russia
snv@kitidis.tsu.ru, yevtushenko@elefot.tsu.ru
[2] American University of Sharjah, PO Box 26666, UAE
kelfakih@aus.edu

Abstract. In this paper, we propose a fault model and a method for deriving complete test suites for nondeterministic FSMs with respect to the separability relation. Two FSMs are separable if there exists an input sequence such that the sets of output responses of these FSMs to the sequence do not intersect. In contrast to the well-known reduction and equivalence relations, the separability relation can be checked when the «all weather conditions» assumption does not hold for a nondeterministic Implementation Under Test (IUT). A (complete) test suite derived from the given (nondeterministic) FSM specification using the separability relation can detect every IUT that is separable from the given specification after applying each test case only once. Two algorithms are proposed for complete test derivation without the explicit enumeration of all possible implementations. The first algorithm can be applied when the set of possible implementations is the set of all complete nondeterministic submachines of a given mutation machine. The second algorithm is applied when the upper bound on the number of states of an IUT is known.

Keywords: separability relation, testing nondeterministic FSMs.

1 Introduction

A number of conformance testing methods have been developed for deriving tests when the system specification and implementation are represented by nondeterministic FSMs [1–16]. Non-determinism occurs due to various reasons such as performance, flexibility, limited controllability, and abstraction [8] [11] [13] [17].

A number of methods have been proposed for test generation against nondeterministic FSM specifications with the guaranteed fault coverage with respect to appropriate fault models. The methods given in [3] and [4] derive test suites with respect to the equivalence relation for a nondeterministic implementation against a nondeterministic specification while the methods given in [6], [8], [9], and [10] derive test suites for a complete deterministic implementation against a nondeterministic specification with respect to the reduction relation. Two FSMs are *equivalent* if they have the same input/output behavior and an FSM T is a *reduction* of FSM S if the input/output behavior of T is a subset of that of S. Hierons [11] presented a test derivation method with respect to the reduction relation when a system implementation

A. Petrenko et al. (Eds.): TestCom/FATES 2007, LNCS 4581, pp. 305–318, 2007.
© IFIP- International Federation for Information Processing 2007

can be nondeterministic. Petrenko and Yevtushenko [15] generalize the work given in [18] and proposed a method that derives a test suite with respect to the reduction and equivalence relations for a nondeterministic implementation against possibly a partial specification.

When deriving test suites with respect to the reduction and equivalence relations with the guaranteed fault coverage the so-called *complete testing* assumption [3–4] (called «*all weather conditions*» by Milner in [19]) is assumed to be satisfied when testing a nondeterministic implementation. According to this assumption, if an input sequence (a test case) is applied a number of times to a nondeterministic implementation, then all possible output sequences of the implementation to this test case can be observed. However, when an Implementation Under Test (IUT) has a limited controllability, as happens, for instance, in remote testing, the complete testing assumption cannot be satisfied. In this case, the only relation that can be used for the preset test derivation with the guaranteed fault coverage [20] [21] is the separability relation defined by Starke in [22]. Two FSMs are *separable* if there is an input sequence, called a *separating sequence*, such that the sets of output responses of these FSMs to the sequence do not intersect, i.e., the sets are disjoint. It is known [15] [21] that test suites derived with respect to the equivalence and reduction relations cannot be used for testing the separability relation. The fact that an FSM with m states is not equivalent (or not a reduction) of an FSM with n states can always be established by an input sequence of length up to mn [15], while there exist two FSMs which can be separated only with an input sequence of exponential length [5].

The separability relation was further studied by Alur et al. in [5] where an algorithm for deriving a separating sequence for two separable states of an FSM with n states is proposed and is shown that the upper bound on the length of a shortest separating sequence is exponential. In [23], it is shown that given FSMs S with n states and T with m states, the length of a shortest separating sequence is at most 2^{mn-1} and this upper bound is reachable. An algorithm is proposed for deriving a shortest separating sequence of the given FSMs. However, experiments with the proposed algorithm show that on average, the length of a shortest separating sequence is less than mn and the existence of a separating sequence significantly depends on the number of nondeterministic transitions in the given FSMs. For all conducted experiments the upper bound 2^{mn-1} on the length of a separating sequence was never reached.

In this paper, we consider the test derivation w.r.t. the reduction relation when «all weather conditions» assumption may not be held for an IUT. A test suite is called complete up to the separability relation if it detects every non-reduction of the specification FSM from a given fault domain that is separable from the given FSM specification. If each test case of the test suite is applied to an IUT of the fault domain once and an IUT is separable from the specification FSM, then the IUT will be detected with such a test suite. However, this test suite can also detect some other implementations which are non-reductions of the specification FSM but are not separable with the specification FSM. Correspondingly, we refine the notions of a fault model and of a complete test suite. We propose a method for deriving a complete test suite without the explicit enumeration of FSMs of the fault domain when the fault domain is the set of all submachines of a given mutation machine (including those which are non-deterministic) and when the fault domain has each implementation FSM up to m states. We also demonstrate that not every test suite that is complete w.r.t. the reduction relation under the complete testing assumption can be used for testing up to the separability relation when each test case is applied to an IUT at most once.

This paper is organized as follows. Section 2 includes all necessary definitions. In Section 3 we refine the notion of a fault model and define a complete test suite w.r.t. the refined fault model. Sections 4 and 5 contain algorithms for building a complete test suite w.r.t. the refined model where the fault domain is the set of all complete (not only deterministic) submachines of a given nondeterministic FSM (a mutation machine) and where the fault domain is the set of all nondeterminisitic FSMs with at most m states. Section 6 concludes this paper.

2 Preliminaries

A *finite state machine* (*FSM*) S is a 5-tuple $\langle S, I, O, h_S, s_1 \rangle$, where S is a finite nonempty set with s_1 as the initial state; I and O are input and output alphabets; and $h_S \subseteq S \times I \times O \times S$ is a behavior relation. The behavior relation defines all possible transitions of the machine. Given a current state s_j and input symbol i, a 4-tuple $(s_j, i, o, s_k) \in h_S$ represents a possible transition from state s_j under the input i to the next state s_k with the output o, usually written as $s_j \xrightarrow{i/o} s_k$. If for each pair $(s, i) \in S \times I$ there exists $(o, s') \in O \times S$ such that $(s, i, o, s') \in h_S$ then FSM S is said to be *complete*; otherwise, FSM S is *partial*. If for each $(s, i, o) \in S \times I \times O$ there is at most one transition $(s, i, o, s') \in h_S$ then FSM S is said to be *observable*. Given FSM $S = \langle S, I, O, h_S, s_1 \rangle$, state s and an input i, state s' is a *successor* of state s under the input i or simply an *i-successor* of state s if there exist $o \in O$ such that the 4-tuple $(s, i, o, s') \in h_S$. Given a set of states $b \subseteq S$ and an input i, the set of states b' is a *successor* of the set b under the input i or simply an *i-successor* of b if b' is the set of all *i-successors* of states of the set b.

In the usual way, the behavior relation h_S is extended to input and output sequences. Given states $s, s' \in S$, input sequence $\alpha = i_1 i_2 \dots i_k \in I^*$ and output sequence $\beta = o_1 o_2 \dots o_k \in O^*$. Transition $(s, \alpha, \beta, s') \in h_S$ if there exist states $s = s_1, s_2, \dots ,$ $s_k, s_{k+1} = s'$ such that $(s_i, i_i, o_i, s_{i+1}) \in h_S$, $i = 1, \dots , k$. As usual, given a defined input sequence α at state s, $h_S^O(s, \alpha)$ denotes the set of all output sequences which FSM S produces at state s under the input sequence α, i.e. $h_s^O(s, \alpha) = \{\beta: \exists s' \in S \ [(s, \alpha, \beta, s') \in h_S]\}$.

Given an FSM $M = \langle M, I, O, h_M, m_1 \rangle$, an FSM $S = \langle S, I, O, h_S, s_1 \rangle$, $S \subseteq M$, $s_1 = m_1$, is a *submachine* of FSM M if $h_S \subseteq h_M$, i.e., if each transition of FSM S is obtained by fixing an appropriate transition of the FSM M. The set of all complete submachines, including those which are non-deterministic, of a complete FSM S is denoted $Sub_{nd}(S)$.

Given FSMs $S = \langle S, I, O, h_S, s_1 \rangle$ and $T = \langle T, I, O, h_T, t_1 \rangle$, the *intersection* $S \cap T$ is defined as the largest connected submachine of the FSM $\langle S \times T, I, O, h, s_1 t_1 \rangle$ where $(st, i, o, s't') \in h \Leftrightarrow (s, i, o, s') \in h_S \ \& \ (t, i, o, t') \in h_T$.

Complete FSMs S and T are *equivalent*, written $S \cong T$, if for each sequence $\alpha \in I^*$ $[h_T{}^O(t_1, \alpha) = h_S{}^O(s_1, \alpha)]$, i.e., the sets of output sequences of FSMs S and T under each input sequence coincide. If there exist sequence $\alpha \in I^*$ $[h_T{}^O(t_1, \alpha) \neq h_S{}^O(s_1, \alpha)]$ then FSMs S and T are *distinguishable*, written $S \ncong T$.

A state t of a complete FSM T is a *reduction* of a state s of a complete FSM S, written $t \leq s$, if for each sequence $\alpha \in I^*$ $[h_T{}^O(t, \alpha) \subseteq h_S{}^O(s, \alpha)]$, i.e., the set of output sequences of FSM S at state s contains the set of output sequences of FSM T at state t under each input sequence. If there exists sequence $\alpha \in I^*$ $[h_T{}^O(t, \alpha) \not\subseteq h_S{}^O(s, \alpha)]$ then state t is not a reduction of state s, written $t \not\leq s$. FSM T is a *reduction* of a FSM S, written $T \leq S$, if the reduction relation holds between the initial states, i.e., for each sequence $\alpha \in I^*$ $[h_T{}^O(t_1, \alpha) \subseteq h_S{}^O(s_1, \alpha)]$. If there exists sequence $\alpha \in I^*$ $[h_T{}^O(t_1, \alpha) \not\subseteq h_S{}^O(s_1, \alpha)]$ then FSM T is not a reduction of FSM S, written $T \not\leq S$.

A state t of a complete FSM T is *r-compatible* with a state s of a complete FSM S [21], written $t \simeq s$, if there exists a complete FSM $B = \langle B, I, O, h_B, b_1 \rangle$ and state $b \in B$ such that b is a reduction of both states t and s. If states t and s are not *r*-compatible then they are *r-distinguishable*, written $t \not\simeq s$. In this case, there exists an *r*-distinguishability finite set W of input sequences such that for each complete FSM $B = \langle B, I, O, h_B, b_1 \rangle$ and each state $b \in B$ there exists $\alpha \in W$ such that $[h_B{}^O(b, \alpha) \not\subseteq h_S{}^O(s, \alpha)]$ or $[h_B{}^O(b, \alpha) \not\subseteq h_T{}^O(t, \alpha)]$. FSMs T and S are *r*-compatible, written $T \simeq S$ (or *r*-distinguishable, written $T \not\simeq S$) if the corresponding relation holds between their initial states.

Complete FSMs T and S are *non-separable*, written $T \sim S$, if for each sequence $\alpha \in I^*$ $[h_T{}^O(t_1, \alpha) \cap h_S{}^O(s_1, \alpha) \neq \varnothing]$, i.e., the sets of output sequences of FSMs T and S under each input sequence intersect. If there exists sequence $\alpha \in I^*$ $[h_T{}^O(t_1, \alpha) \cap h_S{}^O(s_1, \alpha) = \varnothing]$ then FSMs T and S are *separable*, written $T \nsim S$. In the latter case, the sequence α is called a *separating* sequence of FSMs T and S.

3   Fault Model and a Test Suite

When testing w.r.t. the reduction relation the traditional fault model is a triple $\langle S, \leq, \mathfrak{R} \rangle$, where S is a specification FSM, $\leq$ is the reduction relation, fault domain $\mathfrak{R}$ is the set of all possible (faulty and non-faulty) implementation FSMs with the same input and output alphabets as the specification FSM S. As usual, FSMs of the set $\mathfrak{R}$ represent all possible faults which can happen when implementing the specification. An implementation FSM $T \in \mathfrak{R}$ is called *conforming* if $T \leq S$; otherwise, T is a *non-conforming* implementation. Given the specification FSM S, a *test case* is a finite input sequence of S. As usual, a *test suite* is a finite set of test cases. Given an implementation FSM $T \in \mathfrak{R}$ and an output response β of T to a test case α, the FSM T *passes* the test case if β is in the set of output responses of the specification FSM S to α. Otherwise, the FSM T *fails* the test case. Given a test suite, an Implementation Under Test (IUT) *passes* the test suite if the IUT passes each test case.

In this paper, we generalize the traditional fault model $<S, \leq, \mathcal{R}>$ by adding the relation $\nleftrightarrow$ into the fault model. Formally a fault model becomes a 4-tuple $<S, (\leq, \nleftrightarrow), \mathcal{R}>$. This model indicates that a test suite is complete up to the subset of $\mathcal{R}$ that contains all implementations $T \nleftrightarrow S$. That is a test suite is *complete* w.r.t. to the fault model $<S, (\leq, \nleftrightarrow), \mathcal{R}>$ if each non-reduction T of S such that $T \nleftrightarrow S$ can be detected with this test suite.

Here we note that in FSM-based testing when using a traditional fault model it is usually assumed that each non-conforming implementation can be detected with a complete test suite. However, if the assumption of «all weather conditions» fails when testing a non-deterministic implementation, as happens, for example, in the remote testing, then a non-deterministic implementation cannot be tested up to the reduction or up to the equivalence relation. In this paper, we show that in this case, an implementation can be tested up to the separability relation without relying on «all weather conditions» assumption. As usual, we assume that both specification and implementation FSMs are complete. However, we do not require either the specification FSM or an implementation FSM to be observable.

The relation «not a reduction» contains the separability relation, i.e. for nondeterministic FSMs $\nleftrightarrow \subseteq \nleq$. However, as the following example shows a complete test suite w.r.t. the fault models $<S, \leq, \mathcal{R}>$ is not always complete w.r.t. the fault model $<S, (\leq, \nleftrightarrow), \mathcal{R}>$.

Consider the specification FSM S in Figure 1 with states $\{a,b\}$, inputs $\{x, y\}$ and outputs $\{1,2,3,4\}$. States of S are separated by the input y and both states are deterministically reachable from the initial state a. We use the method for test derivation [21] and obtain a test suite $TS = \{xy, yxy, yyy\}$ w.r.t. the fault model $<S, \leq, \mathcal{R}_2>$ where $\mathcal{R}_2$ contains each complete FSM with up to 2 states over the input alphabet $\{x, y\}$. The test suite TS is complete when the assumption of «all weather conditions» holds. However, if the assumption of «all weather conditions» fails then the implementation FSM T in Figure 1 that is separable with S, i.e., is not a reduction of S, can remain undetected with the test suite TS when each test case is applied at most once, i.e., TS is not complete w.r.t. the fault model $<S, (\leq, \nleftrightarrow), \mathcal{R}_2>$. By direct inspection, one can assure that a shortest sequence that separates the initial states a and 1 of S and T has length four while to each input sequence of length up to three the sets of output responses of S and T intersect.

S	a	b	T	1	2
x	$a/0,1,2,3$	$a/1,2$	x	$1/0$ $2/1$	$1/0,1$
y	$b/1,2$	$a/0$ $b/3$	y	$1/1$ $2/0,2$	$1/3$ $2/0$

Fig. 1. FSMs S and T

The reason is that the r-distinguishability relation between states of the specification FSM cannot be used when deriving tests w.r.t. the separability relation. Given two r-distinguishable states s_1 and s_2 of the specification FSM with an r-distinguishability set W and a state t of an implementation FSM, there exists a sequence $\alpha \in W$ such that $t \not\leq_\alpha s_1$ or $t \not\leq_\alpha s_2$. Therefore, if a test suite has two sequences α_1 and α_2 that take the specification FSM to states s_1 and s_2 appended with the r-distinguishability set W then each implementation FSM that is taken with sequences α_1 and α_2 to the same state t will be detected with such a test suite. Unfortunately, the above property does not hold for the separability relation. Given two separable states s_1 and s_2 in the specification FSM with a separating sequence α, a state t of an implementation FSM can be non-separable with both states s_1 and s_2 w.r.t. the sequence α.

Given the specification FSM S and the fault domain $\mathfrak{R}$, a complete test suite w.r.t. the fault model $<S, (\leq, \nleftrightarrow), \mathfrak{R}>$ can be derived by the explicit enumeration of all machines of the set $\mathfrak{R}$. For each $T \in \mathfrak{R}$ that is separable with S, a separating sequence is derived that is used as a test case to detect a wrong implementation FSM T. The set of all test cases is a complete test suite w.r.t. the fault model $<S, (\leq, \nleftrightarrow), \mathfrak{R}>$. Below we include the algorithm given in [23] for deriving a separating sequence for two complete FSMs.

Algorithm 1. Deriving a separating sequence of two complete FSMs
Input: Complete FSMs $S = \langle S, I, O, h_S, s_1 \rangle$ and $T = \langle T, I, O, h_T, t_1 \rangle$
Output: A shortest separating sequence of FSMs S and T (if it exists)

Step 1. Derive the intersection $S \cap T$. If the intersection is a complete FSM then the FSMs S and T are non-separable. END Algorithm 1.

Step 2. If the intersection $S \cap T$ is a partial FSM, then derive a truncated successor tree of the intersection $S \cap T$. The root of this tree, which is at the 0th level, is the initial state (s_1, t_1) of the intersection; the nodes of the tree are labeled with subsets of states of the intersection. Given already derived j tree levels, $j \geq 0$, a non-leaf (intermediate) node of the jth level labeled with a subset P of states of the intersection, and an input i, there is an outgoing edge from this non-leaf node labeled with i to the node labeled with the subset of the i-successors of states of the subset P. A current node *Current*, at the kth level, $k \geq 0$, labeled with the subset P of states, is claimed as a leaf node if one of the following conditions holds:

Rule 1: There exists an input i such that each state of the set P has no i-successors in the intersection $S \cap T$.
Rule 2: There exists a node at a jth level, $j < k$, labeled with subset R of states with the property: for each state (s', t') of R there exists a state (s, t) of P such that $(s', t') \leq (s, t)$.

Step 3. If none of the paths of the truncated tree derived at Step 2 is terminated using Rule 1 then FSMs S and T are non-separable. END Algorithm 1. Otherwise, if there is a leaf node, *Leaf*, labeled with the subset P of pairs of states such that for some input i, each pair of the set P has no i-successors, then a shortest sequence αi where α

labels the path from the root of the tree to *Leaf*, is a shortest separating sequence of FSMs S and T.

Theorem 1. Given FSMs S and T over input alphabet I and output alphabet O, Algorithm 1 returns a shortest separating sequence of FSMs S and T (if a separating sequence exists).

Proof. In order to separate FSMs S and T we need an input sequence α under which the intersection $S \cap T$ enters the set P of states such that there exists some input i that separates each pair of the set P, i.e., in the intersection, each state of the set P has no successors under input i. If none of the paths of the truncated tree derived at Step 2 of Algorithm 1 is terminated according termination Rule 1, then there is no such an input sequence α, and thus, FSMs S and T are non-separable. If there exists a path of the truncated tree derived at Step 2 of Algorithm 1 that is terminated according termination Rule 1 then the sequence α which labels this path takes the intersection $S \cap T$ to the set P such that there exists an input i that separates each pair of the set P, and thus, αi is a separating sequence of FSMs S and T. Given a current node *Current* labeled with a subset P at k^{th} level, let there exist a node at a jth level, $j < k$, labeled with subset R of states with the property: for each state (s', t') of R there exists a state (s, t) of P such that $(s', t') \leq (s, t)$. In this case, each input sequence that separates each pair of states of the set P also separates each pair of states of the set R. Therefore, each path of the truncated tree traversing the node *Current* can be terminated, since a shorter separating sequence can be derived when traversing the node labeled with the set R. □

As usual, the explicit enumeration of all machines in the fault domain $\mathfrak{R}$ can be applied only to small fault domains. Accordingly, in the following section, we propose a method for deriving a complete test suite w.r.t. the fault model $<S, (\leq, \nleftrightarrow), \mathfrak{R}>$ without the explicit enumeration of the machines in $\mathfrak{R}$. This is done by using a nondeterministic FSM called a *mutation machine* MM in [9], to represent, in a compact way, all possible implementations of S. In this case, the fault domain $\mathfrak{R}$ equals the set $Sub_{nd}(MM)$ of all complete submachines of MM. In Section 5, we extend the method to the case when the upper bound on the number of states of an IUT is given.

4 Deriving a Complete Test Suite w.r.t. Fault Model $<S, (\leq, \nleftrightarrow)$, $Sub_{nd}(MM)>$

According to Algorithm 1, in order to derive a complete test suite w.r.t. the fault model $<S, (\leq, \nleftrightarrow), Sub_{nd}(MM)>$ a truncated successor tree of the intersection $S \cap T$ should be derived for each complete submachine T of the FSM MM. Therefore, given a current node *Current* labeled with a subset P and input i, we should have an edge labeled by i not only to the i-successor of P but to all non-empty subsets of the i-successor. Moreover, we cannot terminate a path comparing the label of its node with labels of another path, since now these paths can belong to different submachines of

MM. The above two observations lead us to a method below when deriving a complete test suite w.r.t. the fault model $<S, (\leq, \nrightarrow), Sub_{nd}(MM)>$.

Algorithm 2. Deriving   a   complete   test   suite   w.r.t.   the   fault   model $<S, (\leq, \nrightarrow), Sub_{nd}(MM)>$

Input: Complete FSMs S and MM

Output: A complete test suite TS w.r.t. the fault model $<S, (\leq, \nrightarrow), Sub_{nd}(MM)>$

Step 1. Derive the intersection $S \cap MM$.

Step 2. Derive a truncated successor tree of the intersection $S \cap MM$. The root of this tree, which is at the 0th level, is the initial state (s_1, m_1) of the intersection; the nodes of the tree are labeled with subsets of states of the intersection. Given already derived j tree levels, $j \geq 0$, a non-leaf (intermediate) node of the jth level labeled with a subset P of states of the intersection, and an input i, there is an outgoing edge from this non-leaf node labeled with i to the node labeled with each subset of the i-successor of the subset P. A current node *Current*, at the kth level, $k \geq 0$, labeled with the subset P of states, is claimed as a leaf node if one of the following conditions holds:

> **Rule 1:** There exists an input i such that each state of the set P has no i-successors in the intersection $S \cap MM$.
>
> **Rule 2:** There exists a node at the path from the root to this node at j^{th} level, $j < k$, labeled with subset R of states with the property: for each state (s', m') of R there exists a state (s, m) of P such that $(s', m') \leq (s, m)$.

Step 3. For each path of the tree terminated using Rule 1, include into TS an input sequence that labels the path appended with an input i such that each state of the set P corresponded to the final node of the path has no i-successors in the intersection $S \cap MM$.

For each path of the tree terminated using Rule 2, include into TS an input sequence that labels the path.

Theorem 2. Given FSMs S and MM over the input alphabet I and output alphabet O, Algorithm 2 returns a complete test suite w.r.t. the fault model $<S, (\leq, \nrightarrow), Sub_{nd}(MM)>$.

Proof. According to Algorithm 1, when deriving a separating sequence for two FSMs S and T we use the truncated tree of $S \cap T$. In our case, each FSM T that should be separated with S (if S and T are separable) is a submachine of MM, i.e., $S \cap T$ is a submachine of $S \cap MM$. In order to get an appropriate truncated subtree for each submachine of FSM MM at Step 2 of Algorithm 1, for each non-leaf node *Current* labeled with a subset P and each input i, we add an outgoing edge to each non-empty subset of the i-successor of P. Thus, for each complete submachine T of MM a truncated tree for separating T and S is a subtree of the tree derived by Algorithm 2.   □

Example. As an application example, consider FSMs S in Figure 1 and MM in Figure 2. We apply Algorithm 2 we obtain the intersection (Figure 3) and the truncated successor tree in Figure 4. Therefore, the set $\{xx, xyx, xyyx, xyyy, yxx, yxyx, yxyy, yyxx, yyxyx, yyxyy, yyy\}$ is a complete test suite w.r.t. the fault model $<S, (\leq, \nrightarrow), Sub_{nd}(MM)>$.

MM	1	2
x	1/0,1 1/2,3 2/2,3	1/2,3 2/2,3
y	1/1,2,3 2/1,2	1/0,2 2/2,3

S∩MM	a1	a2	b1	b2
x	a1/0,1,2,3 a2/2,3	a1/2,3 a2/2,3	a1/1,2 a2/2	a1/2 a2/2
y	b1/1,2 b2/1,2	b1/2 b2/2	b1/3	a1/0 b2/3

Fig. 2. FSMs MM and S∩MM

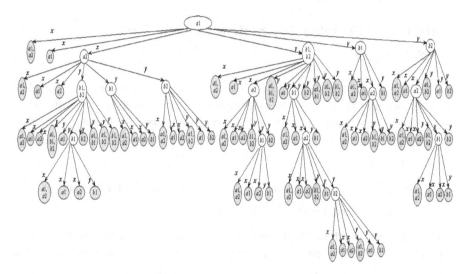

Fig. 3. The truncated successor tree

Sometimes a test suite derived using Algorithm 2 can be shortened by relaxing the conditions of Step 3. For example, given a tail edge $\{sm\} \xrightarrow{i} \{s'm'\}$ of some path terminated using Rule 1 and labeled with an input sequence αi, it can happen that the set of output responses of the intersection $S \cap MM$ to i at state sm coincides with that of the MM at state m. In this case, i is unnecessary for separating new submachines of MM from S and it is enough to include into a test suite the sequence α instead of αi. More analysis is needed for reducing a test suite. This is a part of our future work.

We implemented the above algorithm and performed some experiments with FSMs with small number of states. As our experiments show, the total length of a complete test suite w.r.t. the fault model $<S, (\leq,\nrightarrow), Sub_{nd}(MM)>$ significantly depends on the number of nondeterministic transitions in the specification and mutation machines. Table 1 contains a selected part of conducted experiments for FSMs with 5 states. Each row in the table represents an average test suite length of 100 randomly generated specification FSMs. Each FSM S is a complete nondeterministic FSM with $|S|$ states, $|I|$ inputs, $|O|$ outputs, where for 20 percent of pairs (s, i) there is more than one outgoing transition from state s under input i. Each FSM MM is derived by adding (up to 25 percent) additional transitions to FSM S.

Table 1. A selected part of conducted experiments

$|I|$	$|O|$	Average test suite length
2	2	2432
3	3	15407
3	4	6826

5   Deriving a Complete Test Suite w.r.t. Fault Model $<S, (\leq,\nrightarrow), \mathfrak{R}_m>$

Let $\mathfrak{R}_m$ be the fault domain of a given specification S that contains each complete implementation FSM of S, over the same input and output alphabets of S, with up to a given number m of states. The following theorem can be used for deriving a complete test suite w.r.t. the fault model $<S, (\leq,\nrightarrow), \mathfrak{R}_m>$. This theorem is a corollary of Theorem 2 given in our previous work [23].

Theorem 3. Given the specification FSM S with n states, a test suite $I^{2^{mn-1}}$ is complete w.r.t. the fault model $<S, (\leq,\nrightarrow), \mathfrak{R}_m>$.

In the following, based on the idea of counting states of the specification FSM when deriving a complete test suite w.r.t. the fault model $<S, \leq, \mathfrak{R}_m>$) (a SC-method [15]), we propose a test derivation method for reducing the test suite $I^{2^{mn-1}}$. In this case, unlike the above method given in Algorithm 1, we derive a truncated tree using only the specification FSM S. Before terminating a path at a node labeled with a subset K of states of the specification FSM, we make sure that for each complete FSM T with up to m states the path traverses all possible subsets of the Cartesian product $K \times T$ in the intersection $S \cap T$, i.e., the path should traverse not less than $2^{|K| \cdot m}$ subsets of K. If K contains the initial state, then the initial state of the intersection can be excluded from any subset that labels a non-root node of the tree, i.e., the path should traverse not less than $(2^{|K| \cdot m - 1} + 1)$ subsets of K.

Algorithm 3. Deriving a complete test suite w.r.t. the fault model $<S, (\leq,\nrightarrow), \mathfrak{R}_m>$

Input: Complete FSM S and an upper bound m on the number of states of any FSM implementation of S

Output: A complete test suite TS w.r.t. the fault model $<S, (\leq, \nleftarrow), \mathfrak{R}_m>$

Step 1. Derive a truncated successor tree of the specification FSM S. The root of this tree, which is at the 0th level, is the initial state s_0 of the FSM S; the nodes of the tree are labeled with subsets of states of the FSM S. Given already derived j levels of the tree, $j \geq 0$, a non-leaf (intermediate) node of the jth level labeled with a subset K of states of the FSM S, and an input i, there is an outgoing edge from this non-leaf node labeled with i to the node labeled with the i-successor of the subset K. A current node *Current*, at the kth level, $k \geq 0$, labeled with the subset K of states of S is claimed as a leaf node if the path from the root to this node has $2^{|K| \cdot m}$ nodes labeled with subsets of K and the initial state s_0 is not in K. If the initial state is in K then the node *Current* is claimed as a leaf node if the path from the root to this node traverses $(2^{|K|m - 1} + 1)$ nodes labeled with subsets of K.

Step 2. Include into TS each input sequence which labels the path from the root to a leaf node in the above truncated tree.

Theorem 4. Given the specification FSM S over the input alphabet I and an integer m, Algorithm 3 returns a complete test suite w.r.t. the fault model $<S, (\leq, \nleftarrow), \mathfrak{R}_m>$.

Proof. Given an implementation FSM T, consider the truncated tree $Tree_S$ of the specification FSM S and the truncated tree $Tree_{S \cap T}$ of $S \cap T$. Given a path in the $Tree_S$ to a node labeled with a subset K of states of S, the corresponding path in the tree $Tree_{S \cap T}$ leads to a node that is labeled with a subset P of states of the intersection $S \cap T$ such that the first item of each pair of P is in the set K. The number of such non-empty subsets is $2^{|K| \cdot m} - 1$. Thus, when a path of the $Tree_S$ traverses $2^{|K| \cdot m}$ nodes labeled with subsets of K the corresponding path in the tree $Tree_{S \cap T}$ traverses two nodes labeled with the same subset and can be terminated, according to Algorithm 1. When the initial state of the specification FSM S is in the set K then each subset traversed by the corresponding path in the tree $Tree_{S \cap T}$ does not contain the initial state, i.e., the number of such subsets is $2^{(|K|-1) \cdot m} - 1$. Respectively, when K contains the initial state a path can be terminated if it traverses $(2^{(|K|-1) \cdot m} + 1)$ nodes labeled with subsets of K (counting the initial state of the specification that labels the root of the $Tree_S$), since the corresponding path in the tree $Tree_{S \cap T}$ traverses two nodes labeled with the same subset or with a subset that contains the initial state of the intersection. $\square$

As an example, we consider the specification FSM S in Figure 1 (left hand) and derive a complete test suite w.r.t. the fault model $<S, (\leq, \nleftarrow), \mathfrak{R}_2>$. At Step 2 a current node labeled with the state a is claimed as a leaf node if the path from the root to this node traverses $(2^{m-1} + 1) = 3$ nodes labeled with a. A current node labeled with the state b is claimed as a leaf node if the path from the root to this node traverses $2^m = 4$ nodes labeled with b. Finally, a current node labeled with the subset $\{a, b\}$ is claimed as a leaf node if the path from the root to this node traverses $(2^{2m-1} + 1) = 9$ nodes labeled with a, b and $\{a, b\}$. A complete test suite has the total length 277 (Figure 4).

Here we notice that Algorithm 3 does not return a shortest test suite. Consider, for example, a test case *xyyyyyyy* of the above test suite and the corresponding path of the truncated successor tree $Tree_S$: $a_x a_y b_y \{a,b\}_y \{a,b\}_y \{a,b\}_y \{a,b\}_y \{a,b\}_y \{a,b\}$. By direct inspection, one can assure that if an implementation FSM has states 1 and 2 then the corresponding path in the truncated tree $Tree_{S \cap T}$ will be already terminated after $\{a1\}_x \{a2\}_y \{b1,b2\}_y \{b1\}_y \{b2\}_y \{a,b\}_y$. By using such analyzing, a complete test suite with total length 89 can be derived for the fault model $<S, (\leq, \not\sim), \Re_2>$. Thus, more analysis of termination rules is needed for reducing the length of obtained test suites. Here we recall (Section 3) that a complete test suite of length 11 is derived using the SC-method w.r.t. the fault model $<S, \leq, \Re_2>$ under the assumption of «all weather conditions». According to this condition, each sequence of the test suite should be applied at least eight times to a given IUT since, on average, there are eight different output responses to a test case. Thus, the total length of a test suite complete w.r.t. the fault model $<S, \leq, \Re_2>$ is around 100 and this test suite still does not guarantee the detection of all implementations with up to 2 states, that are separable from the given specification FSM if we lack the necessary controllability and/or observability over an IUT.

However, more rigorous analysis is necessary in order to refine termination rules, since in general, the exponential bound on the length of a test case cannot be reduced [23].

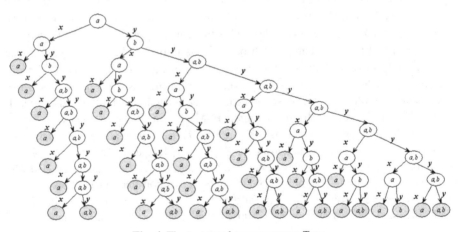

Fig. 4. The truncated successor tree $Tree_S$

6   Conclusion and Future Research Work

In this paper, we have proposed a method for the test derivation against nondeterministic FSMs with respect to the separability relation. This relation can be used without assuming that an implementation under test should satisfy the «all weather conditions» assumption. Refined notions of a fault model and a complete test suite are given. A test suite is called complete up to the separability relation if it detects every implementation that is separable from (i.e., is not a reduction of) the given FSM specification. This complete fault coverage is guaranteed if each test case of the test suite is applied to an IUT only once. The test suite can also detect some implementations that are not reductions of the specification FSM but are non-separable from the specification FSM.

Two algorithms are presented for complete test suite derivation with respect to the separability relation. The first algorithm can be applied when the set of possible implementations is the set of all complete nondeterministic submachines of a given mutation machine. The second algorithm is applied when the upper bound on the number of states of an IUT is known. The proposed algorithms do not return shortest test suites and more work is needed for reducing the length of obtained test suites. Unfortunately, the exponential upper bound on the length of a test case cannot be reduced [23], except of the case when the specification is deterministic or we consider only deterministic implementations. For simplicity of presentation, in this paper, we assume that the specification FSM is complete but the proposed algorithms do not rely on this assumption and thus, can be extended to partial specification FSMs.

References

1. Kloosterman, H.: Test derivation from non-deterministic finite state machines. In: Proceedings of the IFIP Fifth International Workshop on Protocol Test Systems, Canada, pp. 297–308 (1992)
2. Tripathy, P., Naik, K.: Generation of adaptive test cases from nondeterministic Finite State models. IFIP Trans. C: Commun. System C-11, 309–320 (1993)
3. Luo, G., Petrenko, A., Bochmann, G.v.: Selecting test sequences for partially specified nondeterministic finite state machines. In: Proc. 7th International Workshop on Protocol Test Systems (1994)
4. Luo, G., Bochmann, G.v., Petrenko, A.: Test selection based on communicating non-deterministic finite-state machines using a generalized Wp-method. IEEE Transactions on Software Engineering 20(2), 149–161 (1994)
5. Alur, R., Courcoubetis, C., Yannakakis, M.: Distinguishing tests for nondeterministic and probabilistic machines. In: Proc. the 27th ACM Symposium on Theory of Computing, pp. 363–372 (1995)
6. Petrenko, A., Yevtushenko, N., Bochmann, G.v.: Testing deterministic implementations from their nondeterministic specifications. In: Proc. 9th International Workshop on Protocol Test Systems, pp. 125–140 (1996)
7. Boroday, S.Y.: Distinguishing Tests for Non-Deterministic Finite State Machines. In: Proc. IFIP TC6 11th International Workshop on Testing of Communicating Systems, pp. 101–107 (1998)
8. Hierons, R.M.: Adaptive testing of a deterministic implementation against a nondeterministic finite state machine. The Computer Journal 41(5), 349–355 (1998)
9. Koufareva, I., Evtushenko, N., Petrenko, A.: Design of tests for nondeterministic machines with respect to reduction. Automatic Control and Computer Sciences, USA, vol. 3 (1998)
10. Hierons, R.M.: Using candidates to test a deterministic implementation against a non-deterministic finite state machine. The. Computer Journal 46(3), 307–318 (2003)

11. Hierons, R.M.: Testing from a non-deterministic finite state machine using adaptive state counting. IEEE Transactions on Computers 53(10), 1330–1342 (2004)
12. Hierons, R.M., Ural, H.: Concerning the ordering of adaptive test sequences. In: König, H., Heiner, M., Wolisz, A. (eds.) FORTE 2003. LNCS, vol. 2767, pp. 289–302. Springer, Heidelberg (2003)
13. Hwang, I., Kim, T., Hong, S., Lee, J.: Test selection for a nondeterministic FSM. Computer Communications 24, 1213–1223 (2001)
14. Zhang, F., Cheung, T.: Optimal transfer trees and distinguishing trees for testing observable nondeterministic finite-state machines. IEEE Transactions on Software Engineering 29(1), 1–14 (2003)
15. Petrenko, A., Yevtushenko, N.: Conformance tests as checking experiments for partial nondeterministic FSM. In: Proc. 5th International Workshop on Formal Approaches to Testing of Software (2005)
16. Miller, R., Chen, D., Lee, D., Hao, R.: Coping with nondeterminism in network protocol testing. In: Proceedings of the 17th IFIP International Conference on Testing of Communicating Systems, USA (2005)
17. Tanenbaum, A.S.: Computer Networks. Prentice-Hall, NJ (1996)
18. Petrenko, A., Yevtushenko, N.: Testing from partial deterministic FSM specifications. IEEE Trans. on Computers 54(9), 1154–1165 (2005)
19. Milner, R. (ed.): A Calculus of Communication Systems. LNCS, vol. 92. Springer, Heidelberg (1980)
20. Spitsyna, N., Trenkaev, V., El-Fakih, K., Yevtushenko, N.: FSM interoperability testing, Wor. In: Progress: 23rd International Conference on Formal Techniques for Networked and Distributed Systems (2003)
21. Spitsyna, N.: FSM-based test suite derivation strategies for discrete event systems. Ph.D. Thesis, Tomsk State University, pp. 1–158 (2005)
22. Starke, P.: Abstract automata, pp. 3–419. American Elsevier, New York (1972)
23. Spitsyna, N., El-Fakih, K., Yevtushenko, N.: Studying the Separability Relation between Finite State Machines. Submitted to Software Testing, Verification and Reliability (2006)

Learning and Integration of Parameterized Components Through Testing

Muzammil Shahbaz[1], Keqin Li[2], and Roland Groz[2]

[1] France Telecom R&D
Meylan, France
muhammad.muzammilshahbaz@orange-ftgroup.com
[2] LIG, Computer Science Lab
Grenoble Universités, France
{Keqin.Li,Roland.Groz}@imag.fr

Abstract. We investigate the use of parameterized state machine models to drive integration testing, in the case where the models of components are not available beforehand. Therefore, observations from tests are used to learn partial models of components, from which further tests can be derived for integration. We have extended previous algorithms to the case of finite state models with predicates on input parameters and observable non-determinism. We also propose a new strategy where integration tests can be derived from the data collected during the learning process. Our work typically addresses the problem of assembling telecommunication services from black box COTS.

1 Introduction

Model based testing has gained momentum in many industrial fields, in particular in the domain of testing complex systems, e.g., telecom services, which are composed of various components developed independently. It is not uncommon for these components to be collected from different sources as COTS (Commercial-off-the-shelf), their formal models are not always available and no detailed technical corpora is provided with the components. Therefore, engineers find difficulty in providing a required system integration if they have limited knowledge of the behaviors of the components, which they use in the system.

To address this problem, we propose to generate formal models directly from the components through testing. These models are generated as state machine models so that rigorous techniques of model based integration testing could readily be applied. This provides us room to investigate methods of state machine inference from black box components, i.e., the components whose internal structure is unknown. Among various such methods, Angluin's algorithm [1] is well-known that learns a deterministic automata in a polynomial time. This work has yielded positive results in applied research [14], [7] etc., where real problems were put under case-studies. However there remained less explicit emphasis in these works on learning expressive models.

A. Petrenko et al. (Eds.): TestCom/FATES 2007, LNCS 4581, pp. 319–334, 2007.
© IFIP- International Federation for Information Processing 2007

In our particular case of integration of a variety of components, we have observed that the nature of components integration elicit potential interoperability problems due to exchange of data values from arbitrarily complex domains. In this case, learning DFA models for such components would be inadequate and impractical due to the chance of state-explosion and loss of genericity of the model. Therefore, we need to advance from simple state-machine inference to the inference of more expressive models that can maintain the fine granularity of complex systems, i.e., parametric details and also some notion of nondeterminism. Also, as the size of input data is directly proportional to the testing effort, there is a good argument to model expressive forms that can detail the intended behaviors of the component in a compact form and can be learnt through less number of test cases. We have proposed techniques based on Angluin's algorithm to adapt it for more expressive models than DFA, starting from Mealy machines [9] to simple parameterized models [10].

In this paper, we enrich our model to incorporate parameterized predicates on transitions with observable nondeterminism. This model is more expressive compared to the models proposed in the previous works of automata inference [1], [7], [9], [10], [2] in terms of parameterized inputs/outputs, infinite domain of parameter values, predicates on input parameters and observable nondeterminism when interacting with input parameter values. Compared to usual EFSM models [13], [12], we stop short of including variables in the model, because when we learn a black box, we cannot distinguish in its internal structure what would be encoded as (control) state and what would be encoded in variables. All state information in our model is encoded in the state machine structure.

We propose an algorithm to infer such parameterized models based on Angluin's algorithm. We also have significantly improved the algorithm in two ways. The basic algorithm and all its adaptations stated above check for certain concepts in order to make a conjecture of the model. Inspired by [14], we reduced one of these concepts, called *consistency* and hence reduced the number of test cases needed to perform this concept. Furthermore, the algorithm assumes an oracle that provides a counterexample when the conjecture is wrong. In the context of industrial applications where this oracle assumption is quite unrealistic, we propose a technique to find potential counterexamples from the models taking advantage of our integration testing strategy. The counterexamples are provided back to the learning procedure to refine the learned model, thus making it an iterative process [9]. We also consider former approaches, e.g., property-based testing [11] and scenario-based testing [10] and propose a new integration testing technique which is illustrated with the help of an example of integrating two parameterized components. The organization of the paper is as follows. The formal definition of the parameterized model is given in section 2 and its learning algorithm is described in section 3. The integration testing strategy and related discussion is covered in section 4 and finally section 5 concludes the paper.

2 Parameterized Model

A *Parameterized Finite State Machine (PFSM)* M is a tuple $M = (Q, I, O, D_I, D_O, \Gamma, q_0)$, where

- Q is a finite set of states
- I is a finite set of input symbols
- O is a finite set of output symbols
- D_I is a set of input parameter values
- D_O is a set of output parameter values
- q_0 is an initial state
- Γ is a set of transitions

A transition $t \in \Gamma$ is described as: $t = (q, q', i, o, p, f)$, where $q \in Q$ is a source state, $q' \in Q$ is a target state, $i \in I$ is an input symbol, $o \in O$ is an output symbol, $p \subseteq D_I$ is a predicate on input parameter values and $f : p \longrightarrow D_O$ is an output parameter function. We consider that the model is restricted with the following three properties.

Property 1 (Input Enabled). The model is *input enabled*, i.e., $\forall q \in Q, \forall i \in I$ and $\forall x \in D_I, \exists t \in \Gamma$ such that $t = (q, q', i, o, p, f)$, in which $x \in p$.

The machine can be made input enabled by adding loop back transitions on a state for all those inputs (and associated predicate for parameter values) which are not acceptable for that state. Such transitions contain a special symbol Ω in O. Similarly, there exists transitions which do not take input parameter values into account. Such transitions contain a special symbol $\perp$ with the input symbol that expresses the absence of parameter value. For the sake of simplicity, we do not write this symbol while modeling a problem with PFSM.

Property 2 (Input Deterministic). The model is *input deterministic*, i.e., for $t_1, t_2 \in \Gamma$ such that $t_1 = (q_1, q_1', i_1, o_1, p_1, f_1)$, $t_2 = (q_2, q_2', i_2, o_2, p_2, f_2)$ and $t_1 \neq t_2$, if $q_1 = q_2 \wedge i_1 = i_2$ then $p_1 \cap p_2 = \phi$.

Property 3 (Observable). The model is *observable*, i.e., for $t_1, t_2 \in \Gamma$ such that $t_1 = (q_1, q_1', i_1, o_1, p_1, f_1)$, $t_2 = (q_2, q_2', i_2, o_2, p_2, f_2)$ and $t_1 \neq t_2$, if $q_1 = q_2 \wedge i_1 = i_2$ then $o_1 \neq o_2$.

Property 3 ensures that two transitions having same source state and same input symbol would generate different output symbols. This helps us determining the target states that are possibly different for each transition in the learning algorithm.

When M is in state $q \in Q$ and receives an input $i \in I$ along with the parameter value $x \in D_I$, then the target state q', the output o and the output parameter value function f are determined by the functions δ, λ and σ respectively, which are described as follows:

- $\delta : Q \times I \times D_I \longrightarrow Q$ is a target state function
- $\lambda : Q \times I \times D_I \longrightarrow O$ is an output function
- $\sigma : Q \times I \longrightarrow D_O{}^{D_I}$ is an output parameter function. $D_O{}^{D_I}$ is the set of all functions from D_I to D_O.

The properties 1 and 2 ensure that δ and λ are mappings. For an input symbol sequence $\omega = i_1, \ldots, i_k$ and an input parameter value sequence $\alpha = x_1, \ldots, x_k$, where each $i_j \in I, x_j \in D_I, 1 \leq j \leq k$, we define a parameterized input sequence, i.e., the association of ω and α as $\omega \otimes \alpha = i_1(x_1), \ldots, i_k(x_k)$, where each x_j is associated with i_j and $|\omega| = |\alpha|$. The association of output symbol sequence and output parameter value sequence is defined analogously. Then, for the state $q_1 \in Q$, when applying a parameterized input sequence $\omega \otimes \alpha$, M moves successively from q_1 to the states $q_{j+1} = \delta(q_j, i_j, x_j), \forall 1 \leq j \leq k$. We extend the functions from input symbols to parameterized input sequences as $\delta(q_1, \omega, \alpha) = q_{k+1}$ to denote the final state q_{k+1} and $\lambda(q_1, \omega, \alpha) = o_1(y_1), \ldots, o_k(y_k)$, where each $o_j = \lambda(q_j, i_j, x_j), y_j = \sigma(q_j, i_j)(x_j), \forall 1 \leq j \leq k$, to denote the complete parameterized output sequence, when applying $\omega \otimes \alpha$ on q_1.

An example of PFSM model is given in Figure 1, in which $Q = \{q_0, q_1, q_2, q_3, q_4, q_5\}$, $I = \{a, u\}$, $O = \{s, t\}$, $D_I = D_O = \mathbb{Z}$, the set of integers.

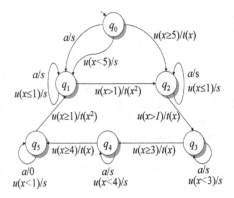

Fig. 1. Example of PFSM Model

3   Algorithm

Assume an unknown PFSM $M = (Q, I, O, D_I, D_O, \Gamma, q_0)$ with known input symbols I and input parameter domain D_I can be used to model a component C. For any parameterized input sequence or a test case $\omega \otimes \alpha, (\omega \in I^*, \alpha \in D_I^*)$ for a component, we assume that $\lambda(q_0, \omega, \alpha)$ can be known from testing. We also assume that C can be reset to its initial state before each test. The key part of the learning algorithm is using observation table. We define the structure of the table and related definitions in the section below and then the algorithm in the subsequent section.

3.1   Observation Table

The observation table is used to generate test cases for an unknown component, to organize the result of each test case and finally to make a PFSM conjecture when certain properties on the table are satisfied. The rows and columns of the table are labelled by input strings which are associated with the input parameter values in order to construct test cases. The result of a test case is organized in the cells of the table in the form of a pair of input parameter value sequence and parameterized output. After the table conforms to the properties, a conjecture is made where some rows of the table are regarded as states and transitions are derived from the observations recorded in the table. We shall describe the basic structure and properties of the table in this section and rest of the explanation regarding construction of test cases, organization of outputs and making a conjecture out of the table will be explained in the next section.

Structure. Let $\mathcal{U} = \{\omega \otimes \alpha | \omega \in I^+, \alpha \in D_I^+\} \cup \{\omega | \omega \in I^*\}$ be the set of parameterized input sequences and input symbol sequences. We define $IS(u), u \in \mathcal{U}$, an input symbol sequence from u such that if $u = \omega$ or $u = \omega \otimes \alpha, \omega \in I^*, \alpha \in D_I^*$, then $IS(u) = \omega$. Also, $pref(\gamma)^k$ is the prefix of some sequence $\gamma \in I^* \cup O^* \cup D_I^* \cup D_O^*$ of length $k \leq |\gamma|$. For example, for $\gamma = i_1, ..., i_n$, where every $i_j \in I, 1 \leq j \leq n, pref(\gamma)^k = i_1, ..., i_k, 1 \leq k \leq n$. Similarly, $suff(\gamma)^k$ is the suffix of γ of length $k \leq |\gamma|$. Let $\mathcal{P} = \{(\alpha, \varpi \otimes \beta) | \alpha \in D_I^+, \varpi \in O^+, \beta \in D_O^+\}$ be the set of the pair of input parameter value sequence and parameterized output sequence.

Table 1. Example of an Observation Table

		E	
		a	u
S	ϵ	$(\perp, s \otimes \perp)$	$(1, s \otimes \perp), (5, t \otimes 5)$
R	a	$((\perp, \perp), s \otimes \perp)$	$((\perp, 1), s \otimes \perp), ((\perp, 5), t \otimes 25)$
	u	$((1, \perp), s \otimes \perp)$	$((5, 1), s \otimes \perp), ((1, 5), t \otimes 25)$

An observation table is denoted as (S, R, E, T). S and R are nonempty finite sets of input strings and make the rows of the table. S *is used to identify potential states in the conjecture* and R is used to satisfy properties on the table. In PFSM, states are not only determined by input symbol sequence but also by parameter value sequence, thus formally, $S \subseteq \mathcal{U}$ is a set of input symbol sequences and parameterized input sequences, and $R \subseteq \mathcal{U}$ extends the rows such that for all $r \in R$, there exists $s \in S, e \in E$ and $IS(r) = IS(s) \cdot e$. Whenever, a sequence s is added to S, R is extended in the following way: i) if $s = \omega, \omega \in I^*$ an input symbol sequence, then R will be extended by $\omega \cdot i$, for all $i \in I$. ii) if $s = \omega \otimes \alpha, \omega \in I^+, \alpha \in D_I^+$ a parameterized input sequence, then R will be extended by $\omega \cdot i \otimes \alpha \cdot x$, for all $i \in I$, where x is selected from D_I. The selection policy for parameter values is up to the system's specific requirements, however a general idea is given in section 4.

$E \subseteq I^+$ is a nonempty finite set of input symbol sequences that make the columns of the table and *separate the different states of the conjecture*. The elements of $(S \cup R) \times E$ are used to construct test cases in the algorithm which are associated with parameter value sequences from D_I^+, and their results (observations) are organized in the table with the help of a function T mapping from $(S \cup R) \times E$ to $2^{\mathcal{P}}$. For example, a parameterized output of a test case derived from $s \in S \cup R$, $e \in E$ and associated with parameter value sequence $\alpha \in D_I{}^+$ will be organized in $T(s, e)$ in the form of a pair of input parameter value sequence and parameterized output sequence, i.e., $(\alpha, \varpi \otimes \beta), \alpha \in D_I{}^+, \varpi \in O^+, \beta \in D_O{}^+$. The observations from $T(s, e)$ are used to *identify potential transitions in the conjecture and label them with input/output and parameter values*. Table 1 is an example of an observation table, in which S contains only one row ϵ and R contains two rows a and u, whereas E contains two columns a and u respectively.

Properties. Each test case driven from $s \in S \cup R, e \in E$ may generate different parameterized output sequences depending upon the selection of different input parameter value sequences from $D_I{}^+$. Thus, there may exist $(\alpha_1, \varpi_1 \otimes \beta_1), (\alpha_2, \varpi_2 \otimes \beta_2) \in T(s, e)$ such that $\varpi_1 \neq \varpi_2$, i.e, $T(s, e)$ contains pairs in which the output sequences are different. Let $\eta(T(s, e))$ be the number of different output sequences contained by $T(s, e)$, then we can divide $T(s, e)$ into $\eta(T(s, e))$ *distinguishing subsets*, i.e., $T(s, e) = \bigcup\limits_{k=1}^{\eta(T(s,e))} d_k(s, e)$, where in each $d_k(s, e) = \{(\alpha_1{}^{(k)}, \varpi_1{}^{(k)} \otimes \beta_1{}^{(k)}), \dots, (\alpha_m{}^{(k)}, \varpi_m{}^{(k)} \otimes \beta_m{}^{(k)})\} \subseteq T(s, e), m = |d_k(s, e)|$, $\varpi_1 = \dots = \varpi_m$, the output sequences are same. Let $OS(d_k(s, e)) = \varpi_1{}^{(k)} = \dots = \varpi_m{}^{(k)}$ be the output sequence from $d_k(s, e)$, then for any $d_1(s, e), d_2(s, e) \subset T(s, e), OS(d_1(s, e)) \neq OS(d_2(s, e))$ and $d_1(s, e) \cap d_2(s, e) = \emptyset$. For every $d_k(s, e)$, we define $\rho(d_k(s, e)) = \{\alpha_1{}^{(k)}, \dots, \alpha_m{}^{(k)}\}$, a set of *distinguishing parameter value sequence* from $d_k(s, e)$, and $PS(d_k(s, e)) = \{(suff(\alpha_1{}^{(k)})^{|e|}, \beta_1{}^{(k)}), \dots, (suff(\alpha_m{}^{(k)})^{|e|}, \beta_m{}^{(k)})\}$, the set of pairs of i/o parameter value sequences from $d_k(s, e)$.

Since $T(s, e), s \in S \cup R, e \in E$ represents a possible transition in the conjecture, if $T(s, e)$ contains many distinguishing subsets then each subset may represent a different transition. Therefore, we call such s a disputed row. Formally, $s \in S$ is *disputed* iff for any $e \in E$, $\eta(T(s, e)) > 1$, i.e., $T(s, e)$ contains more than one *distinguishing subsets*. The table must contain additional rows to treat disputed rows. A disputed row s is *treated* iff for every distinguishing subset $d_k(s, e) \subset T(s, e), 1 \leq k \leq \eta(T(s, e))$, there exists $t \in S \cup R$ such that $t = IS(s) \cdot e \otimes \alpha, \alpha \in \rho(d_k(s \cdot e))$. The table is called *dispute − free* iff all the disputed rows $s \in S$ are *treated*.

For any $s_1, s_2 \in S \cup R$, s_1 and s_2 are comparable with the help of the following definitions.

- s_1 and s_2 are *compatible*, denoted by $s_1 \equiv s_2$, iff $\forall e \in E, \forall(\alpha_1, \varpi_1 \otimes \beta_1) \in T(s_1, e), \forall(\alpha_2, \varpi_2 \otimes \beta_2) \in T(s_2, e)$, if $suff(\alpha_1)^{|e|} = suff(\alpha_2)^{|e|}$, then $\varpi_1 \otimes \beta_1 = \varpi_2 \otimes \beta_2$. This means that common input parameters produce the same output parameters.
- s_1 and s_2 are *balanced*, denoted by $s_1 \leftrightarrow s_2$, iff $\forall e \in E, \forall(\alpha_1, \varpi_1 \otimes \beta_1) \in T(s_1, e), \exists(\alpha_2, \varpi_2 \otimes \beta_2) \in T(s_2, e)$ such that $suff(\alpha_1)^{|e|} = suff(\alpha_2)^{|e|}$ and $\forall(\alpha_2, \varpi_2 \otimes \beta_2) \in T(s_2, e), \exists(\alpha_1, \varpi_1 \otimes \beta_1) \in T(s_1, e)$ such that $suff(\alpha_2)^{|e|} =$

$suff(\alpha_1)^{|e|}$. This means any input parameter combination on one has been tested on the other.

– s_1 and s_2 are *equivalent*, denoted by $s_1 \cong s_2$, iff $s_1 \leftrightarrow s_2$ and $s_1 \equiv s_2$, i.e., s_1 and s_2 are *balanced* and they remain *compatible*.

A table is *balanced* iff for every $s, t \in S \cup R$ such that $s \equiv t$, $s \leftrightarrow t$. The table is called *closed* iff for each $t \in R$, there exists $s \in S$ such that $s \cong t$. A *closed* table makes sure that no row in R is different from the rows in S that gives out the potential states of the conjecture.

3.2 Algorithm

The algorithm starts by initializing (S, R, E, T) with $E = I$ and $S = R = \emptyset$, i.e., each input symbol makes one column and there are no rows initially. The first step is to add ϵ to S, where ϵ is an empty string. Thus, R will be extended by adding $\epsilon \cdot i$, for all $i \in I$. Table 1 shows the extensions of S, R and E, where $I = \{a, u\}$.

The test cases are constructed by the elements of $(S \cup R) \times E$. Since $S \cup R$ contains the input symbol sequences as well as the parameterized input sequences, the test cases in each case are constructed in the following way:

i) if $s = \omega \in S \cup R$ an input symbol sequence and $e \in E$, then a test case is constructed as $\omega \cdot e \otimes \alpha_1 \cdot \alpha_2$, where α_1 and α_2 are selected from $D_I{}^*$ such that $|\omega| = |\alpha_1|$ and $|e| = |\alpha_2|$.

ii) If $s = (\omega \otimes \alpha_1) \in S \cup R$ a parameterized input sequence and $e \in E$, then a test case is constructed as $\omega \cdot e \otimes \alpha_1 \cdot \alpha_2$, where α_2 will be selected from $D_I{}^+$ such that $|e| = |\alpha_2|$.

The result of each test case is organized in the table by just filling the cells with output sequences, and does not lead to the extension of rows or columns. Let $\omega \otimes \alpha$ be a test case, where $\omega \in I^+, \alpha \in D_I{}^+$, generating a parameterized output sequence $\lambda(q_0, \omega, \alpha) = \varpi \otimes \beta, \varpi \in O^+, \beta \in D_O{}^+$, then the table will be filled as follows:

i) if there exists $s = \omega_1 \in S \cup R, e = \omega_2 \in E$ such that $\omega_1 \cdot \omega_2$ is a prefix of ω or

ii) if there exists $s = \omega_1 \otimes \alpha_1 \in S \cup R$ and $e = \omega_2 \in E$ such that $\omega_1 \cdot \omega_2$ is a prefix of ω and α_1 is a prefix of α,

then there is a prefix $\alpha_p = pref(\alpha)^{|\omega_1 \cdot \omega_2|}, \beta_p = pref(\beta)^{|\omega_1 \cdot \omega_2|}, \varpi_p = pref(\varpi)^{|\omega_1 \cdot \omega_2|}$ and $T(s, e)$ will be appended by $(\alpha_p, \varpi' \otimes \beta')$, where $\varpi' = suff(\varpi_p)^{|\omega_2|}, \beta = suff(\beta_p)^{|\omega_2|}$.

The table is made *balanced* after every test case performed. Whenever it is not *balanced*, find $s, t \in S \cup R, e \in E, (\alpha_1, \varpi_1 \otimes \beta_1) \in T(s, e)$ such that $s \equiv t$ and there does not exist $(\alpha_2, \varpi_2 \otimes \beta_2) \in T(t, e)$ where $suff(\alpha_1)^{|e|} = suff(\alpha_2)^{|e|}$, then construct test case $IS(t) \cdot e \otimes pref(\alpha)^{|\alpha| - |e|} \cdot suff(\alpha_1)^{|e|}$ where α is selected from $\rho(d_k(t, e))$, for any $d_k(t, e) \subseteq T(t, e), 1 \leq k \leq \eta(T(s, e))$.

The table is made *dispute − free* after balancing. Let $s \in S$ be *disputed* then find $e \in E$ such that $\eta(T(t, e)) > 1$. Then, for every distinguishing subset $d_k(t, e) \subset T(t, e), 1 \leq k \leq \eta(T(t, e))$, add $IS(s) \cdot e \otimes \alpha$ to R where α is selected from $\rho(d_k(t, e))$. Remove the original row $s \cdot e \in S \cup R$ if it exists. Construct additional test cases for the missing elements of the table.

When the table is made *balanced* and *dispute* − *free*, it is made *closed*. Whenever it is not *closed*, find $t \in R$ such that $s \not\cong t, \forall s \in S$ and move t to S and extend R accordingly. Construct additional test cases for the missing elements of the table.

When table is *balanced*, *dispute* − *free* and *closed*, a PFSM conjecture M' is made from the table in the following way:

- Each $s \in S$ is a state of the conjecture
- $\epsilon \in S$ is the initial state

    For each $s \in S, i \in I$, there exists $\eta(T(s,i))$ transitions. Thus, each distinguishing subset $d_k(s,i) \subseteq T(s,i), 1 \le k \le \eta(T(s,i))$, defines one transition $\{s, s', i, o, p, f\}$, in which $p = \{suff(\alpha)^1, \forall \alpha \in \rho(d_k(s,i))\}$, $f = \sigma(s,i) = PS(d_k(s,i))$ and s', i are determined by $\delta(s,i,x), \lambda(s,i,x), \forall x \in p$, resp., in the following way:
- $\delta(s,i,x) = t \in S | t \cong (IS(s) \cdot i \otimes \alpha) \in S \cup R, \alpha \in (\rho(d_k(s,i)))^*$
- $\lambda(s,i,x) = OS(d_k(s,i))$

The termination of the algorithm is guaranteed by the finite space of states and transitions of the black box component modeled as PFSM. The operations which keep the algorithm extending the table are two, i.e., disputed row treatment and making the table closed.

A row is disputed if a row (or state) has more than one outputs (or possible transitions) for the same input symbol but for different set of parameter values. If a state in the actual component has m different transitions for an input symbol and a parameter value for each transition has been tested during the process, then there will be at most m rows added in the table for such state and input symbol.

A table is not closed when a row r in R is not equivalent to any row in S. Then by definition, r will be moved to S and will represent a state of the conjecture. If there are n states in the actual component, then there will be at most $n - 1$ moves from R to S, since there is initially one row in S and there cannot be more than n.

As to balancing the table, it is nothing more than recording output sequences in the existing table for those input parameter values that are not recorded previously. The number of test cases required for balancing the table is calculated as follows. Let $m_{r,e}$ is the number of different input parameter values recorded in $T(r,e), r \in S \cup R, e \in E$, and n_e is the number of different input parameter values recorded in $T(s,e), \forall s \in S \cup R$. Then, the number of test cases required for balancing each $T(r,e)$ is $n_e - m_{r,e}$.

3.3 Illustration

We illustrate the learning algorithm of PFSM model on the example given in Figure 1. The summary of the algorithm is given below.

Input: I, D_I
Output: Conjecture M'
begin

Initialize (S, R, E, T) by $E = I$, $S = \epsilon$, $R = \epsilon \cdot i, \forall i \in I$;

Construct the test cases from $(S \cup R) \times E$;

Organize result in the table accordingly ;

while *table is not balanced* **or** *not dispute $-$ free* **or** *not closed* **do**

Make the table *balanced* such that for every $s, t \in S \cup R | s \equiv t$,
$s \leftrightarrow t$;

Make the table *dispute $-$ free* such that for all $s \in S, e \in E$, where
$\eta(T(s, e)) > 1$, s is treated ;

Make the table *closed* such that for every $t \in R$, there exists $s \in S$
such that $s \cong t$;

end

Make a conjecture M' from the table.

end

Algorithm 1. Summary of the Learning Algorithm

We start by initializing (S, R, E, T) with the input symbols from $I = \{a, u\}$ and construct test cases to fill the table, shown in Table 1. In the test cases, we associate parameter values 1 and 5 and balance the table accordingly. Thus, the row ϵ becomes disputed, since $\eta(T(\epsilon, u)) > 1$.

Table 2. Table is not *closed*

	a	u
ϵ	$(\perp, s \otimes \perp)$	$(1, s \otimes \perp), (5, t \otimes 5)$
a	$((\perp, \perp), s \otimes \perp)$	$((\perp, 1), s \otimes \perp), ((\perp, 5), t \otimes 25)$
$u \otimes 1$	$((1, \perp), s \otimes \perp)$	$((1, 1), s \otimes \perp), ((1, 5), t \otimes 25)$
$u \otimes 5$	$((5, \perp), s \otimes \perp)$	$((5, 1), s \otimes \perp), ((5, 5), t \otimes 5)$

Table 3. Table is not *dispute $-$ free*

	a	u
ϵ	$(\perp, s \otimes \perp)$	$(1, s \otimes \perp), (5, t \otimes 5)$
a	$((\perp, \perp), s \otimes \perp)$	$((\perp, 1), s \otimes \perp), ((\perp, 5), t \otimes 25)$
$u \otimes 1$	$((1, \perp), s \otimes \perp)$	$((1, 1), s \otimes \perp), ((1, 5), t \otimes 25)$
$u \otimes 5$	$((5, \perp), s \otimes \perp)$	$((5, 1), s \otimes \perp), ((5, 5), t \otimes 5)$
aa	$((\perp, \perp, \perp), s \otimes \perp)$	$((\perp, \perp, 1), s \otimes \perp), ((\perp, \perp, 5), t \otimes 25)$
au	$((\perp, 5, \perp), s \otimes \perp)$	$((\perp, 5, 1), s \otimes \perp), ((\perp, 1, 5), t \otimes 25)$

We add two parameterized sequences $u \otimes 1$ and $u \otimes 5$ to R and refill the table by constructing test cases for new rows and balance it respectively, shown in Table 2. The table is not closed, since row a in R is not equivalent to any row in S (that contains only one row ϵ and $a \not\cong \epsilon$). Thus, we move a to S and extend R accordingly, shown in table 3. Balancing the table makes the row a disputed, as $\eta(T(a, u)) > 1$. Hence, we add two more parameterized sequences in R and construct test cases to fill new rows. Table 4 is *balanced, dispute $-$ free* and

Table 4. Table is *balanced, dispute − free* and *closed*

	a	u
ϵ	$(\perp, s \otimes \perp)$	$(1, s \otimes \perp), (5, t \otimes 5)$
a	$((\perp, \perp), s \otimes \perp)$	$((\perp, 1), s \otimes \perp), ((\perp, 5), t \otimes 25)$
$u \otimes 1$	$((1, \perp), s \otimes \perp)$	$((1, 1), s \otimes \perp), ((1, 5), t \otimes 25)$
$u \otimes 5$	$((5, \perp), s \otimes \perp)$	$((5, 1), s \otimes \perp), ((5, 5), t \otimes 5)$
aa	$((\perp, \perp, \perp), s \otimes \perp)$	$((\perp, \perp, 1), s \otimes \perp), ((\perp, \perp, 5), t \otimes 25)$
$(a, u) \otimes (\perp, 1)$	$((\perp, 1, \perp), s \otimes \perp)$	$((\perp, 1, 1), s \otimes \perp), ((\perp, 1, 5), t \otimes 25)$
$(a, u) \otimes (\perp, 5)$	$((\perp, 5, \perp), s \otimes \perp)$	$((\perp, 5, 1), s \otimes \perp), ((\perp, 5, 5), t \otimes 5)$

closed. Figure 2 shows the conjecture from the current table. Note that we can use arbitrary input parameter values every time we construct test cases, whereas in the example, only 1 and 5 are used for sake of simplicity. However, using many different parameter values is more likely to reveal interesting information.

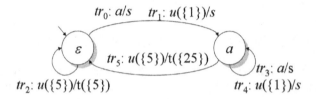

Fig. 2. The first conjecture of the example

3.4 Dealing with Counterexamples

The original learning algorithm for DFA [1], its improvements [14], [7] and its adaptations to more expressive models [9], [2], [10] performs an additional concept on the observation table, i.e., the table must be *consistent* before making the conjecture. The consistency concept can be described informally in the following way. If there are two equivalent rows $s, t \in S$, then all the subsequent rows in $S \cup R$, which extend s, t with some input symbol $i \in I$, must also be equivalent. In other words, the two apparently similar states (i.e., rows in S) must have same successive states for all inputs implied on those states. If the table is found not consistent, then the corresponding input sequence (which makes the successive states different) is added to E. This means that rows are extended with longer input sequences and then new test cases are constructed to fill the table. In this way, two apparently similar states in S become different.

In the learning algorithm of PFSM, we do not perform this concept because any two rows in S remain inequivalent during the whole process. Therefore inconsistency does not occur in the first iteration of the learning process. If a conjecture made from the table is not correct and there is a counterexample (an input sequence) that rejects the conjecture (the output sequence differs from the conjecture when applying counterexample to the component), then the

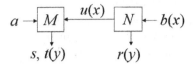

Fig. 3. Example of a Composed System

counterexample is fixed back into the table in order to refine the conjecture, which is considered as the next iteration of the learning process. In all above-mentioned algorithms, a counterexample is fixed by adding all its prefixes in S and hence new test cases are constructed for new rows. That is where the inconsistency may occur while adding prefixes in S.

This concept can be avoided altogether if the method of fixing a counterexample in the table is modified in such a way that instead of adding all prefixes in S, we only add the relevant sequence in the table that results in difference between the conjecture and the actual component. Furthermore, this addition will not be reflected in S, so that no two rows in S become equivalent. A general idea is discussed in [3], inspired by [14], applied on DFA algorithm. However, we deal differently in our case which is described below.

Let $c = \omega \otimes \alpha, \omega \in I^+, \alpha \in D_I^+$ be a counterexample for the current conjecture. Then c will be fixed in the observation table as follows:

If there exists $s \in S \cup R$ such that $IS(s)$ is the longest prefix of ω then add $e = suff(IS(c))^{|IS(c)|-|IS(s)|}$ in E, if it is not already present. In case where $s = \omega_1 \otimes \alpha_1 \in S \cup R$ a parameterized input sequence and α_1 is not a prefix of α then add $\omega_1 \otimes pref(\alpha)^{|\omega_1|}$ in R. Organize $\lambda(q_0, \omega, \alpha)$ in the table and make it *balanced, dispute − free* and *closed* for a new PFSM conjecture.

We have observed that fixing the counterexample in this way actually gives the same result as fixing the inconsistency in other algorithms. In other algorithms, E is extended only when inconsistency is found, which is reflected after fixing the counterexample. In our explanation, we extend E immediately while fixing the counterexample which keeps the rows in S inequivalent.

4   Integration Testing

In [9], we described the overall testing procedure in which the model is Mealy machine, with adaptations to a restricted form of PFSM in [10]. We suppose that we are provided with a set of components and the architecture of communication linking them. That is, we know for each component its interfaces. Each interface is a set of input and output symbol types and the types of associated parameters. Interface of two components can be pairwise connected, provided they are complementary (inputs and outputs correspond, and parameter types match). In an integrated architecture, non-connected interfaces will be considered as external interfaces to the environment. An example of a composed system of two components M and N is shown in Figure 3.

In order to associate PFSM models to components, we must provide a mapping from interfaces and parameter domains to sets I, O, D_I, D_O for each component. In this mapping, we may omit unrelevant parameters: some expertise may be needed there to identify which parts of the system are of interest. Some high-level description of the integration (e.g. with component diagrams, use cases...) could help in identifying relevant elements. We assume that through this mapping, each machine can be modeled with a PFSM, i.e, all state information will be captured in finite state. We also assume that typical parameter values to be tested are provided for each input: those could be provided by scenarios (esp. for external interfaces) or, failing that, chosen randomly. And for parameters considered not relevant, there should be some mechanism to assign them a value (either a default value, or some value linked to the values of other parameters, e.g. observed values in similar type).

We first learn each component in isolation, using algorithm 1 up to the first conjecture: we call this "unit testing". Thus, we get a PFSM model for each component. Actually, when the conjecture is made, some transitions will be labelled as "unchecked" as will be explained in section 4.2. From that point, we proceed to integration testing, where we connect the actual components using the specified architecture. We also connect the models of components: for this, since PFSM are a restricted form of EFSM, we use the IF tool-set [4] to compute interaction sequences. When we execute a test case, we submit external input symbols along with external input parameter values to the integrated system, observe the external output symbols and the external output parameter values. At the same time, by observing the internal interfaces, we also obtain the input and output sequences of the components. By using the mapping to the inputs and output of the models, and running the corresponding sequences on the integrated model, we can detect any discrepancy between the observed behaviors of components and that of their models. Those discrepancies can then be used as counterexamples to refine the models.

In order to choose integration tests, we can first use some information provided as scenarios or properties of the system, as described below in section 4.1. In any case, we shall be able to use the information from unit testing to derive systematic integration test cases, as described in section 4.2. Additionally, random walk on the model could provide a cheap test generation strategy: it could also be related to a coverage of the "unchecked" transitions.

4.1   Test Generation by Scenario or Model Checking

In component integration, the integrator may have a number of test scenarios for the global interaction of the system with its environment. Additionally, sample parameter values are provided for all external interfaces of the system. For each test scenario, a test case is constructed, in which the input parameter values are selected according to the ranges specified in the test scenario. In executing the test case, we check two properties:

- Whether the test scenario has been respected. If the test scenario has not been respected, an error has been detected in the system of components.

– Whether the observed behaviors conform to the models of components. If there is a discrepancy between the observed behavior of one component and its model, we go back to the unit testing procedure to refine the model with the input sequence as counterexample.

Another source for test cases could come from property checking. If some properties are specified for the system, then we can model-check those properties on the composed model. Any counterexample for the property could then be run on the system to check whether the actual system also includes a violation of the property. This combination of model-checking with learned models has been quite extensively studied in [5]. If no specific property is provided, we can still check for generic properties. In particular, we could check for livelocks, since our unit testing cannot guarantee that the models do not livelock when integrated (deadlocks are a different matter since we make our models input-enabled).

4.2 Test Generation Using Information from Learning Procedure

In the unit testing procedure, in the step of making a conjecture, the set of states is taken from S. When we want to define a transition from a state s for an input symbol and a set of parameter value, we try to identify the corresponding sequence s' in $S \cup R$, through observation recorded in T. If s' is in S, the next state of the transition is that sequence. If s' is in R, we find the sequence $t \in S$ which is equivalent to s'.

In the first case, since the sequences s and s' are all in S, they are not equivalent to each other. So, we are sure that in the real model of the component, the state reached by s and the state reached by s' are different, and there must exist such a transition from the state reached by s to the state reached by s'.

In the latter case, we cannot distinguish the state reached by s' and the state reached by t using the current set E of separating sequences. So, in the conjecture, we assume these two states are the same, and there is a transition from the state reached by s to the state reached by t.

But this conjecture may be wrong. In the real model of the component, the state reached by s' and the state reached by t can be different. These two states can be distinguished by certain sequence. From the point of view of identifying counterexamples for the conjecture, in the integration testing procedure, we should try to separate these states by executing long sequences from them. Based on this observation, we propose the following integration testing technique.

In making a conjecture in the unit testing procedure, for $s \in S$, $i \in I$, $\alpha \in (\rho(d_k(s, i)))^*$, if $t = (IS(s) \cdot i \otimes \alpha) \in R$ then we label the transition as *unchecked*, and we record the sequence t with it and refer to it as the *hidden sequence*.

Our test generation strategy for integration testing will be specifically targeted at covering unchecked transitions. For each unchecked transition, we extend its hidden sequence with several parameterized input sequences whose lengths are limited by a predefined threshold k to obtain a group of sequences. From all these sequences obtained, we remove those sequences which have been executed in unit testing, and those sequences in which there is not any interaction with another component. Those sequences are local to a given component, and should be extended to a global test sequence. Therefore, we take rest of the sequences as test purposes, and obtain a group of test cases which contain external inputs/outputs only using the method described in [8]: basically we search the composed model for global sequences whose projections on the local component match the test purpose. Actually, in a single search, we may compute the test cases for several components. By executing these test cases, we may identify counterexamples.

In the example of a composed system shown in Figure 3, component M has $I_M = \{a, u\}$ and $O_M = \{s, t\}$, and component N has $I_N = \{b\}$ and $O_N = \{u, r\}$, respectively. The PFSM model of component M is shown in Figure 1. After unit testing, the first conjecture $M^{(1)}$ of component M is learnt, shown in Figure 2. The PFSM model of component N is shown in Figure 4. It is learnt exactly in its unit testing.

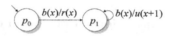

Fig. 4. PFSM Model of Component N

In $M^{(1)}$, among the 6 transitions, transitions tr_1, tr_2, tr_3, tr_4 and tr_5 are unchecked. For unchecked transition tr_2, its hidden sequence is $u(5)$. We extend it to obtain $u(5){\cdot}a{\cdot}a$, $u(5){\cdot}a{\cdot}u(5)$, $u(5){\cdot}u(5){\cdot}a$, and $u(5){\cdot}u(5){\cdot}u(5)$. Among them, using $u(5) \cdot a \cdot u(5)$ as test purpose, we obtain a test case $b(4)/r(4) \cdot b(4)/t(5) \cdot a/s \cdot b(4)/t(25)$.

In executing this test case, the expected behavior of component M is $u(5)/t(5){\cdot}a/s{\cdot}u(5)/t(25)$, and the observed behavior is $u(5)/t(5){\cdot}a/s{\cdot}u(5)/t(5)$. This means that a counterexample $u(5) \cdot a \cdot u(5)$ is identified.

Table 5. Table is *balanced*, *dispute − free* and *closed* after fixing the counterexample

	a	u	au
ϵ	$(\perp, s \otimes \perp)$	$(\perp, s \otimes \perp), (5, t \otimes 5)$	$((\perp, 5), (s, t) \otimes (\perp, 25)), ((\perp, 1), (s, s) \otimes (\perp, \perp))$
a	$((\perp, \perp), s \otimes \perp)$	$((\perp, 1), s \otimes \perp), ((\perp, 5), t \otimes 25)$	$((\perp, \perp, 5), (s, t) \otimes (\perp, 25)), ((\perp, \perp, 1), (s, s) \otimes (\perp, \perp))$
$u \otimes 5$	$((5, \perp), s \otimes \perp)$	$((5, 1), s \otimes \perp), ((5, 5), t \otimes 5)$	$((5, \perp, 5), (s, t) \otimes (\perp, 5)), ((5, \perp, 1), (s, s) \otimes (\perp, \perp))$
$u \otimes 1$	$((1, \perp), s \otimes \perp)$	$((1, 1), s \otimes \perp), ((1, 5), t \otimes 25)$	$((1, \perp, 5), (s, t) \otimes (\perp, 25)), ((1, \perp, 1), (s, s) \otimes (\perp, \perp))$
aa	$((\perp, \perp, \perp), s \otimes \perp)$	$((\perp, \perp, 1), s \otimes \perp), ((\perp, \perp, 5), t \otimes 25)$	$((\perp, \perp, \perp, 5), (s, t) \otimes (\perp, 25)), ((\perp, \perp, \perp, 1), (s, s) \otimes (\perp, \perp))$
$(a, u) \otimes (\perp, 1)$	$((\perp, 1, \perp), s \otimes \perp)$	$((\perp, 1, 1), s \otimes \perp), ((\perp, 1, 5), t \otimes 25)$	$((\perp, 1, \perp, 5), (s, t) \otimes (\perp, 25)), ((\perp, 1, \perp, 1), (s, s) \otimes (\perp, \perp))$
$(a, u) \otimes (\perp, 5)$	$((\perp, 5, \perp), s \otimes \perp)$	$((\perp, 5, 1), s \otimes \perp), ((\perp, 5, 5), t \otimes 5)$	$((\perp, 5, \perp, 5), (s, t) \otimes (\perp, 5)), ((\perp, 5, \perp, 1), (s, s) \otimes (\perp, \perp))$
$(u, u) \otimes (5, 1)$	$((5, 1, \perp), s \otimes \perp)$	$((5, 1, 1), s \otimes \perp), ((5, 1, 5), t \otimes 5)$	$((5, 1, \perp, 5), (s, t) \otimes (\perp, 5)), ((5, 1, \perp, 1), (s, s) \otimes (\perp, \perp))$
$(u, u) \otimes (5, 5)$	$((5, 5, \perp), s \otimes \perp)$	$((5, 5, 1), s \otimes \perp), ((5, 5, 5), t \otimes 5)$	$((5, 5, \perp, 5), (s, t) \otimes (\perp, 5)), ((5, 5, \perp, 1), (s, s) \otimes (\perp, \perp))$
$(u, a) \otimes (5, \perp)$	$((5, \perp, \perp), s \otimes \perp)$	$((5, \perp, 1), s \otimes \perp), ((5, \perp, 5), t \otimes 5)$	$((5, \perp, \perp, 5), (s, t) \otimes (\perp, 5)), ((5, \perp, \perp, 1), (s, s) \otimes (\perp, \perp))$
$(a, a, u) \otimes (\perp, \perp, 5)$	$((\perp, \perp, 5, \perp), s \otimes \perp)$	$((\perp, \perp, 5, 1), s \otimes \perp), ((\perp, \perp, 5, 5), t \otimes 5)$	$((\perp, \perp, 5, \perp, 5), (s, t) \otimes (\perp, 5)), ((\perp, \perp, 5, \perp, 1), (s, s) \otimes (\perp, \perp))$
$(a, a, u) \otimes (\perp, 1, 1)$	$((\perp, 1, 1, \perp), s \otimes \perp)$	$((\perp, 1, 1, 1), s \otimes \perp), ((\perp, 1, 1, 5), t \otimes 25)$	$((\perp, 1, 1, 5), (s, t) \otimes (\perp, 25)), ((\perp, 1, 1, 1, 1), (s, s) \otimes (\perp, \perp))$
$(u, a, u) \otimes (5, \perp, 1)$	$((5, \perp, 1, \perp), s \otimes \perp)$	$((5, \perp, 1, 1), s \otimes \perp), ((5, \perp, 1, 5), t \otimes 5)$	$((5, \perp, 1, 5), (s, t) \otimes (\perp, 5)), ((5, \perp, 1, 1, 1), (s, s) \otimes (\perp, \perp))$
$(u, a, u) \otimes (5, \perp, 5)$	$((5, \perp, 5, \perp), s \otimes \perp)$	$((5, \perp, 5, 1), s \otimes \perp), ((5, \perp, 5, 5), t \otimes 5)$	$((5, \perp, 5, 5), (s, t) \otimes (\perp, 5)), ((5, \perp, 5, 1), (s, s) \otimes (\perp, \perp))$

Going back to unit testing, we fix the counterexample in the observation table by adding $a \cdot u$ in E and then making the table *balanced*, *dispute − free* and *closed*, we obtain a new conjecture $M^{(2)}$ for component M. The table is shown in Table 5 and Figure 5 is the conjecture. The new conjecture then will again be put under integration testing with component N and new global test sequences will be generated according to the process described above. This may identify new counterexamples or end the integration process if no discrepancy is found [9]. In the former case, the conjecture will then be refined again through fixing new counterexamples in the table.

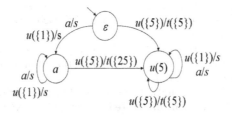

Fig. 5. Conjecture $M^{(2)}$ of Component M

5 Conclusion

We have presented an approach that makes it possible to use model-based testing techniques, in particular test generation for integration testing, in the absence of initial models. We extend previous work done in this direction [7], [9], [5] to deal with arbitrary data values, avoiding the complexity of expanding into DFA or FSM models. The model is richer than the models used by [10] or [2]. We use an incremental testing approach where new interoperability tests can be derived to check systematically the models derived from previous observations. From those tests, refined models of the system can be built, or faults in the system can be identified, as explained in [9].

We are currently working on a tool, called RALT (Rich Automata Learning and Testing), to run the approach on case studies to be provided by France Telecom. We have already implemented the learning algorithms for DFA [1], for Mealy machine [9] and for simple parameterized machine [10], and need to interface to actual test drivers, so that we can compare them all.

We also consider research perspectives to deal with even more complex models. In particular, we could try to move closer to EFSM models by incorporating variables. To circumvent the hidden nature of state structure in black boxes, we could either rely on additional structure information provided by the integrator (moving from black to some kind of grey box) or use some heuristics to differentiate control states from variables. Other direction is to consider sufficient information (e.g., parts of source code) and derive complex models, as performed in [6], [15]. We are also investigating other types of test generation strategies for integration testing.

References

1. Angluin, D.: Learning regular sets from queries and counterexamples. Information and Computation 2, 87–106 (1987)
2. Berg, T., Jonsson, B., Raffelt, H.: Regular inference for state machines with parameters. In: Baresi, L., Heckel, R. (eds.) FASE 2006 and ETAPS 2006. LNCS, vol. 3922, pp. 107–121. Springer, Heidelberg (2006)

3. Berg, T., Raffelt, H.: Model checking. In: Model-Based Testing of Reactive Systems, pp. 557–603 (2004)
4. Bozga, M., Graf, S., Ober, I., Ober, I., Sifakis, J.: The IF toolset. In: Bernardo, M., Corradini, F. (eds.) Formal Methods for the Design of Real-Time Systems. LNCS, vol. 3185, pp. 237–267. Springer, Heidelberg (2004)
5. Elkind, E., Genest, B., Peled, D., Qu, H.: Grey-box checking. In: Najm, E., Pradat-Peyre, J.F., Donzeau-Gouge, V.V. (eds.) FORTE 2006. LNCS, vol. 4229, pp. 420–435. Springer, Heidelberg (2006)
6. Ernst, M.D., Perkins, J.H., Guo, P.J., McCamant, S., Pacheco, C., Tschantz, M.S., Xiao, C.: The Daikon system for dynamic detection of likely invariants. Science of Computer Programming (2006)
7. Hungar, H., Niese, O., Steffen, B.: Domain-specific optimization in automata learning. In: Hunt Jr., W.A., Somenzi, F. (eds.) CAV 2003. LNCS, vol. 2725, pp. 315–327. Springer, Heidelberg (2003)
8. Koné, O., Castanet, R.: Test generation for interworking systems. Computer Communications 23(7), 642–652 (2000)
9. Li, K., Groz, R., Shahbaz, M.: Integration testing of components guided by incremental state machine learning. In: TAIC PART, pp. 59–70. IEEE Computer Society, Washington (2006)
10. Li, K., Groz, R., Shahbaz, M.: Integration testing of distributed components based on learning parameterized i/o models. In: Najm, E., Pradat-Peyre, J.F., Donzeau-Gouge, V.V. (eds.) FORTE 2006. LNCS, vol. 4229, pp. 436–450. Springer, Heidelberg (2006)
11. Peled, D., Vardi, M.Y., Yannakakis, M.: Black box checking. In: FORTE. IFIP Conference Proceedings, vol. 156, pp. 225–240. Kluwer, Dordrecht (1999)
12. Petrenko, A., Boroday, S., Groz, R.: Confirming configurations in EFSM testing. IEEE Trans. Softw. Eng. 30(1), 29–42 (2004)
13. Ramalingom, T., Thulasiraman, K., Das, A.: Context independent unique state identification sequences for testing communication protocols modelled as extended finite state machines. Computer Communications 26(14), 1622–1633 (2003)
14. Rivest, R.L., Schapire, R.E.: Inference of finite automata using homing sequences. In: Machine Learning: From Theory to Applications, pp. 51–73 (1993)
15. Walkinshaw, N., Bogdanov, K., Holcombe, M.: Identifying state transitions and their functions in source code. In: TAIC PART, pp. 49–58. IEEE Computer Society, Washington (2006)

An EFSM-Based Passive Fault Detection Approach

Hasan Ural and Zhi Xu

School of Information Technology and Engineering (SITE)
University of Ottawa, Ottawa, Ontario, Canada, K1N 6N5
{ural,zxu061}@site.uottawa.ca

Abstract. Extended Finite State Machine (EFSM)-based passive fault detection involves modeling the system under test (SUT) as an EFSM M, monitoring the input/output behaviors of the SUT, and determining whether these behaviors relate to faults within the SUT. We propose a new approach for EFSM-based passive fault detection which randomly selects a state in M and checks whether there is a trace in M starting from this state which is compatible with the observed behaviors. If a compatible trace is found, we determine that observed behaviors are not sufficient to declare the SUT to be faulty; otherwise, we check another unchecked state. If all the states have been checked and no compatible trace is found, we declare that the SUT is faulty. We use a Hybrid method in our approach which combines the use of both Interval Refinement and Simplex methods to improve the performance of passive fault detection.

1 Introduction

Passive fault detection is a fundamental part of passive testing which determines whether a system under test (SUT) is faulty by observing the input/output (I/O) behaviors of the SUT without interfering with its normal operations [10]. Compared with active fault detection, in which a tester has complete control over the inputs and devises a test sequence to reveal possible faults of the SUT, passive fault detection is more applicable under circumstances where the control is impractical or impossible, such as network fault management [10].

In Extended Finite State Machine (EFSM)-based passive fault detection, the specification of an SUT N is modeled as an EFSM M, N is treated as a blackbox, and the observed I/O behaviors of N is represented as a sequence E of observed I/O events. Determining whether N is faulty with respect to M is then based on the existence of traces in M that are compatible with E, i.e., a trace in M is compatible with E if E maps to a sequence of consecutive transitions of M starting at a state s of M. If the number of traces in M compatible with E is zero, then E is sufficient to determine that N is faulty. Otherwise, E is declared to be insufficient to determine whether N is faulty, i.e., there is at least one trace in M compatible with E and E needs to be augmented with additional I/O events of N to continue with passive fault detection.

A. Petrenko et al. (Eds.): TestCom/FATES 2007, LNCS 4581, pp. 335–350, 2007.
© IFIP- International Federation for Information Processing 2007

Usually, EFSM-based passive fault detection approaches are derived from Finite State Machine-based passive fault detection approaches. The FSM-based fault detection approach in [9] checks the observed sequence of I/O events one-by-one from the beginning, and reduces the size of the set S' of possible current states by eliminating impossible states until either S' is empty (N is faulty) or there is at least one state in S' (no fault is detected). The approach in [9] has been applied for passive fault detection in FSM-based systems [22, 23]. This approach has been extended to systems specified in the EFSM model by [7, 10, 11, 21] and adopted to systems specified in the Communicating Finite State Machine (CFSM) model by [14, 15, 16, 17, 18]. Another approach to EFSM-based passive fault detection focuses on characterizing specifications of an SUT in terms of invariants [3, 4, 5, 6].

This paper proposes a new approach for EFSM-based passive fault detection which is summarized as follows: assume that the subset S_0 of states of M contains all possible starting states of E. Randomly pick a state s in S_0 and determine whether there exists a trace in M that starts at s and is compatible with E. If such a trace is found, then stop and declare that E is not sufficient to determine whether N is faulty. In this case, the starting state and the current state of N can be determined readily using this trace. Otherwise, continue to check other states in S_0. After checking all the states in S_0, if no trace in M is found to be compatible with E, then N will be declared faulty.

The proposed approach provides information about possible starting state and possible trace compatible with E at the end of passive fault detection. Such information cannot be provided by the existing approaches derived from [9] unless a post-processing is performed or a backward checking approach is taken for exploring the information about possible starting state and possible trace [1, 2]. In addition, the proposed approach utilizes a Hybrid method to evaluate constraints in predicates associated with transitions in an EFSM which combines the use of both Interval Refinement [8, 19] and Simplex [13] methods for performance improvement during passive fault detection. We show that using only the Interval Refinement method has a similar performance to the Hybrid method but suffers from inaccuracy whereas using only the Simplex method has the same accuracy as the Hybrid method but suffers from poor performance.

The rest of the paper is organized as follows. Section 2 gives preliminaries needed for our discussion, including definitions and notations used in our presentation. Section 3 presents the proposed approach for EFSM-based passive fault detection in detail. Section 4 provides experimental evaluations. Section 5 concludes this paper with some final remarks and directions for future research.

2   Preliminaries

The proposed approach for EFSM-based passive fault detection is based on the specification of SUT N given as a Simplified Extended Finite State Machine (SEFSM) and the sequence of I/O behaviors a tester observes during the execution of N given as a sequence E of observed I/O events.

A *Simplified Extended Finite State Machine* (SEFSM) M is $(S, E_m, \bar{x}, T)$:

1. $S = \{s_1, \ldots, s_n\}$ is a finite set of states;
2. E_m is a finite set of I/O events. $e(\bar{y}) \in E_m$ is an input or output event, and $\bar{y} = (y_1, y_2, \ldots, y_p)$ is a vector of parameters of the I/O event e, called *local variables*;
3. $\bar{x} = (x_1, \ldots, x_r)$ is a vector of *global variables* which are accessible within all transitions;
4. T is a finite set of transitions.

The difference between $\bar{y}$ and $\bar{x}$ is that $\bar{y}$ is observable from SUT N while $\bar{x}$ is unobservable. Note that all variables are integers. An example SEFSM is shown in Figure 1.

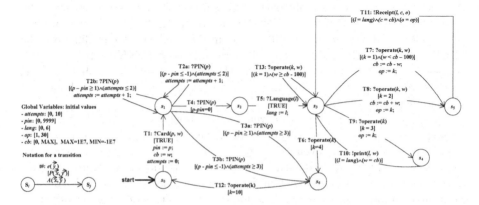

Fig. 1. The SEFSM *ATM* for an Automatic Teller Machine (ATM) system

A *transition* $t \in T$ in an SEFSM is $(s_i, s_j, e(\bar{y}), P(\bar{x}, \bar{y}), A(\bar{x}, \bar{y}))$:

1. s_i is the starting state of t;
2. s_j is the ending state of t;
3. $e(\bar{y}) \in E_m$ is an input event prefixed with "?" or output event prefixed with "!" that can be observed once t is activated;
4. $P(\bar{x}, \bar{y})$ is a predicate expressing the conditions to be satisfied for the activation of t which consists of conjunctive terms, each of which is defined as a *constraint*, connected by "$\wedge$" (and) operators;
5. $A(\bar{x}, \bar{y})$ is an action consisting of a sequence of assignment statements, each updating a global or local variable as a function of elements of $\bar{x}$ and $\bar{y}$.

Examples of an I/O event, predicate, and action are: "!display(y)" is an I/O event "display" which outputs the value of y, "$(3 \times x_1 + (-1) \times x_2 \geq 0) \wedge (1 \times x_1 + 4 \times y_2 \leq 4)$" is a predicate, and "$x_3 := 3 \times x_1 + (-1) \times x_2 + (-5); x_1 := x_3;$" is an action, respectively.

Because $\bar{y}$ is observable from N while $\bar{x}$ is unobservable, the I/O events with global variables as parameters must be modified. For example, if x is a global variable, an input event "?read(x)" will be transformed to "?read(a) $x:=a;$" where a is a local variable and the action "$x:=a;$" assigns the value of a to x; similarly, an output event "!display(x)" will be transformed to "!display(a) [$a = x$]" where the predicate "[$a = x$]" guarantees the output value is equal to the value of x.

In this paper, a constraint cs is represented by $\sum_{i=1}^{k} a_i x_i = I$ (a_i is a coefficient, x_i is a global variable, I is an interval) after replacing the local variables of $\bar{y}$ by the actual values of the parameters observed during the execution of N. For example, the constraint "$3 \times x_1 + (-1) \times x_2 \geq 0$" is represented by the expression "$3 \times x_1 + (-1) \times x_2 = [0, \text{MAX}]$". MAX is defined as 1×10^7 and MIN is defined as -1×10^7 in this paper.

Note that an event-driven extended finite state machine (EEFSM) model is used in [10]. The differences between EEFSM and SEFSM models are as follows: the SEFSM model simplifies the structure of predicates in transitions by eliminating the "or" operator in EEFSM. Therefore, in SEFSM, a transition is executable if and only if all the constraints in the predicate are evaluated to be TRUE. Also, in actions associated with transitions in EEFSM, [10] only considered the assignment statements where the left hand side is a global variable, whereas we consider both global and local variables to be on the left hand side of assignment statements.

The sequence E of observed I/O events represents a sequence of I/O behaviors a tester observed during the execution of N, i.e., $e_1 e_2 \ldots e_n$. Like an I/O event in E_m, an observed I/O event e_i, $1 \leq i \leq n$, in E is also categorized as an observed input event prefixed with "?" or an observed output event prefixed with "!". Different from the I/O event in E_m, an observed I/O event in E contains determined values instead of symbols for variables. For example, "?read(3)" is an observed I/O event in E while "?read(y)" is an I/O event in E_m.

A *configuration* depicts a possible status of the SUT N during EFSM-based passive fault detection. A configuration c is a quadruple $(\#, s, [\bar{x}], CS(\bar{x}))$ where

1. $\#$ is the number of observed I/O events that have been checked to reach the configuration;
2. s is the possible current state of N;
3. $[\bar{x}]$ is a vector of *intervals* which represents the ranges of possible values which the variables in $\bar{x}$ can take;
4. $CS(\bar{x})$ records the constraints on variables in $\bar{x}$. These constraints are obtained from both predicates and actions. As $CS(\bar{x})$ contains only global variables, we shall henceforth use CS as the abbreviation of $CS(\bar{x})$.

For example, $c = (3, s_6, \{x_1 = [0,5], x_2 = [1,2]\}, \{x_1 + x_2 \geq 0; 3x_1 - x_2 \leq 9; \})$ is a configuration. (see Figure 2) According to configuration c in Figure 2, 3 observed I/O events have been checked; the current possible state of N is s_6; the value of x_1 is greater or equal to 0 and less than or equal to 5, and the value of x_2 is greater or equal to 1 and less than or equal to 2; the values of x_1 and x_2 must satisfy two constraints "$x_1 + x_2 \geq 0$" and "$3x_1 - x_2 \leq 9$" at the same time.

#:	3
s:	s_6
$[\vec{x}]$:	$x_1 = [0, 5], x_2 = [1, 2]$
$CS(\vec{x})$:	$x_1 + x_2 \geq 0; 3x_1 - x_2 \leq 9;$

Fig. 2. A configuration c

A *trace* represents the sequence of status of the SUT N during EFSM-based passive fault detection. *Trace-Tree* records all the traces that have been checked during EFSM-based passive fault detection.

1. A trace *trace* is a sequence of configurations, which are connected by transitions;
2. A Trace-Tree *Tree* for s consists of all the traces starting from a state $s \in S_0$. Each node in *Tree* represents a configuration and each edge stands for a transition between two configurations. Every trace $trace_i$ of length k, from s to a leaf in *Tree*, is compatible with a prefix of E ($e_1 e_2 \ldots e_k, k \leq |E|$);
3. A trace in M compatible with E, henceforth called *compatible trace of E*, is defined as a trace in Trace-Tree for s with length equal to $|E|$.

3   The Proposed Approach

Given a specification SEFSM M of an SUT N, a sequence E of observed I/O events, and $S_0 \subseteq S$, the proposed approach proceeds as follows:

1. Pick an unchecked state s from S_0;
2. Build a Trace-Tree for s by finding all the possible traces starting from state $s \in S_0$;
3. If a compatible trace of E is found, declare this trace as a compatible trace of E; if no compatible trace of E can be found in Trace-Tree for s, go to (1);
4. If all states in S_0 have been checked and no compatible trace of E is found, declare that "N is faulty".

3.1   Algorithm Main

In algorithm *Main*, we randomly select a state s from $S_0 \subseteq S$ of SEFSM M and try to find a compatible trace of E starting from s. If *trace* is found to be a compatible trace of E, this algorithm will terminate and declare *trace* as a compatible trace of E; if all the states in S_0 have been checked and no compatible trace of E is found, the algorithm will report "N is faulty".

3.2   Algorithm Search_Trace_Tree

Algorithm *Search_Trace_Tree* searches for a compatible trace of E starting from a state s using the data structures for configuration and Trace-Tree.

Algorithm 1. Algorithm Main

1: **Given:** an SEFSM M,
2:    a sequence E of observed I/O events, and
3:    $S_0 = \{ s_1, s_2, \cdots, s_n \}$
4: **Return:** "N is faulty", or "$trace$ is a compatible trace of E"
5: **Begin:**
6:    **while** $(S_0 \neq \emptyset)$
7:       randomly select a state s from S_0;
8:       $S_0 \leftarrow S_0 \setminus \{ s \}$;
9:       $trace \leftarrow$ **Search_Trace_Tree**(M, s, E);{search for a compatible trace of E}
10:       **If**($trace \neq$ NULL)
11:          return ("$trace$ is a compatible trace of E");
12:       **endwhile**
13:       return ("N is faulty");   {no compatible trace of E is found}
14: **End**

3.3   Algorithm Check_Trace and the Hybrid Method

A trace consists of a sequence of configurations which represents the sequence of changes in the status of N through E. Algorithm **Check_Trace**(M, **trace**, E, **Tree**) checks if there is a trace compatible with E. It first initializes the current configuration $c_{current}$ to the first configuration from $trace$, sets the current possible state s to the state in $c_{current}$ and gets the observed I/O event e to be considered from E. Then, all transitions in M starting from s (i.e., set T_s of transitions) are checked one by one. Those transitions passing both control portion and data portion fault detection will be considered as executable transitions corresponding to the observed I/O event e. As there may be more than one executable transition, algorithm *Check_Trace* picks the first one of them to

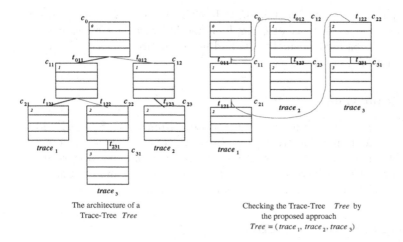

The architecture of a
Trace-Tree *Tree*

Checking the Trace-Tree   *Tree* by
the proposed approach
Tree = (*trace* $_1$, *trace* $_2$, *trace* $_3$)

Fig. 3. The architecture of a Trace-Tree and its representation during passive fault detection

---

Algorithm 2. Algorithm *Search_Trace_Tree*(M, s, E)

1: **Given:** an SEFSM M,
2:    a state $s \in S_0$, and
3:    a sequence E of observed I/O events
4: **Return:** a compatible trace *trace*, or NULL
5: **Begin:**
6:    *Tree* ← NULL;   {initialize the Trace-Tree *Tree*}
7:    *trace* ← NULL;   {initialize the trace *trace*}
8:    $[\bar{x}]_0$ ← set the initial intervals of the global variables in M;
9:    c_0 ← (0, s, $[\bar{x}]_0$, $\emptyset$); {create the initial configuration $c_0 = (\#, s, [\bar{x}], CS)$}
10:    *trace*.add(c_0);   {add c_0 as the first configuration in this trace}
11:    *Tree*.add(*trace*);
12:    **while** (*Tree* $\neq \emptyset$)
13:       *trace* ← *Tree*.get(0); {get the first trace in *Tree*}
14:       succ ← **Check_Trace(M, trace, E, Tree)**; {check if this trace is compatible with E}
15:       **if** (succ = TRUE) {if *trace* is compatible with E}
16:          return (*trace*);
17:       **else**
18:          *Tree*.delete(*trace*); {delete *trace* from *Tree*}
19:    **endwhile**
20:    return (NULL); {no trace compatible with E has been found}
21: **End**

---

continue checking and adds all other transitions as branches into the Trace-Tree *Tree*. The procedure of checking a Trace-Tree is described in Figure 3. In Figure 3, *Tree* consists of three traces. For example, when checking configuration c_{11}, there exist two executable transitions, t_{121} and t_{122}. For each executable transition, a new configuration will be built. For c_{21}, which corresponds to the first executable transition t_{121}, we add c_{21} to the end of *trace*$_1$; for c_{22}, we build a new trace, *trace*$_3$, and set c_{22} as the starting configuration of *trace*$_3$. Then we continue checking *trace*$_1$ with c_{21}. *trace*$_3$ will be checked if and only if *trace*$_1$ and *trace*$_2$ are determined not compatible with E. Whenever a compatible trace of E is found, algorithm *Check_Trace* returns this trace.

When searching for executable transitions within algorithm *Check_Trace*, two steps are applied to a transition $t \in T_s$: In the first step, which corresponds to function *control_portion_checking* in algorithm *Check_Trace*, we compare the I/O event associated with transition t with the observed I/O event e (in E) by the prefix symbol, event name and possibly the number of parameters. If this comparison produces a mismatch, we stop processing transition t. Otherwise, we continue with the second step, which corresponds to the data portion fault detection, where we replace the local variables of $\bar{y}$ in predicate $t.P(\bar{x}, \bar{y})$ by the actual values of the parameters of the observed I/O event e and then transform the predicate into a list of constraints stored in *newCS*. After the replacement, the data portion fault detection problem is reduced to a *Constraint Satisfaction*

Problem (CSP) which is defined as follows: given (1) a configuration c, in which $c.[\bar{x}]$ contains a vector of intervals representing the ranges of possible values of global variables and $c.CS$ stores existing constraints on $\bar{x}$; and (2) a set $newCS$ of new constraints, which is generated from $t.predicate(\bar{x}, \bar{y})$, determine if there exists at least one combination of values, called *solution*, in $c.[\bar{x}]$ that satisfies the existing constraints in $c.CS$ and new constraints in $newCS$ simultaneously. If there exists a solution, the predicate $t.predicate(\bar{x}, \bar{y})$ will be considered consistent with the configuration c. If no solution exists, it means that an inconsistency has been detected.

To solve this CSP, the Interval Refinement method can be used, as done in [10]. However, because of the dependency problem, the results of the Interval Refinement method may not be accurate, i.e., some transitions may falsely be reported as executable. For example: assume a configuration c with $c.[\bar{x}] : x_1 = [1, 2], x_2 = [1, 2]$, x_1 and x_2 are integers; $c.CS$: $\{cs : x_1 - x_2 = 0;\}$, and check two transitions t_1 with a constraint cs_1 in its predicate as: $x_1 + x_2 = 3$; t_2 with a constraint cs_2 in its predicate as: $x_1 + x_2 \leq 4$. By applying the Interval Refinement method, both transition t_1 and t_2 will be judged as executable. However, t_1 is not executable because x_1 and x_2 are integers and there is no solution for both cs and cs_1 at the same time. To guarantee the correctness of results, the Simplex method can be used instead of the Interval Refinement method, as done in [11]. Although the Simplex method is accurate, it is slower than the Interval Refinement method. Another difference between these two methods is that, in the Interval Refinement method, the intervals are narrowed; while in the Simplex method, the intervals will be untouched.

To combine the advantages of both the Interval Refinement and Simplex methods, we propose a Hybrid method, which is as accurate as the Simplex method and as efficient as the Interval Refinement method. The proposed Hybrid method uses both of these two methods judiciously as follows: given the set T_s of transitions, the current configuration $c_{current}$, and an observed I/O event e, first the Interval Refinement method, together with function $control\_portion\_checking$, is used to decide which transitions in T_s are executable. If no transition in T_s is evaluated to be executable, the current trace will be determined not compatible with E. If more than one transition is evaluated to be executable, the Simplex method will be applied to check the correctness of the Interval Refinement method in declaring these transitions executable. If only one transition is evaluated to be executable by the Interval Refinement method, the Simplex method will not be applied because this transition will be evaluated by the Simplex method implicitly by checking the last configuration of this trace. That is, at the end of a trace, before the trace is determined to be compatible with E, the Simplex method is applied to confirm that there exists no inconsistency in the last configuration of this trace. For example, consider a trace $trace$ $(c_1 c_2 \ldots c_k, k \leq |E|$) in the Trace-Tree $Tree$. If c_k is checked by the Simplex method and no inconsistency is found, $trace$ is guaranteed to be compatible with a prefix of E $(e_1 e_2 \ldots e_k, k \leq |E|)$ because c_k contains all the constraints within the configurations from c_1 to

c_{k-1}. Therefore, if no inconsistency found in the last configuration of *trace* by the Simplex method, the transitions associated with *trace* are all executable.

After evaluating all the transitions in T_s, we continue to perform actions by function ***action***(t_c, *e*, *c*) on the configurations in C with their corresponding transitions in T_s. After performing actions, we add the first configuration in C to the end of *trace* and continue to check *trace* starting from this configuration. Other configurations in C will be considered as the initial configuration of new branches, which are represented as new traces in the Trace-Tree.

In function ***Interval_Refinement***(*c*.[$\bar{x}$], *c.CS*, *newCS*), the interval arithmetic operations are applied to narrow the intervals of variables in constraints [19]. During refinement, if the interval of a variable is empty, an inconsistency is detected and function *Interval_Refinement* returns FALSE. Otherwise, *c*.[$\bar{x}$] is updated based on the new constraints *newCS* and *newCS* is added into the set *c.CS*.

In function ***Simplex***(*c*.[$\bar{x}$], *c.CS*, $\emptyset$), we adopt an open source tool *lp_solve* which is a free linear programming solver based on the revised Simplex method and the Branch-and-bound method [11, 12]. If no solution exists, function *Simplex* returns FALSE. Both *c*.[$\bar{x}$] and *c.CS* are unchanged within function *Simplex*.

The worst case computational complexities of Interval Refinement and Simplex methods are exponential. [10, 11, 20] show that the average complexities of both methods in practice are polynomial. However, because the Simplex method is more complex than the Interval Refinement method, the speed of the Simplex method is slower than that of the Interval Refinement method. However, the use of the Simplex method in conjunction with the Interval Refinement method does not adversely affect the efficiency of the Hybrid method because the frequency of applying the Simplex method in the Hybrid method is very low; and the Interval Refinement method narrows the intervals which helps reduce the cost of applying the Simplex method.

3.4   Function *action*

When a transition has been evaluated to be executable, a new configuration will be constructed to record the status of SUT N after this transition. The construction of a new configuration depends on the *action* part, $A(\bar{x},\bar{y})$, in the transition which consists of a sequence of assignment statements. Given a configuration c in the set of configurations built for all executable transitions, an observed I/O event e and a transition(t_c corresponding to c, function *action*(t_c, e, c) performs the actions associated with t_c, and builds a new configuration c_{next} which stands for the status of SUT N after t_c. The details of algorithm *action* are presented as follows: In the first step, we replace the local variables in the right hand expression (RHE) of an assignment statement by their values in e which gives an $RHE = \sum_{i=1}^{k} a_i x_i$. After the replacement, RHE without local variables is used to update the value of the left hand variable (LHV) in the configuration c. If LHV is a local variable, we use the value of RHE in the assignment statement to replace the existing value of this local variable. If LHV

Algorithm 3. Algorithm *Check_Trace(M, trace, E, Tree)*

1: **Given:** an SEFSM M,
2:   a trace *trace*,
3:   a sequence E of observed I/O events, and
4:   a Trace-Tree *Tree*,
5: **Return:** FALSE, or   {*trace* is not a compatible trace of E}
6:       TRUE   {*trace* is a compatible trace of E}
7: **Begin:**
8:   $c_{current} \leftarrow trace.get(0)$;{get the first configuration}
9:   **while** ($c_{current} \neq$ NULL and $c_{current}.\# \neq E.\#$){ if there is an observed I/O event to be checked}
10:     $s \leftarrow c_{current}.s_c$;
11:     $T_s \leftarrow$ all transitions in M starting at s;
12:     $e \leftarrow E.get(c_{current}.\# + 1)$;{ get the observed I/O event e}
13:     $C \leftarrow \emptyset$;
14:     **for** each transition t in T_s {evaluate transitions}
15:       $c \leftarrow c_{current}$;
16:       **if** (*control_portion_checking*(c, t, e) = FALSE)
17:         end the for loop; {the control portion is inconsistent}
18:       **else**   {the data portion fault detection commences}
19:         $newCS \leftarrow replace(t.P(\bar{x}, \bar{y}), e)$;{eliminate local variables}
20:         **if** (***Interval_Refinement***($c.[\bar{x}]$, $c.CS$, $newCS$) = FALSE)
21:           end the for loop; {the data portion is inconsistent}
22:         **else**
23:           $C \leftarrow C \cup \{ c \}$; {c is modified and needs to be added to C}
24:     **endfor**
25:     **if** ($C = \emptyset$) return (FALSE); {if no executable transition is found}
26:     **else**
27:       **if** ($|C| > 1$ or $c_{current}.\# + 1 = |E|$) {checking by the Simplex method}
28:         **for** each configuration c in C
29:           **if** (***Simplex***($c.[\bar{x}]$, $c.CS$, $\emptyset$) = FALSE) $C \leftarrow C \setminus \{ c \}$;
30:         **endfor**
31:       **else**
32:         continue;
33:       **if** ($C = \emptyset$) return (FALSE); {if no configuration in C is consistent}
34:       **else**
35:         **for** each configuration c in C
36:           $c \leftarrow$ ***action***(t_c, e, c) ; {perform actions associated with t_c which is the executable transition corresponding to c}
37:           **if** (c = NULL)
38:             end the for loop;
39:           **else**
40:             **if** (c is the first configuration in C)
41:               add c to *trace*;
42:               $c_{current} \leftarrow c$;
43:             **else**
44:               build a new trace *branch_trace*;
45:               add c to *branch_trace*; {create a new branch}
46:               add *branch_trace* to *tree*;
47:         **endfor**
48:   **endwhile**
49:   return (TRUE); {a trace compatible with E is found}
50: **End**

Algorithm 4. Algorithm $action(t_c, e, c)$

1: **Given:** a transition t_c,
2:     an observed I/O events e, and
3:     the current configuration c
4: **Return:** new configuration c_{next}, or {the configuration after transition t_c}
5:             NULL {construction failed}
6: **Begin:**
7:     $local\_var \leftarrow$ set the values of the set of local variables according to e;
8:     $c_{next} \leftarrow c$;
9:     $assignments \leftarrow t_c.A(\bar{x}, \bar{y})$; {put the assignments in $t_c.A(\bar{x}, \bar{y})$ into a vector}
10:    **while**($assignments$ is not an empty sequence)
11:        $a \leftarrow$ remove(a, $assignments$); {pick the first assignment}
12:        replace the local variables in a using $local\_var$; {the first step}
13:        if ($a.LHV$ is a local variable) {the second step}
14:            $q \leftarrow$ find the index of variable $a.LHV$ in $local\_vars$;
15:            $local\_vars[q] \leftarrow a.RHE$; {replace by the value of RHE}
16:        else
17:            $q \leftarrow$ find the index of variable $a.LHV$ in $c.[\bar{x}]$;
18:            $[x_q] \leftarrow R(a.RHE)_{[\bar{x}]}$; {update the interval of $a.LHV$ in $[\bar{x}]$}
19:            if ($a.LHV$ appears in $a.RHE$)
20:                **for** every constraint cs in $c_{next}.CS$ that contains $a.LHV$
21:                    replace the $a.LHV$ in cs by $(a.LHV - \sum_{i=1, i \neq q}^{k} a_i x_i)/a_q$;
22:                **endfor**
23:            else {if $a.LHV$ does not appear in $a.RHE$}
24:                **for** every constraint cs in $c_{next}.CS$ that contains $a.LHV$
25:                    replace the variable $a.LHV$ in cs with $[x_q]$;
26:                    change a to a new constraint cs';
27:                    $c_{next}.CS \leftarrow c_{next}.CS \cup cs'$; {add this new constraint}
28:                **endfor**
29:        **endwhile**
30:        return (c_{next});
31: **End**

is a global variable, we first replace the interval of LHV in $c.[\bar{x}]$ by the value of interval $R(RHE)_{[\bar{x}]}$, then update the constraints containing LHV in $c.CS$. If LHV appears in RHE, for every constraint cs in $c.CS$ that contains LHV, we replace LHV in cs by $(a.LHV - \sum_{i=1, i \neq q}^{k} a_i x_i)/a_q$. If LHV does not appear in RHE, for every constraint cs in $c.CS$ that contains LHV, we replace the occurrences of LHV with $[x_q]$ and add the assignment to $c.CS$ as a new constraint. For example, the assignment "$x_1 := x_2 + x_3 - 3$" can be added as a constraint "$x_2 + x_3 - x_1 = 3$". Note that in [10], in the situation where LHV does not appear in RHE, all the constraints in $c.CS$ containing LHV will be discarded. However, those discarded constraints may contain constraints on not only LHV but also other global variables. Considering this, we keep those constraints and replace LHV in them by the interval of LHV in $[\bar{x}]$.

3.5   Optimization on Constraints

In algorithm *Check_Trace* and function *action*, evaluating and storing constraints are complex and time consuming. In order to reduce the complexity, we optimize the constraint related operations as follows: First, the values of global variables are represented by intervals. For a variable $x_i = [\underline{x_i}, \overline{x_i}]$, if its lower bound is equal to its higher bound (i.e. $\underline{x_i} = \overline{x_i}$), [10] considers the value of variable x_i as a *determined value*. Whenever the value of a global variable is determined, [10] replaces this variable in constraints with its determined value. For example, given the variable $x_1 = [1, 1]$ and a constraint cs: $x_1 + x_2 - x_3 = $ [-1, 5], x_1 in cs can be replaced by 1. Therefore, the new constraint after replacement would be cs: $x_2 - x_3 = $ [-2, 4]. We adopt this replacement strategy in our approach.

Second, consider the situation in which a new constraint cs contains a single variable in the expression, for example $x_1 \leq 8$. It would be unnecessary to check cs with former constraints in $c.CS$ and keep it in $c.CS$. Instead, we use cs to directly narrow the interval of this variable in $[\bar{x}]$. For example, given the existing interval of x_1 in $[\bar{x}]$ as $x_1 = [0, 20]$, and a new constraint cs as $x_1 \leq 8$, the narrowed interval is $x_1 = [0, 8]$. If the narrowed interval is not empty, we use the narrowed interval to replace the existing interval in $[\bar{x}]$. Otherwise, we report that an inconsistency is found.

Third, when searching for a compatible trace of E, a transition t in M may be encountered more than once, i.e. the observed I/O event e_i and e_k ($i \neq k$) in E may correspond to the same transition t in M. In this case, we may have two constraints cs_1: and cs_2: $cs_1 : \sum_{i=1}^{k} a_i x_i = I_1$ and $cs_2 : \sum_{i=1}^{k} b_i x_i = I_2$ such that $\forall i, 1 \leq i \leq k$, $a_i = z \times b_i$ where z is a constant. We will call cs_1 and cs_2 *similar*. For example, $x_1 + x_2 = [1, 2]$ and $3x_1 + 3x_2 = [0, 9]$ are similar. Then, given a new constraint cs, if there is a constraint within $c_{current}.CS$ that is similar to cs, we can reduce the number of constraints that need to be checked by the Hybrid method. In order to determine whether there is a constraint cs' in current CS that is similar to cs, we apply the following algorithm (called *Similarity_Checking*) before checking cs with function *Interval_Refinement*. If a constraint cs' similar to cs is found, we replace the interval of constraint $cs'.I$ by $(cs'.I \times z) \cap cs.I$. Thus, by applying algorithm *Similarity_Checking*, we can reduce the number of constraints that need to be checked by the Hybrid method.

4   Experiments

We made an experimental comparison of Interval Refinement, Simplex and Hybrid methods for EFSM-based passive fault detection on the ATM system of Figure 1. Within the SEFSM *ATM*, there are five global variables, i.e. $\bar{x} = ($ *attempts, pin, lang, op, cb*); seven states, i.e. $S_0 = \{s_0, s_1, \ldots, s_6\}$; and fifteen transitions. Local variables are defined within transitions. Each global variable is assigned an interval standing for its initial values. S_0 is determined by the

Algorithm 5. Algorithm *Similarity_Checking*

1: **Given:** a new constraint cs ($\sum_{i=1}^{k} a_i x_i = I_1$), and
2:         a set of existing constraints CS
3: **Return:** FALSE, or   {inconsistency detected}
4:         TRUE   {no inconsistency detected}
5: **Begin:**
6:   **for** each constraint cs' in CS {cs':$\sum_{i=1}^{k} b_i x_i = I_2$}
7:     **if** (cs and cs' are similar)
8:       $z \leftarrow a_i/b_i$;
9:       $cs.I \leftarrow (cs'.I \times z) \cap cs.I$;
10:      **if** ($cs.I = \emptyset$) return (FALSE); {cs is inconsistent with cs'}
11:      **else**
12:        $cs'.I \leftarrow cs.I$;return (TRUE); {cs is consistent with cs'}
13:   **endfor**
14:   return (TRUE);   {no inconsistency is found by cs}
15: **End**

tester according to the specific application at hand. In this experiment, S_0 is chosen to be equal to S.

In the experiment, we considered two cases. In Case I, called *correct implementation*, there is at least one trace in M that is compatible with E and this compatible trace is expected to be reported. In this case, we randomly generate a sequence E_s of observed I/O events ($|E_s| = 1000$) based on the SEFSM ATM and starting from state s_0. Within E_s, we randomly select five sequences with lengths of 20, 50, 100, 200, and 500 observed I/O events.

In Case II, called *faulty implementation*, there is no trace in M that is compatible with E and "faulty" is expected to be reported. First, we create a faulty specification ATM' from ATM by altering the next state, expanding a constraint in the predicate, or narrowing a constraint in the predicate of a randomly selected transition. Then, we randomly generate a sequence E_s of observed I/O events ($|E_s| = 1000$) based on the SEFSM ATM' and starting from state s_0. Within E_s, we randomly select ten sequences containing the altered transition with length of 30 observed I/O events.

We compared three implementations. The first implementation is the Hybrid method; the second implementation replaces the Hybrid method by the Interval Refinement method so that a transition is checked only by the Interval Refinement method (the same as in [10]); the third implementation replaces the Hybrid method by the Simplex method so that a transition is checked only by the Simplex method (the same as in [11]).

According to the results, in Case I, all three implementations successfully find the corresponding traces. In Case II with next state fault and expanded constraint fault, all three implementations report fault correctly. But, the fault with narrowed constraint cannot be detected by all the three implementations because an observed I/O event generated by narrowed constraint will certainly satisfy the original constraint.

Figure 4, left, compares the efficiency of these three implementations in terms of the average time cost. According to the results, the Interval Refinement method requires the least amount of time; the Hybrid method requires a little bit more time than the Interval Refinement method; and the Simplex method is the most expensive in terms of time. As the length of sequence E of observed I/O events increases, the time consumed for these three methods all increases.

Moreover, to compare the rate of increase of time costs, along with the increase of $|E|$, we compute the average rate of time costs between (1) Simplex method and Interval Refinement method (Simplex/IR); (2) Hybrid method and Interval Refinement method (Hybrid/IR). In Figure 4, right, we see that the time costs of the Interval Refinement method and Hybrid method are quite similar and, with the increase in the length of E, the difference between these two methods is not noticeable. We also see that the time cost of the Simplex method is much more than that of the Interval Refinement method, and as the length of E increases, the disparity between these two methods also increases.

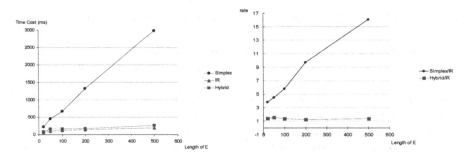

Fig. 4. The results of Case I by applying the Hybrid method, Interval Refinement method, and Simplex method (left) and rates of time cost of three methods (right)

5   Conclusions

In this paper, we have proposed an approach for EFSM-based passive fault detection which provides information about possible starting state and possible trace at the end of passive fault detection; and utilizes a Hybrid method which combines the use of both Interval Refinement and Simplex methods for performance improvement during passive fault detection. Through experiments, we show that, compared with using only the Interval Refinement or only the Simplex method, the Hybrid method guarantees the correctness of results with a reasonable time cost.

In future research, some model checking techniques can be adopted in the proposed approach for EFSM-based passive fault detection to help exploring the Trace-Tree. Also, it would be interesting to see how our proposed approach can help solving the problems of fault location and fault identification.

Acknowledgments

This work is supported in part by the Natural Science and Engineering Research Council of Canada under grant RGPIN 976 and CITO/OCE of the Government of Ontario. The authors wish to thank Dr. Fan Zhang for many useful discussions.

References

1. Alcalde, B., Cavalli, A., Chen, D., Khuu, D., Lee, D.: Network protocol system passive testing for faulty management - a backward checking approach. In: de Frutos-Escrig, D., Núñez, M. (eds.) FORTE 2004. LNCS, vol. 3235, pp. 150–166. Springer, Heidelberg (2004)
2. Alcalde, B., Cavalli, A.: Parallel passive testing of system protocols c towards a real-time exhaustive approach. In: International Conference on Network, International Conference on Systems and International Conference on Mobile Communications and Learning Technologies (ICN/ICONS/MCL 06), pp. 42–42 (2006)
3. Arnedo, J.A., Cavalli, A., Nunez, M.: Fast testing of critical properties through passive testing. In: Hogrefe, D., Wiles, A. (eds.) TestCom 2003. LNCS, vol. 2644, pp. 295–310. Springer, Heidelberg (2003)
4. Bayse, E., Cavalli, A., Nunez, M., Zaidi, F.: A passive testing approach based on invariants: Application to the wap. Computer Networks and ISDN Systems 48, 247–266 (2005)
5. Cavalli, A., Gervy, C., Prokopenko, S.: New approaches for passive testing using an extended finite state machine specification. Information and Software Technology 45, 837–852 (2003)
6. Cavalli, A., Vieira, D.: An enhanced passive testing approach for network protocols. In: International Conference on Networking, International Conference on Systems and International Conference on Mobile Communications and Learning Technologies (ICN/ICONS/MCL06), pp. 169–169 (2006)
7. Chen, D., Wu, J., Chu, T.: An enhanced passive testing tool for network protocols. In: International Conference on Computer Networks and Mobile Computing (ICCNMC 03), pp. 513–516 (2003)
8. Hansen, E., Walster, G.: Global Optimization Using Interval Analysis, 2nd edn. New York: Marcel Dekker Inc (2004)
9. Lee, D., Netravali, A.N., Sabnani, K.K., Sugla, B., John, A.: Passive testing and applications to network management. In: IEEE International Conference on Network Protocols (ICNP97), pp. 113–122 (1997)
10. Lee, D., Chen, D., Hao, R., Miller, R.E., Wu, J., Yin, X.: A formal approach for passive testing of protocol data portions. In: IEEE International Conference on Network Protocols (ICNP02), pp. 122–131 (2002)
11. Lee, D., Chen, D., Hao, R., Miller, R.E., Wu, J., Yin, X.: Network protocol system monitoring c a formal approach with passive testing. IEEE/ACM Transactions on Networking 14, 424–437 (2006)
12. LP_Solve: Tool   lp_solve,   version   5.5.0.9   (2007) http://lpsolve.sourceforge.net/5.5/
13. Marriott, K., Stuckey, P.: Programming with Constraints: An Introduction. MIT Press, Cambridge, Mass (1998)
14. Miller, R.: Passive testing of networks using a cfsm specification. In: IEEE International Performance, Computing and Communications Conference (IPCCC98), pp. 111–116 (1998)

15. Miller, R., Arisha, K.: On fault location in networks by passive testing. In: IEEE International Performance, Computing and Communications Conference (IPCCC00), pp. 281–287 (2000)
16. Miller, R., Arisha, K.: Fault identification in networks by passive testing. In: 34th Annual Simulation Symposium, pp. 277–284 (2001)
17. Miller, R., Arisha, K.: Fault identification in networks using a cfsm model by passive testing. Technical report, UMIACS (2001)
18. Miller, R., Arisha, K.: Fault coverage in networks by passive testing. In: International Conference on Internet Computing, pp. 413–419 (2001)
19. Moore, R.: Interval Analysis. Prentice-Hall Inc, Englewood Cliffs, N.J (1966)
20. Spielman, D., Teng, S.: Smoothed analysis: Why the simplex algorithm usually takes polynomial time. Journal of the ACM 51, 385–463 (2004)
21. Tabourier, M., Cavalli, A.: Passive testing and application to the gsm-map protocol. Information and Software Technology 41, 813–821 (1999)
22. Wu, J., Zhao, Y., Yin, X.: From active to passive: Progress in testing of internet routing protocols. In: IFIP FORTE01, pp. 101–118 (2001)
23. Zhao, Y., Yin, X., Wu, J.: Online test system, an application of passive testing in routing protocols test. In: 9th IEEE International Conference on Networks, pp. 190–195 (2001)

Test Data Variance as a Test Quality Measure: Exemplified for TTCN-3[*]

Diana Vega[1], Ina Schieferdecker[1,2], and George Din[2]

[1] Technical University Berlin, Franklinstr. 28/29, D-10623 Berlin
{vega,ina}@cs.tu-berlin.de
[2] Fraunhofer FOKUS, Kaiserin-Augusta-Allee 31, D-10589 Berlin
{schieferdecker,din}@fokus.fraunhofer.de

Abstract. Test effectiveness is a central quality aspect of a test specification which reflects its ability to demonstrate system quality levels and to discover system faults. A well-known approach for its estimatation is to determine coverage metrics for the system code or system model. However, often these are not available as such but the system interface only, which basically define structural aspects of the stimuli and responses to the system.

Therefore, this paper focuses on the idea of using test data variance analysis as another analytical approach to determine test quality. It presents a method for the quantitative evaluation of structural and semantical variance of test data. Test variance is defined as the test data distribution over the system interface data domain. It is expected that the more the test data varies, the better the system is tested by a given test suite. The paper instantiates this method for black-box test specifications written in TTCN-3 and the structural analysis of send templates. Distance metrics and similarity relations are used to determine the data variance.

1 Introduction

Today's test specifications used in industry and for standardised test suites are usually complex (several hundred test cases, several thousand lines of test code, etc.). As they are hard to evaluate and assess, test quality aspects are constantly subject of discussions. Various test metrics have been developed already measuring selected aspects [1,2,3]. Therefore [4] provided a framework for the different quality aspects of test specifications: it proposed a quality model for test specifications based on the ISO/IEC 9126 [5] quality model. The concrete quality analysis for *Testing and Test Control Notation* (TTCN-3) test suites however concentrated on internal quality aspects only — to analyse potentials of test suite reuse and maintenance.

However, TTCN-3 [6] being standardized by *European Telecommunications Standards Institute* (ETSI) allows not only to specify tests abstractly, but also to make them executable by compilation and execution together with additional

---

[*] This work has been supported by the *Krupp von Bohlen und Halbach - Stiftung.*

A. Petrenko et al. (Eds.): TestCom/FATES 2007, LNCS 4581, pp. 351–364, 2007.

run-time components (such as an *SUT adapter*). By that not only the internal, but also the external quality is of interest.

Test effectiveness is the external quality aspect of a test specification which reflects its ability to demonstrate system quality levels and to discover system faults — in other words, its ability to fulfill a given set of test purposes. According to [4], test effectiveness is divided into

- the *suitability* aspect which is characterised by *test coverage*. Coverage constitutes a measure for test completeness and can be measured on different levels, e.g. the degree to which the test specification covers system requirements, system model, system code and alike,
- the *test correctness* aspect which reflects the correctness of a test specification with respect to the system specification or the set of test purposes, and
- finally the *fault-revealing capability* on the capability of a test specification to actually reveal faults.

In practice, both system model and system code are not always available to the testers, for example when testing third-party components, integrated with off-the-shelf components or tested on system and acceptance level. Hence, the test correctness and fault-revealing capabilities are hard and if not impossible to determine. In contrast, system interfaces as such are available (often also provided in terms of interface specifications and/or documentations) test coverage for the system interfaces could be analysed — despite the fact, that a more thorough analysis would be possible if more information in terms of system model and system code would be available. In the latter case, for white-box (structural) testing code coverage metrics and for black-box (functional) testing system model coverage metrics are in use. Traditionally, code metrics have been used only. With the advances of model-based system development, system model coverage metrics have been adapted from code coverage metrics by using state coverage (the counterpart for statement coverage), transition coverage (the counterpart for branch coverage) and alike.

In this paper we investigate the typical, but less comfortable situation where only the system interface is given. The system interface defines the input data to the system (the stimuli) and the output data to the system (the reactions). These are either provided in form of data structures (e.g. accessing the system information directly), of messages (e.g. accessing the system via asynchronous communication means) or by means of operations (e.g. accessing the system via synchronous operation invocations).

This paper describes an approach to analyse system interface coverage by means of test data variance. It is expected that the more the test data variates, the better the system is tested by the given test suite. We concentrate on asynchronous test data only by analysing type and send value templates only[1].

---

[1] Without loss of generality, we do not consider signatures and signature templates. The approach however can be extended to handle the case of synchronous communication.

Please note that although we consider external test quality, we are using analysis methods that do not execute the test suite itself. We use such an approach of analysing the abstract test suite itself as every test execution involves also the SUT. Hence because of the SUT capabilities and quality, a test execution may reveal selected errors only, may allow to execute a small subset of test cases only and alike - although the test suite might principally be able to reveal more errors. Therefore, we are aiming at determining the *error revealing potential* of a test suite instead.

The paper is structured as follows: after reviewing related work in Section 2, the principal approach is explained in Section 3 and the data variance computation method is presented in Section 4. Distance metrics for TTCN-3 types are discussed in Section 5 and further aspects of test data variance are discussed in Section 6. An example is given in Section 7 and details of our implementation are highlighted in Section 8. Concluding remarks in Section 9 complete the paper.

2   Related Work

Independent of a test specification language, the *test data adequacy criterion* remains among the most important factors for the effectiveness of a test. According to [7], a test data adequacy criterion is a formalism to decide if a software has been tested enough by test data. The same author introduces in [8] a theoretical model for the notion of adequacy in the context of test effectiveness. Many other test adequacy criteria are defined in the literature [9] such as the well-known control-flow criteria, but also more complicated ones such as the Modified Condition/Decision Coverage (MC/DC) or Reinforced Condition/Decision Coverage (RC/DC) criteria introduced by Kapoor [10].

Approaches to study how good the test data is selected include the notion of *distance* between programs, where programs are the tested systems. A test data set is considered to be *adequate* if it distinguishes the tested program from other programs which are *sufficiently far* from it, i.e. produce different input-output behavior. Close programs producing same results are considered the same [8]. A similar concept of *Adaptive Random Testing* is provided in [11] where random test inputs rely on the notion of *distance* between the test values. The authors define the object distance in the context of object-oriented programs. In addition, they propose a method to compute the distance and use it to generalize their approach.

Another approach of test data coverage called *statistical coverage* is presented in [12]. The concept of statistical coverage derives from statistical testing and requires continues testing until it is unlikely that new coverage items will appear. The proposed statistical coverage method uses a binomial distribution to compute the probability of finding new coverage item and an associated confidence interval, with the assumption that software test runs are independent of each other.

All these approaches intend to study *test data variance* as a measure of test data values spread over the input domain. Given the very large number of possible inputs (e.g. almost all types have a theoretically unlimited number of values) that could be feed to a program to be tested, the goal is to minimize the number of test cases (in software testing, test data applied to the same system behaviour) in a test suite while keeping test effectiveness as high as possible.

3   The Principal Approach

The basic idea for test data variance of black-box tests is to analyse the coverage of test inputs with respect to the system interface and its structure as depicted in Figure 1.

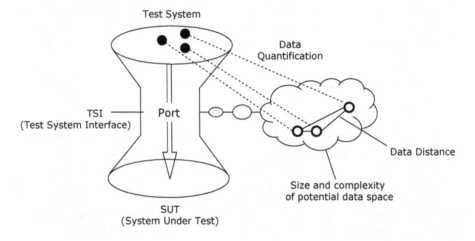

Fig. 1. The principle test data variance approach

In order to get a conceptual framework for presenting our approach, we use TTCN-3 terminology [6]. The *system under test (SUT)* is represented with its interfaces and in relation to the test system only: the *test system interface (TSI)* consists of a number of *ports*. Every port can be of different *port type*. A port type defines the kind of a port to be *message-based* or *signature-based* and to be *unidirectional* or *bidirectional*. For every *test case*, the TSI is defined (explicitly or implicitly) within the *system* clause. By that, one test suite represented by a (set of) TTCN-3 modules can test different TSIs (even of potentially different SUTs, although that is not recommended).

A high test coverage with respect to a given TSI requires that the test data has to fulfill the following criteria:

- Every port and every type transportable to the SUT via that port (incl. every type element in case of structured types) have to be "touched" by the test data.
- The test data has to be representative with respect to a given type. Representative data can be identified either semantically or structurally, i.e. data can vary with respect to qualitative or quantitative similarity.

While quantitative similarity is by use of distance measures easier to derive, qualitative similarity is assumed to provide better results. In particular, the partitioning method can be used to provide a qualitative characterization of an input space. In this paper, we provide a computation method open to both the qualitative and quantitative similarity of test data (see Section 4), but concentrate later on quantitative similarity for TTCN-3 (see Section 5) only as this can be derived purely from the TSI. For the sake of simplicity, we assume in the following

- that all test cases have the *same TSI*,
- that the TSI consists of *one port* only,
- that this port is a *message-based* port, and
- which can transport *data to the SUT*[2].

4   The Data Variance Computation Method

Let us assume a set of test data of a given type being sent over a given port to the SUT, from which we select two (i.e. two TTCN-3 templates resolving to concrete values). Their distance is calculated by use of type specific *distance* metrics. By use of a type specific *distance threshold*, the similarity or dissimilarity of the test data is being determined. The type coverage is finally determined by the number of subsets of similar test data, meaning that a set of dissimilar test data covers a type better than one with similar data.

For that, we consider basically *types T*. As in structured types however elements/fields can be optional, a type extended with omit are being considered in those cases:

$$T' = T \cup \{omit\}$$

Without loss of generality, we consider subsequently T' for types. We partition types into subtypes by considering a value v of T' and the set of values being logically or numerically nearby v:

$partition_{T'} : T' \rightarrow \Pi_{T'}$
for $v \neq omit : partition_{T'}(v) = \{v' \in T : similar(v, v')\}$
$partition_{T'}(omit) = \{omit\}$

The *similarity* of values is a Boolean relation which can be used to determine qualitative or quantitative similarity. For the moment, we restrict ourselves to quantifiable similarity. For that, we use the *distance* between values and a *distance threshold* so that any two values are considered similar whenever their distance is smaller than the distance threshold[3]:

---

[2] This is in fact not a limitation, but a precondition that test data can be sent to the SUT via that port.

[3] Please note that this gives us a dynamic classification of values into sets of similar values — depending on the chosen values, the set of values considered similar will differ. This is different to the qualitative approach of equivalence classes [13], where equivalence classes (also representing data partitions) are statically defined.

$similar_{T'} : T' \times T' \to \mathcal{B}$

for $v_1, v_2 \neq omit : similar_{T'}(v_1, v_2) = \{ \begin{smallmatrix} true \text{ for } distance_{T'}(v_1,v_2)<threshold_{T'} \\ false \text{ otherwise} \end{smallmatrix}$

$similar_{T'}(v, omit) = \{ \begin{smallmatrix} true \text{ for } v=omit \\ false \text{ otherwise} \end{smallmatrix}$

The distance of a type is defined as a float value in between 0 and 1:

$distance_{T'} : T' \times T' \to [0..1]$

for $v_1, v_2 \neq omit : distance_{T'}(v_1, v_2) = \{ \begin{smallmatrix} >0 \text{ as defined in Section 5 for } v_1 \neq v_2 \\ 0 \text{ otherwise} \end{smallmatrix}$

$distance_{T'}(v, omit) = \{ \begin{smallmatrix} 0 \text{ for } v=omit \\ 1 \text{ otherwise} \end{smallmatrix}$

Finally, we determine the *coverage* of a type T' for a given set of values of that type:

$coverage_{T'} : \Pi_{T'} \to [0..1]$

$coverage_{T'}(V) = \#partitions_{T'}(V) \star threshold_{T'}$

where the *number of partitions* determines the number of varying data of a value set for a given type:

$\#partitions_{T'} : \Pi_{T'} \to \mathcal{N}$

$\#partitions_{T'}(\emptyset) = 0$

for $V \neq \emptyset : \#partitions_{T'}(V) = 1 + \#partitions_{T'}(V')$

with $V' = V \setminus partition_{T'}(v)$ for a selected $v \in V$

That completes the data variance computation method, which allows us to determine type coverage based on qualitative or quantitative notion of test data variance.

5   Distance Metrics for TTCN-3 Values

This section defines the distance measures to derive the quantitative similarity of TTCN-3 values being sent to the SUT. The TTCN-3 type system consists of *basic* and *structured* types. Their distance definitions are given in Table 1 and Table 2[4]. Please remember that the distance for *omit* has been already defined in Section 4: it is maximum, i.e. 1, between *omit* and any other concrete value and minimum, i.e. 0, between *omit* and *omit*.

As defined in Section 4, every type T has an associated $threshold_T$, which is the basis to determine data similarity out of the data distance. In spite of our pure quantitative data analysis, our analysis of selected test suites (i.e. for *Session Initiation Protocol* (SIP), *IP Multimedia Subsystem* (IMS), *SS7 MTP3-User Adaptation Layer* (M3UA) and *Internet Protocol version 6* (IPv6)) indicated that a uniform threshold of $\frac{1}{3}$ is a good basis for representing the data variance requirements: $\frac{1}{3}$ means that there should be three representative values

---

[4] We left out the objid type as it is often used together with ASN.1 specifications only.

Table 1. Distance Metrics for Values of Basic TTCN-3 Types

Basic Type	Distance based on	Definition of distance d for values x and y
Integer	One-Dimensional Euclidian Distance	$d(x,y) = \frac{|x-y|}{sizeof(Integer)}$
Float	One-Dimensional Euclidian Distance	$d(x,y) = \frac{|x-y|}{sizeof(Float)}$
Boolean	Inequality	$d(x,y) = \{ \begin{smallmatrix} 0 \text{ for } x=y \\ 1 \text{ otherwise} \end{smallmatrix}$
Bitstring	Hamming Distance	number of positions for which the bits are different (the shorter bitstring is extended into the longer bitstring by filling it with leading '0'B) divided by the longer length: $d(x,y) = \frac{\eth(x,y)}{maxlength(x,y)}$ with $\eth(x,y)$ = number of i where $x_i \neq y_i$
Hexstring	Hamming Distance	same but with leading '0'H
Octetstring	Hamming Distance	same but with leading '0'O
Charstring	Hamming Distance	same but with leading " " (spaces)
Universal Charstring	Hamming Distance	same but with leading " " (spaces)

Table 2. Distance Metrics for Values of Structured TTCN-3 Types

Structured Type	Distance based on	Definition of distance d for values x and y
Record	N-Dimensional Euclidian Distance	$d(x,y) = \frac{\sqrt{\sum_{i=1}^{n}(d(x_i,y_i))^2}}{n}$
Record of	Hamming Distance	$d(x,y) = \frac{\sum_{i=1}^{n}\eth(x,y)}{maxlength(x,y)}$ with $\eth(x,y)$ = number of i where $d(x_i,y_i) > \frac{1}{3}$ and where the record sequence is extended into the longer record sequence by filling it with leading *omit*
Set	N-Dimensional Euclidian Distance	same as for record
Set of	Hamming Distance	same as for record of
Enumerated	Inequality	$d(x,y) = \frac{|\mathfrak{n}(x)-\mathfrak{n}(y)|}{n}$ where $\mathfrak{n}$ is the sequentially numbered index of the enumeration
Union	Distance defined above	$d(x,y) = d(\mathfrak{v}(x),\mathfrak{v}(y)) = \{ \begin{smallmatrix} 1 \text{ for } \mathfrak{v}(x)=\mathfrak{v}(y) \\ 0 \text{ otherwise} \end{smallmatrix}$

such as from the "beginning", "middle" and "end" of a type. Only for the case of *Boolean* and two-value enumerations this threshold should even be reduced to $\frac{1}{2}$. However note that, if we consider an optional field of these types (and hence T' instead) we take $threshold_{T'} = \frac{1}{3}$ as in this case *omit* constitutes an own similarity class of the data being sent to the SUT.

6   Distance Metrics for TTCN-3 Templates

In general, test data to be analysed with respect to their type coverage are not just concrete values, but templates that are sent over the same port to the SUT. These templates constitute a *template subset* to be considered, where the following aspects complicate the analysis:

- *global and local templates*: Templates can be defined in global or local scope. For the latter case, a call flow analysis would be adequate in order to derive the template subset precisely. For the moment, we analyse all templates independently of their scope.
- *template parameters*: Template fields may use parameters directly or parameters within more complicated expressions. In both cases, a symbolic analysis is needed to derive the limitations for the template fields. For the moment, we use for parameterized fields the maximum distance as they have the potential to spread the field type completely.
- *modified templates*: Template modifications are used to change the setting of template fields of global or local templates. The updated field can be defined in terms of concrete values or more complex expressions, which may reference parameters, functions and alike. Currently, we use distances for concrete values only and a maximum distance in all other cases.
- *inline templates*: In this case, the send template is formed directly in the send statement where the values may take any form of expression such as function calls, variables reference and alike. A precise analysis of inline templates requires a combination of call flow analysis with symbolic computation. As for modified templates, we use distances for concrete values only and a maximum distance in all other cases.

These template aspects make the analysis of data variance tricky and demonstrate why the quality of a real TTCN-3 test suite is hard to assess. Our current solution overestimates the coverage of a test suite. However, as tool development is progressing the provided measures will become more and more precise.

7   An Example

In this section, we show how to apply the introduced concepts to small TTCN-3 examples. In the listing below we define a simple TTCN-3 record type R that contains two fields: one of type integer of range (1..100) and an optional boolean field. Based on this type definition, several templates are defined $r1, r2 \ldots, r4$.

Listing 1.1. TTCN-3 Example

1

```
type record R {
        integer  i  (1..100),
        boolean  b  optional
}
```

6

```
template R r1:=  {1 , true}
template R r2:=  {10,true}
template R r3:=  {90,omit}
template R r4:=  {35,false}
```

Assuming that all these templates are used as SUT stimuli over a port that carries R, they form a template subset of interest. The next step is to determine each distance $d(r_i, r_j), i \neq j$ — recursively for the fields as given in Table 3 and Table 4 according to the formulas given in Table 1 and then for the complete record (see Table 5) according to the formulas in Table 2.

Table 3. Distances for boolean record field in the example

	true	false	omit
true	0	0,5	1
false	0,5	0	1
omit	1	1	0

Table 4. Distances for integer record field in the example

	1	10	90	35
1	0	0,09	0,89	0,34
10	0,09	0	0,8	0,25
90	0,89	0,8	0	0,45
35	0,34	0,25	0,45	0

We see that the fields themselves are well covered (and, indeed, we have seen this immediately because of the simplicity of the example). Looking however at the records in Table 5, it shows that R is not well covered.

The records $r1$, $r2$ and $r4$ are similar (and are hence considered stimulating the same system behavior — they are considered belonging to the same similarity class). The separation of the templates into similarity classes is made by comparing the distances between them and a threshold value. The threshold of $\frac{1}{3}$ separates R into three similarity classes: r1,r2 and r4 form one similarity class, r3 a second, but the third is missing. It is not so obvious that although the field types are covered, the record type R is not. The situation can be resolved by

Table 5. Distances for records in the example

	r1	r2	r3	r4
r1	0	0,05	0,67	0,3
r2	0,05	0	0,64	0,28
r3	0,67	0,64	0	0,57
r4	0,3	0,28	0,57	0

Table 6. Distances for added record

	r1	r2	r3	r4
r5	0,49	0,44	0,5	0,41

adding e.g. r5 which represents the third similarity class. The distances between r5 and the other templates are computed in Table 6.

Listing 1.2. Extended TTCN-3 Example

```
:
template R r1:=  {1, true}
template R r2:=  {10,true}
template R r3:=  {90,omit}                                        4
template R r4:=  {35,false}
template R r5:=  {99,true} // added to cover R completely
```

Whenever representatives for similarity classes are missing, approaches for test refactoring [14] and/or pattern-driven test generation [15] can help here to improve the quality of a test suite. Once the analysis has shown that selected interface aspects are not covered, additional templates (to be sent by additional test cases) could be proposed for inclusion into the test suite.

8   Implementation

In order to compute the test data variance and system interface coverage, there is a clear requirement for a TTCN-3 tool to automatically compute the variance measures.

Our implementation is based on the TTworkbench [16] product, an Eclipse-based IDE that offers an environment for specifying and executing TTCN-3 tests. The main reason for selecting this tool, is that it provides a metamodel for the TTCN-3 language which is technically realized by using the *Eclipse Modelling Framework* (EMF) provided by Eclipse. EMF is a Java framework and code generation facility which helps turning models rapidly into efficient, correct, and easily customizable Java code.

The generated Java classes provide an interface useful to traverse every TTCN-3 test suite loaded and access every element of it by creating an associated metamodel instance. The plug-in based structure of the tool allows adding new features by plugging them into the core platform. The incorporated TTCN-3 metamodel is a central test repository that can be used to present the test suite in various formats: in the core language format (the CLEditor is used to edit TTCN-3 code in its textual format) or in the graphical format (for which a GFT Editor is provided).

Our work on the automated template distance collector follows up an earlier work [2] where TTCN-3 test quality indicators are derivedfrom a statical analysis of a TTCN-3 test suite. It is designed as a plug-in whose invocation triggers a) the access to the metamodel instance and b) the traversal of elements of interest as shown in Figure 2. The most significant steps are:

- visit test cases
- identify templates used in send operations
- recursively traverse pairs of templates in order to measure their distance

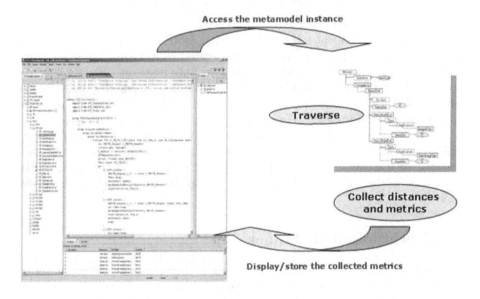

Fig. 2. The principle of the implementation

In the tool terminology, a TTCN-3 project contains all TTCN-3 modules composing an overall test suite. Given a project identifier, the metamodel loader engine is able to load all modules and build a tree-like structure having as root the main module. The code snippet in Figure 3 shows how to extract the *runs on* component name from each test case declaration using the provided EMF API. This one will be used furthermore to find its definition and extract the list of ports used in send direction.

```
Iterator it = testcaseDeclarationListPerModule.iterator();
while(it.hasNext()) {
    mev = (MuTTCNEMFModuleElementView) it.next();

    ValueInterpreter valueInterpreter = ValueInterpreter.create(this.repository.getId(),
                                               (Element)mev.getObject());
    if (valueInterpreter.getInput() instanceof ConstantDeclarationImpl)
    {
        ConstantDeclarationImpl currentTestcase =
                            (ConstantDeclarationImpl)valueInterpreter.getInput();
        if (currentTestcase.getTheType() instanceof FunctionTypeImpl ){
            FunctionTypeImpl currentTestcaseTheType =
                            (FunctionTypeImpl)currentTestcase.getTheType();
            if ( currentTestcaseTheType.getTheToType() instanceof TestcaseBehaviorTypeImpl){
                TestcaseBehaviorTypeImpl testcaseBehaviorTypeImpl =
                            (TestcaseBehaviorTypeImpl)currentTestcaseTheType.getTheToType();
                String runsOnComponentName =
                        testcaseBehaviorTypeImpl.getTheFromType().getTheName().getTheName();

                if(runsOnComponentName.equals(component))
                    testcasesRunsOnComponentList.add(valueInterpreter);
            }
        }
    }
}
```

Fig. 3. Metamodel traversal - code snippet

Every TTCN-3 element has a corresponding EMF element that could be accessed and modified handling only EMF generated Java classes. While searching for simple EMF element definitions translates into accessing directly the tree nodes and getting the needed information, obtaining in parallel the values of two templates whose distance is to be measured, introduces a much more increased degree of complexity. For example, for structured type based templates, it is required to design a visitor that traverses recursively and in parallel every child from each template until values in leaves are reached. Then, the distance formula for templates of basic types is applied to the leaves belonging to the same level in the tree hierarchy and returned to the upper level in a recursive process.

9   Conclusions and Outlook

In this paper we investigate test data variance as a way to assess the test coverage for the system interface quantitatively. We and others consider test data variance as an import factor of test effectiveness. This paper defines a *principal method for deriving test data variance* based on notions of qualitative or quantitative data similarity. This method is then exemplified for TTCN-3 and for *quantitative data similarity*.

We define distance metrics for basic and structured types. A threshold based weighting process of distance evaluation leads to an empirical assessment of data similarity: it is false when the values are different "enough". Different values are counted into separate *similarity classes* representing partitions of a given type where similar values belong to the same partition. The number of present similarity classes of a type in a test suite defines finally the coverage for that type. With the aggregation of the coverage of all stimuli types, we obtain the overall test suite coverage.

Although the approach is in its beginning, it demonstrated already the value of coverage analysis for system interfaces. In future work, the empirical analysis will become more precise with the addition of a dedicated call flow analysis and symbolic execution. In addition, the simplifying assumptions given in Section 3 for defining the data variance computation method are easy to leverage and are being leveraged already in our tool

- by slicing a test suite into sets of test cases with the same test system interface (TSI) and considering the TSIs individually,
- by considering every port of a TSI individually, and
- by expanding the notion of distance, similarity and coverage to signatures and their parameters,

Finally we plan to detail the system interface analysis by extending toward semantical aspects like condition, assertions and behaviours of the interface usage. By all that, we foresee an application of the test data variance analysis in test generation approaches as a control and stopping criteria.

References

1. Sneed, H.M.: Measuring the Effectiveness of Software Testing. In: Beydeda, S., Gruhn, V., Mayer, J., Reussner, R., Schweiggert, F. (eds.) Proceedings of SOQUA 2004 and TECOS 2004. Lecture Notes in Informatics, vol. 58, Gesellschaft für Informatik (2004)
2. Vega, D.E., Schieferdecker, I.: Towards quality of TTCN-3 tests. In: Gotzhein, R., Reed, R. (eds.) SAM 2006. LNCS, vol. 4320, Springer, Heidelberg (2006)
3. Zeiss, B., Neukirchen, H., Grabowski, J., Evans, D., Baker, P.: Refactoring and Metrics for TTCN-3 Test Suites. In: Gotzhein, R., Reed, R. (eds.) SAM 2006. LNCS, vol. 4320, Springer, Heidelberg (2006)
4. Zeiß, B., Vega, D., Schieferdecker, I., Neukirchen, H., Grabowski, J.: Applying the ISO 9126 Quality Model to Test Specifications Exemplified for TTCN-3 Test Specifications. In: Software Engineering 2007 (SE 2007). Lecture Notes in Informatics, Copyright Gesellschaft für Informatik, Köllen Verlag, Bonn (2007)
5. ISO/IEC: ISO/IEC Standard No. 9126: Software engineering – Product quality; Parts 1–4. International Organization for Standardization (ISO) / International Electrotechnical Commission (IEC), Geneva, Switzerland (2001-2004)
6. ETSI: ETSI Standard ES 201 873-1 V3.2.1 (2007-03): The Testing and Test Control Notation version 3; Part 1: TTCN-3 Core Language. European Telecommunications Standards Institute (ETSI), Sophia-Antipolis, France (2007)
7. Weyuker, E.J.: The evaluation of program-based software test data adequacy criteria. Commun. ACM 31, 668–675 (1988)
8. Davis, M., Weyuker, E.: Metric space-based test-base adequacy criteria. Comput. J. 31, 17–24 (1988)
9. Weiss, S.N.: Comparing test data adequacy criteria. SIGSOFT Softw. Eng. Notes 14, 42–49 (1989)
10. Vilkomir, S.A., Bowen, J.P.: Reinforced condition/decision coverage (RC/DC): A new criterion for software testing. In: ZB, pp. 291–308 (2002)

11. Ciupa, I., Leitner, A., Oriol, M., Meyer, B.: Object distance and its application to adaptive random testing of object-oriented programs. In: RT '06: Proceedings of the 1st international workshop on Random testing, pp. 55–63. ACM Press, New York (2006)
12. Howden, W.E.: Systems testing and statistical test data coverage. In: COMPSAC '97: Proceedings of the 21st International Computer Software and Applications Conference, pp. 500–504. IEEE Computer Society, Washington, DC, USA (1997)
13. Grochtmann, M., Grimm, K.: Classification trees for partition testing. Software Testing, Verification and Reliability 3, 63–82 (1993)
14. Zeiss, B., Neukirchen, H., Grabowski, J., Evans, D., Baker, P.: Refactoring for TTCN-3 Test Suites. In: Gotzhein, R., Reed, R. (eds.) SAM 2006. LNCS, vol. 4320, Springer, Heidelberg (2006)
15. Vouffo-Feudjio, A., Schieferdecker, I.: Test patterns with TTCN-3. In: Grabowski, J., Nielsen, B. (eds.) FATES 2004. LNCS, vol. 3395, pp. 170–179. Springer, Heidelberg (2005)
16. TestingTechnologies: TTworkbench: an Eclipse based TTCN-3 IDE (2007) http://www.testingtech.de/products/ttwb_intro.php

Model-Based Testing of Optimizing Compilers

Sergey Zelenov and Sophia Zelenova

Institute for System Programming of Russian Academy of Sciences
{zelenov, sophia}@ispras.ru
http://www.unitesk.com

Abstract. We describe a test development method, named OTK[1], that is aimed at optimizing compiler testing. The OTK method is based on constructing a model of optimizer's input data. The method allows developing tests targeted to testing a chosen optimizer. A formal data model is constructed on the basis of an abstract informal description of an algorithm of the optimizer under test. In the paper, we consider in detail the process of analyzing an optimization algorithm and building a formal model. We also consider in outline the other part of the method, test selection and test running. The OTK method has been successfully applied in several case studies, including test development for several different optimizing compilers for modern architectures.

Keywords: model based testing, compiler testing, formalization of requirements, formal data model, test data generation.

1 Introduction

High level programming languages are the main instruments in software development. Translation of source text written in a high level programming language into executable form is performed by software that is traditionally called "compiler".

Compiler defects break execution of entities resulting from translation: their behavior differs from what is specified in the language specification. Defects in executable entities induced by erroneous compiler are hard to detect and find a workaround, thus correctness of executables obtained from an incorrect compiler is always a doubt. Validation and verification of a compiler is an important activity for dissemination of a compiler in industry.

Validation and verification of compilers is always a very complicated. The main source of difficulties is complexity of input and output: the input is a program with a furcated syntax structure and rich set of context constraints imposed by the language specification, the output is an executable in machine or intermediate language and possesses similar or even higher degree of complexity.

The usual way to cope with complications of compiler validation and verification is a decomposition the validation and verification task into several subtasks that in total cover whole functionality of the compiler.

---

[1] OTK stands for "Optimizer Testing Kit".

A. Petrenko et al. (Eds.): TestCom/FATES 2007, LNCS 4581, pp. 365–377, 2007.

Typical compiler includes the following set of functions:

1. analysis of syntax correctness and parsing of input text;
2. semantic check of input;
3. optimization of the internal representation;
4. generation of the output.

There are many papers concerning validation and verification of the first and second functions of compiler. Papers [7,11,18,24] describe various approaches to validation and verification of parsers. Papers [2,5] describe approaches to validation and verification of semantic checkers.

Nowadays the main function of a compiler is optimization, which allows prodicing faster executable programs. So, the main subtask of the compiler validation and verification is the validation and verification of optimizers.

Papers [6,21,22] describes theoretical studies that use various logical calculi for compiler verification.

Papers [12,8,15,13] contain ideas on creation of oracles that check preservation of program semantics during optimizations. The common shortcoming of these methods is that they do not offer any approach for selection of compiler input data.

Study [9] describes an approach to automation of code generator testing. Specifications developed in XASM language were used for automated filtering tests and obtaining reference results. But this approach to test selection is not systematic and very ineffective.

We use testing [3] based on formal specifications and models [17] as the primary tool for compiler validation and verification.

In this paper we present the OTK method of automated test generation for optimizing compiler testing. The method is based on constructing a model of input data of an optimizer under test. The OTK method allows constructing data models and developing generators of tests targeted to testing a chosen optimizer.

The OTK method consists of the following phases:

1. requirements elicitation;
2. formalization of requirements;
3. automated tests generation;
4. tests execution.

In this paper we zero in on the first and second phases. The third and fourth phases have been described in details in [10].

The remainder of the paper is organized as follows. In Section 2 we describe the OTK method. In Section 3 we present practical applications of the OTK method. In Section 4 the paper is concluded.

2   The OTK Method

The OTK method was developed during joint project of ISP RAS and Intel on testing a set of optimizer units of Intel C++/Fortran compiler in 2001–2003.

Most of compilers perform optimization on some internal representation that is built during parsing and semantics analysis. Straightforward approach to verification of the optimization is to build internal representation of some piece of source code and then optimize it.

The problem is that since internal representation is encapsulated in implementation part of a compiler, then it is very uncertain and therefore tests are difficult to build and are not portable even between different versions of the same compiler. Another problem is that test developers may not have an access to the interface of optimizer units, which are working with internal representation of program code[2].

More practical approach to verification is to use purposely built source code. This approach is easier to implement and is more generic. The OTK method implements this approach.

The OTK method is based on UniTESK approach [4,20] to model-based testing and consists of the following phases.

The first phase is requirements elicitation: analytics study an algorithm of the optimization under test, identify input data requirements and categorize them. The result of the phase is a requirements diagram that contains precisely formulated input data requirements, classified into several groups with established links between them. The diagram is used on the following phases.

The first phase is described in Subsection 2.1.

The second phase is formalization of requirements. Elicited input data requirements get specified using appropriate formal notation. Such specification is called formal data model.

The second phase is described in Subsection 2.2.

The third phase is automated tests generation from the formal data model.

The fourth phase is tests execution that results in test reports that contain information about observed compiler behavior.

The third and fourth phases are described in outline in Subsection 2.3. Details may be found in [10].

Reports analysis, defects identification and corrections is beyond the scope of validation and verification. These issues are not discussed here.

2.1   Process of Analyzing an Optimization Algorithm

The first phase of the OTK method is requirements elicitation. An input data requirements are elicited from an abstract description of the optimization algorithm.

An optimization algorithm is formulated using *entities* of some appropriate abstract representation of an input data, for example, control flow graph, data flow graph, symbol table, etc. In order to perform transformations, an optimizer searches for combinations of entities that match some *patterns*, for example, presence of loops in a routine, presence of some specific statements in the loop, presence of common subexpressions, presence of some specific data dependences between statements. Patterns contains entities significant for the algorithm of the optimization. The goal of this phase is to build a UML-like diagram of these entities.

---

[2] In the project of ISP RAS and Intel we have no access to the interface of optimizers under test due to Intel security policy. The only information available was that an optimization algorithm operates similar to the one described in certain section of the Muchnick's book [14].

Here we proceed with step-by-step detailed description of the process of analyzing an optimization algorithm.

First, one should represent the text of the algorithm under consideration in "if–then" form.

Next, one should mark all branch conditions in this text, i.e. all parts of the text that are located between "if" and "then" words. These branch conditions are *patterns* that the algorithm deals with.

Next, one should mark all entities in all patterns.

Example: Induction-Variable Optimizations Algorithm. Let us consider the induction-variable (IV) optimizations (see [14]). An *induction variable* is a variable whose successive values form an arithmetic progression over some part of the program, usually a loop. There are three important transformations that apply to induction variables:

- *strength reduction* that replaces expensive operations, such as multiplications and divisions, by less expensive ones, such as additions and subtractions;
- *induction-variable removal*, when we may remove an induction variable that serve no useful purpose in the program;
- *linear-function test replacement*, when a variable is used only in the loop-closing test and may be replaced by another induction variable in that context.

For simplicity we consider only the principal part of the algorithm, identifying induction variables. Fig. 7 in Appendix presents the "if–then" form of this algorithm. Patterns are printed in italic. Entities in the patterns are underlined. ▷

Next, one should write out a list of all marked entities. Besides, one should add to this list a principal entity that is a common context where the algorithm is applied. For each entity in the list, one should create some unique identifier.

Example: List of IV-related Entities. A principal entity for the algorithm presented in Fig. 7 is a *loop body*. The list of all entities with corresponding identifiers is shown in Table 1. ▷

Next, one should write out a list of all marked patterns. For each pattern, one should create its graphical representation (a diagram of the pattern) as follows.

- The diagram should contain all entities that the pattern has.
- An entity in the diagram is presented in the form of boxed identifier that corresponds to the entity.
- If an entity in the pattern has some properties, then these properties should be reflected in the diagram under the box of the entity by the label of the form "`<property_identifier> : <value>`".
- If two entities in the pattern are related to each other in some way, then this relation should be reflected in the diagram as an arrow link between corresponding boxes. An arrow should be labeled by the identifier of the corresponding relation. All links fall into two categories:

Table 1. List of IV-related entities

Entity	Identifier
variable	Var
instruction of the form $i = i + c$ or $i = c + i$	Inc
loop constant	Const
subexpression	Expr
induction variable	IndVar
basic IV	BIV
dependent IV	DIV
temporary dependent IV	TIV
assignment	Asgn
loop body	Loop

- *aggregation* that means that one entity contains another;
- *reference* that means that entities are related in some another way.

Any arrow that corresponds to aggregation is marked by a bullet point in the beginning of the arrow.

- Any relation between two entities has cardinality that is reflected by the following labels near the end of the corresponding arrow:

  - without label – "beginning" entity relates to exactly one "end" entity;
  - "0..1" – "beginning" entity relates to 0 or 1 "end" entity;
  - "0..n" – "beginning" entity relates to 0 or more "end" entities;
  - "1..n" – "beginning" entity relates to 1 or more "end" entities.

Example: Diagrams of IV-related Patterns. The algorithm presented in Fig. 7 provides us with the following list of patterns:

1. a variable i is modified by exactly one instruction of the form $i = i + c$ or $i = c + i$, where c is a loop constant, and the instruction is unconditionally executable;
2. a variable i is modified by two or more instructions of the form $i = i + c_n$ or $i = c_n + i$, where all c_n are loop constants, and all the instructions are unconditionally executable;
3.1. a subexpression has any of the forms $\{i * c, c * i, i + c, c + i, i - c, c - i, -i\}$, where i is a basic IV, c is a loop constant;
3.2.1. a subexpression has any of the forms $\{i * c, c * i, i + c, c + i, i - c, c - i, -i\}$, where i is a dependent IV, c is a loop constant, and the subexpression is located after modification of i;
3.2.2. a subexpression has any of the forms $\{i * c, c * i, i + c, c + i, i - c, c - i, -i\}$, where i is a temporary dependent IV[3], c is a loop constant;
4. a subexpression described in the patterns 3.1, 3.2.1, 3.2.2 is assigned to a variable k, and all assignments to k are unconditionally executable;

---

[3] Any temporary variable is always defined before use.

5. there are two or more cases described in the pattern 4 of modification of one variable k.

The corresponding diagrams are presented in Fig. 1.

The property **uncond** reflects that corresponding instruction is unconditionally executable, the property **kind** keeps the information about form of a subexpression, the property **afterIV** reflects that a subexpression is located after modification of used induction variable.

Links **iv** in the patterns 3.1 and 3.2.1 are references since one induction variable (basic or dependent) may be used in several different subexpressions. The other links in the patterns are aggregations. ▷

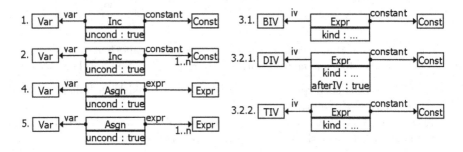

Fig. 1. Diagrams of IV-related patterns

Next, one should improve the diagrams of the patterns, i.e. make the information presented on the patterns more exact: Some entities, links or properties in the diagrams may be renamed or added. The source for such an improvement are those parts of the algorithm that have not been considered yet, i.e. "then" clauses.

Example: Improved Diagrams of IV-related Patterns. "Then" clauses of the items 1 and 2 of the algorithm presented in Fig. 7 say that the variables are in fact basic IVs, "then" clause of the item 3.2.2 says that the temporary dependent IV is related to some subexpression, "then" clauses of the items 4 and 5 say that the variables are in fact dependent IVs.

The corresponding improved diagrams are presented in Fig. 2. ▷

Next, one should check if some entities may be specialized. An entity should be specialized if it has different sets of properties and/or links in the patterns. In this case, the initial entity is called a *generalized entity*.

One should reflect the information about generalization and specialization in a special diagram of generalization. Any entity may occur in the diagram of generalization no more then once. Each specialized entity linked to its generalized entity by a special kind of arrow with big white end. A generalized entity possesses only those properties and links that are common for several entities in the patterns. A specialized entity possesses all properties and links of its generalized entity, and besides, it has some additional properties and links.

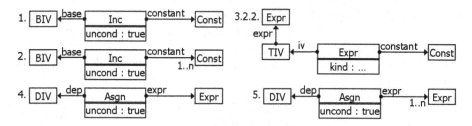

Fig. 2. Improved diagrams of IV-related patterns

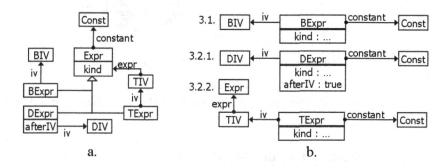

Fig. 3. Generalization of the Expr entity

Next, one should improve the initial diagrams of patterns: Rename the generalized entities to corresponding specialized entities.

Note that not all generalized entities can be renamed during such an improvement. If after the improvement some pattern contains a generalized entity, then this entity may be in fact any of its specialized entity.

Example: Generalization of the Expr Entity. Occurrences of the Expr entity in the patterns 3.1, 3.2.1 and 3.2.2 have different sets of properties and links. Thus, this entity should be specialized. The diagram of generalization is presented in Fig. 3.a.

Now we should improve the diagrams of patterns: We rename the generalized entities Expr in the diagrams of the patterns 3.1, 3.2.1 (Fig. 1), and 3.2.2 (Fig. 2) to specialized entities BExpr, DExpr, and TExpr correspondingly. Note that diagrams of the patterns 4 and 5 can not be improved, since these patterns have no information that may be used for specialization of the Expr entity.

The improved diagrams of the patterns 3.1, 3.2.1 and 3.2.2 are presented in Fig. 3.b. ▷

Finally, one should construct a UML-like data model diagram. Any entity may occur in the data model diagram no more then once. The data model diagram should contain all entities, properties and links that are presented in all the finally obtained diagrams of the patterns. Besides, the data model diagram contains the principal entity that should be linked to some other entities by means of aggregation links.

Example: IV-related Data Model Diagram. A principal entity for the algorithm presented in Fig. 7 is `Loop`. It may contain several `Inc` entities and several `Asgn` entities.

Fig. 4 shows the corresponding data model diagram for the algorithm under consideration. ▷

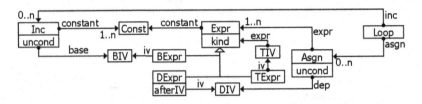

Fig. 4. IV-related data model diagram

The obtained data model diagram is a result of the first phase of the OTK method.

2.2   Formalization of Requirements

The second phase of the OTK method is formalization of requirements. A formal data model is constructed on the basis of the data model diagram elicited on the first phase.

We consider a model representation of a test program as an attributed tree. The role of nodes is played by entities, the role of edges from parents to children is played by aggregation links, the role of attributes is played by properties and reference links.

A formal data model is specified using TreeDL[4] language [19] as follows.

- Each entity is specified using the TreeDL-term "node".
- A generalized entity is specified as an "abstract node", a specialised entity is specified as a derived node.
- A property of an entity is specified as an "attribute" of corresponding node.
- An aggregation link of an entity is specified as a "child" of corresponding node.
- An reference link of an entity is specified as a "attribute late" of corresponding node.
- The cardinality of properties and links is specified using the following modifiers:
  - without modifiers – exactly one element;
  - "?" – 0 or 1 element;
  - "*" – 0 or more elements;
  - "+" – 1 or more elements.

---

[4] TreeDL stands for "Tree Description Language".

Example: IV-related Formal Data Model. Fig. 5 demonstrates a formal data model for the algorithm presented in Fig. 7. ▷

```
node Loop : <OtkNode> {              node DIV : <OtkNode> {
    child Asgn* asgn;                }
    child Inc* inc;                  node Const : <OtkNode> {
}                                    }
node Inc : <OtkNode> {               abstract node Expr : <OtkNode> {
    attribute <boolean> uncond;          attribute <int> kind;
    child BaseIV base;                   child Const constant;
    child Const+ constant;           }
}                                    node BExpr : Expr {
node Asgn : <OtkNode> {                  attribute late BIV iv;
    attribute <boolean> uncond;      }
    child DepIV dep;                 node DExpr : Expr {
    child Expr+ expr;                    attribute <boolean> afterIV;
}                                        attribute late DIV iv;
node TIV : <OtkNode> {               }
    child Expr expr;                 node TExpr : Expr {
}                                        child TIV iv;
node BIV : <OtkNode> {               }
}
```

Fig. 5. IV-related formal data model

2.3 Automated Tests Generation and Tests Execution

Here we proceed with brief description of the third and fourth phases of the OTK method. Detailed description may be found in [10].

The third phase of the OTK method is automated tests generation from the formal data model.

A test coverage criterion is formulated in terms of the data model. A goal of test generation is to cover various combinations of model entities. Tests should contain both combinations that match some of the patterns and combinations that unmatch the patterns in some way. Practice shows that such an approach allows to achieve high level of code coverage of the optimizer under test.

Test program generator is constructed as a structured system of generators of separate data model elements. Such generators in their turn are constructed from generators of subelements, and so on. For example, generator of assignments is usually constructed from two generators of subexpressions and generator of dependent induction variables. All these generators work with model representation of test program structure. The text of test programs appears after applying special mapper component transforming model representation into textual and constructed also on the base of data model structure.

The OTK method is supplied by a tool kit for data model formal description and for developing all required components of a test generator [16].

The fourth phase of the OTK method is automated tests execution.

In the OTK method, an oracle for back-end testing automatically checks preservation of program semantics during back-end pass. To perform this, a mapper should map a model structure to a program with functional semantics being fully described by program's output trace. For such programs, the problem of checking program semantics preservation during optimizer pass is reduced to comparison of output trace of an optimized program with some reference trace.

Checking optimizer correctness is organized as comparison of traces generated by program compiled with optimization and without it.

Example: IV-related Test Program. Fig. 6 shows an example of a test program generated with OTK from the formal data model presented in Fig. 5. The program consists of one loop with several statements, each of which is modification of some induction variable. Some of the statements are located within if-statements that reflect conditionally executable instructions. The program takes several parameters that are used as induction variables, which are modified by the assignments inside the loop and then are printed in the trace. Traces of optimized and nonoptimized programs' executions with several arrays of parameters are compared to find differences in their behavior. Each difference detected is further analyzed for being caused by a bug in an optimizer unit. ▷

```
void f_0(int i_0, int i_1, int s_0, int s_1) {
    int k;
    for( k = 0; k < 100; k++ ) {
        if( cond_asgn() ) {
            s_0 = i_0 - 7;
            s_0 = 7 - i_1;
        }
        if( cond_asgn() ) {
            s_1 = -i_0;
            s_1 = s_0 * 7;
        }
        if( cond_inc() ) {
            i_0 = i_0 + 7;
            i_0 = i_0 + 7;
        }
        i_1 = i_1 + 7;
        i_1 = i_1 + 7;
    }
    printf( "%d %d %d %d\n", i_0, i_1, s_0, s_1 );
}
```

Fig. 6. Example of generated test program for IV optimization

3   Practical Applications

The OTK method was used in several case studies.

During joint project of ISP and Intel in 2001–2003, the OTK method has been applied in testing several optimizing compilers for modern architectures, namely, in GCC, Open64, Intel C++/Fortran compiler.

Test sets developed with the help of the OTK method and targeted to the following compiler's components have been used for testing:

- Common subexpression elimination;
- Jump optimizations;
- Loop fusion optimization;
- Induction variable optimization;

- Linear loop transformations;
- Loop carried dependence detection;
- Register allocation;
- Loop rerolling;
- Subscripts dependence detection;
- Separable and coupled subscripts detection.

Desctiptions of these components may be found in [14,1].

As a result of test execution, several bugs in compilers under test have been found. In the case of GCC testing, we have achieved about 90% of code coverage of the units under test.

During joint project of ISP and Intel in 2004, the OTK method has been successfully applied in testing exception handling mechanism[5] in Intel C++ compiler.

During joint project of ISP and DaimlerChrysler AG in 2005, the OTK method has been successfully applied in testing optimizers of graphical models [23].

Obtained practical results prove effectiveness of the OTK method.

4 Conclusion

This paper presents the OTK method that implements model-based testing approach to optimizing compiler testing. The OTK method supports test development phases starting on requirements elicitation from an algorithm of the optimization under test and ending on automated tests generation and test execution. The process of analyzing an optimization algorithm and building a formal data model is considered in details.

The OTK method is supplied by a tool kit that supports creating formal data models and developing test generators. A generator developed with the help of the OTK allows automatic generating sets of tests that meet a chosen coverage criteria and are targeted to an optimizer under test.

The OTK may be also used in test development for processors of complex structured text.

The OTK method was used in several case studies including commercial compiler testing projects. Obtained practical results prove effectiveness of the OTK method.

References

1. Allen, R., Kennedy, K.: Optimizing Compilers for Modern Architectures. Morgan Kaufmann Publishers, San Francisco (2002)
2. Arkhipova, M.V.: Semantic analyzer tests generation. Numerical Methods and Programming, vol. 7, pp. 55–70 (in Russian) (2006) http://num-meth.srcc.msu.su/english/zhurnal/tom_2006/v7r206.html

---

[5] Checking correctness in this case has been organized as comparison of traces generated by program compiled with compiler under test and compiled by GCC.

3. Beizer, B.: Software Testing Techniques. van Nostrand Reinhold (1990)
4. Bourdonov, I.B., Kossatchev, A.S., Kuliamin, V.V., Petrenko, A.K.: UniTesK Test Suite Architecture. In: Eriksson, L.-H., Lindsay, P.A. (eds.) FME 2002. LNCS, vol. 2391, pp. 77–88. Springer, Heidelberg (2002)
5. Duncan, A.G., Hutchison, J.S.: Using Attributed Grammars to Test Designs and Implementation. In: Proceedings of the 5th international conference on Software engineering, Piscataway, NJ, USA, pp. 170–178. IEEE Press, New York (1981)
6. Hannan, J., Pfenning, F.: Compiler Verification in LF. In: Proc. 7th Annual IEEE Symposium on Logic in Computer Science, pp. 407–418 (1992)
7. Harm, J., Lämmel, R.: Two-dimensional Approximation Coverage. Informatica Journal, 24(3) (2000)
8. Jaramillo, C., Gupta, R., Soffa, M.L.: Comparison Checking: An Approach to Avoid Debugging of Optimized Code. In: Nierstrasz, O., Lemoine, M. (eds.) Software Engineering - ESEC/FSE '99. LNCS, vol. 1687, pp. 268–284. Springer, Heidelberg (1999)
9. Kalinov, A., Kossatchev, A., Posypkin, M., Shishkov, V.: Using ASM Specification for automatic test suite generation for mpC parallel programming language compiler. In: Proc. 4th International Workshop on Action Semantic, AS', BRICS note series NS-02-8, pp. 99–109 (2002)
10. Kossatchev, A.S., Petrenko, A.K., Zelenov, S.V., Zelenova, S.A.: Application of Model-Based Approach for Automated Testing of Optimizing Compilers. In: Proceedings of the International Workshop on Program Understanding. Novosibirsk, pp. 81–88 (2003)
11. Lämmel, R.: Grammar testing. In: Proc. of Fundamental Approaches Software Engineering, vol. 2029, pp. 201–216 (2001)
12. McKeeman, W.: Differential testing for software. Digital Technical Journal 10(1), 100–107 (1998)
13. McNerney, T.S.: Verifying the Correctness of Compiler Transformations on Basic Blocks using Abstract Interpretation. In: Symposium on Partial Evaluation and Semantics-Based Program Manipulation, pp. 106–115 (1991)
14. Muchnick, S.: Advanced Compiler Design and Implementation. Morgan Kaufmann Publishers, San Francisco (1997)
15. Necula, G.: Translation Validation for an Optimizing Compiler. In: Proc. ACM SIGPLAN Conference on Programming Language Design and Implementation, pp. 83–95 (2000)
16. OTK: Optimizer Testing Kit. http://www.unitesk.com/content/category/9/17/35/
17. Petrenko, A.K.: Specification Based Testing: Towards Practice. In: Bjørner, D., Broy, M., Zamulin, A.V. (eds.) PSI 2001. LNCS, vol. 2244, pp. 287–300. Springer, Heidelberg (2001)
18. Purdom, P.: A Sentence Generator For Testing Parsers. BIT 2, 336–375 (1972)
19. TreeDL: Tree Description Language. http://treedl.sourceforge.net/treedl/treedl_en.html
20. UniTESK Technology Web-site. http://www.unitesk.com/
21. Wand, M., Wang, Zh.: Conditional Lambda-Theories and the Verification of Static Properties of Programs. In: Proc. 5th IEEE Symposium on Logic in Computer Science, pp. 321–332 (1990)
22. Wand, M.: Compiler Correctness for Parallel Languages. In: Conference on Functional Programming Languages and Computer Architecture (FPCA), pp. 120–134 (1995)

23. Zelenov, S.V., Silakov, D.V., Petrenko, A.K., Conrad, M., Fey, I.: Automatic Test Generation for Model-Based Code Generators. In: Proc. 2nd International Symposium on Leveraging Applications of Formal Methods, Verification and Validation, ISoLA (2006)

24. Zelenov, S., Zelenova, S.: Automated Generation of Positive and Negative Tests for Parsers. In: Grieskamp, W., Weise, C. (eds.) FATES 2005. LNCS, vol. 3997, pp. 187–202. Springer, Heidelberg (2006)

Appendix

Identifying basic IVs.
We sequentially inspect all variables in all instructions in the body of a loop.

1. If *a variable* i *is modified by exactly one instruction of the form* $i = i + c$ *or* $i = c + i$, *where* c *is a loop constant, and the instruction is unconditionally executable*, then i is a basic IV.

2. If *a variable* i *is modified by two or more instructions of the form* $i = i + c_n$ *or* $i = c_n + i$, *where all* c_n *are loop constants, and all the instructions are unconditionally executable*, then i is replaced by corresponding quantity of different basic IVs.

Identifying dependent IVs.
We repetitively inspect all subexpressions in all instructions in the body of a loop.

3. If *a subexpression has any of the forms* $\{i * c, c * i, i + c, c + i, i - c, c - i, -i\}$, *where* i *is an IV*, c *is a loop constant*, then in the following cases we define new temporary dependent IV j whose value is equal to the subexpression, and we replace the subexpression by j:

   3.1. if i *is a basic IV*, then j depends on i;

   3.2. if

      3.2.1. i *is a dependent IV* or

      3.2.2. i *is a temporary dependent IV*,

      and the *subexpression is located after modification* of i in the body of the loop, then j and i depends on the same basic IV.

4. If *a subexpression described in the item 3 is assigned to a variable* k, *and all assignments to* k *are unconditionally executable*, then we does not define a temporary IV for the subexpression, but we state that k is a dependent IV.

5. If *there are two or more cases described in the item 3 of modification of one variable* k, then k is replaced by corresponding quantity of different dependent IVs.

Fig. 7. The "if–then" form of the principal part of the IV optimizations algorithm (identifying IV) with marked patterns (printed in italic) and marked entities in the patterns (underlined)

Author Index

Lecture Notes in Computer Science

For information about Vols. 1–4450

please contact your bookseller or Springer

Vol. 4499: Y.Q. Shi (Ed.), Transactions on Data Hiding and Multimedia Security II. IX, 117 pages. 2007.

Vol. 4497: S.B. Cooper, B. Löwe, A. Sorbi (Eds.), Computation and Logic in the Real World. XVIII, 826 pages. 2007.

Vol. 4496: N.T. Nguyen, A. Grzech, R.J. Howlett, L.C. Jain (Eds.), Agent and Multi-Agent Systems: Technologies and Applications. XXI, 1046 pages. 2007. (Sublibrary LNAI).

Vol. 4495: J. Krogstie, A. Opdahl, G. Sindre (Eds.), Advanced Information Systems Engineering. XVI, 606 pages. 2007.

Vol. 4494: H. Jin, O.F. Rana, Y. Pan, V.K. Prasanna (Eds.), Algorithms and Architectures for Parallel Processing. XIV, 508 pages. 2007.

Vol. 4493: D. Liu, S. Fei, Z. Hou, H. Zhang, C. Sun (Eds.), Advances in Neural Networks – ISNN 2007, Part III. XXVI, 1215 pages. 2007.

Vol. 4492: D. Liu, S. Fei, Z. Hou, H. Zhang, C. Sun (Eds.), Advances in Neural Networks – ISNN 2007, Part II. XXVII, 1321 pages. 2007.

Vol. 4491: D. Liu, S. Fei, Z.-G. Hou, H. Zhang, C. Sun (Eds.), Advances in Neural Networks – ISNN 2007, Part I. LIV, 1365 pages. 2007.

Vol. 4490: Y. Shi, G.D. van Albada, J. Dongarra, P.M.A. Sloot (Eds.), Computational Science – ICCS 2007, Part IV. XXXVII, 1211 pages. 2007.

Vol. 4489: Y. Shi, G.D. van Albada, J. Dongarra, P.M.A. Sloot (Eds.), Computational Science – ICCS 2007, Part III. XXXVII, 1257 pages. 2007.

Vol. 4488: Y. Shi, G.D. van Albada, J. Dongarra, P.M.A. Sloot (Eds.), Computational Science – ICCS 2007, Part II. XXXV, 1251 pages. 2007.

Vol. 4487: Y. Shi, G.D. van Albada, J. Dongarra, P.M.A. Sloot (Eds.), Computational Science – ICCS 2007, Part I. LXXXI, 1275 pages. 2007.

Vol. 4486: M. Bernardo, J. Hillston (Eds.), Formal Methods for Performance Evaluation. VII, 469 pages. 2007.

Vol. 4485: F. Sgallari, A. Murli, N. Paragios (Eds.), Scale Space and Variational Methods in Computer Vision. XV, 931 pages. 2007.

Vol. 4484: J.-Y. Cai, S.B. Cooper, H. Zhu (Eds.), Theory and Applications of Models of Computation. XIII, 772 pages. 2007.

Vol. 4483: C. Baral, G. Brewka, J. Schlipf (Eds.), Logic Programming and Nonmonotonic Reasoning. IX, 327 pages. 2007. (Sublibrary LNAI).

Vol. 4482: A. An, J. Stefanowski, S. Ramanna, C.J. Butz, W. Pedrycz, G. Wang (Eds.), Rough Sets, Fuzzy Sets, Data Mining and Granular Computing. XIV, 585 pages. 2007. (Sublibrary LNAI).

Vol. 4481: J. Yao, P. Lingras, W.-Z. Wu, M. Szczuka, N.J. Cercone, D. Ślęzak (Eds.), Rough Sets and Knowledge Technology. XIV, 576 pages. 2007. (Sublibrary LNAI).

Vol. 4480: A. LaMarca, M. Langheinrich, K.N. Truong (Eds.), Pervasive Computing. XIII, 369 pages. 2007.

Vol. 4479: I.F. Akyildiz, R. Sivakumar, E. Ekici, J.C.d. Oliveira, J. McNair (Eds.), NETWORKING 2007. Ad Hoc and Sensor Networks, Wireless Networks, Next Generation Internet. XXVII, 1252 pages. 2007.

Vol. 4478: J. Martí, J.M. Benedí, A.M. Mendonça, J. Serrat (Eds.), Pattern Recognition and Image Analysis, Part II. XXVII, 657 pages. 2007.

Vol. 4477: J. Martí, J.M. Benedí, A.M. Mendonça, J. Serrat (Eds.), Pattern Recognition and Image Analysis, Part I. XXVII, 625 pages. 2007.

Vol. 4476: V. Gorodetsky, C. Zhang, V.A. Skormin, L. Cao (Eds.), Autonomous Intelligent Systems: Multi-Agents and Data Mining. XIII, 323 pages. 2007. (Sublibrary LNAI).

Vol. 4475: P. Crescenzi, G. Prencipe, G. Pucci (Eds.), Fun with Algorithms. X, 273 pages. 2007.

Vol. 4474: G. Prencipe, S. Zaks (Eds.), Structural Information and Communication Complexity. XI, 342 pages. 2007.

Vol. 4472: M. Haindl, J. Kittler, F. Roli (Eds.), Multiple Classifier Systems. XI, 524 pages. 2007.

Vol. 4471: P. Cesar, K. Chorianopoulos, J.F. Jensen (Eds.), Interactive TV: a Shared Experience. XIII, 236 pages. 2007.

Vol. 4470: Q. Wang, D. Pfahl, D.M. Raffo (Eds.), Software Process Dynamics and Agility. XI, 346 pages. 2007.

Vol. 4469: K.-C. Hui, Z. Pan, R.C.-k. Chung, C.C.L. Wang, X. Jin, S. Göbel, E.C.-L. Li (Eds.), Technologies for E-Learning and Digital Entertainment. XVIII, 974 pages. 2007.

Vol. 4468: M.M. Bonsangue, E.B. Johnsen (Eds.), Formal Methods for Open Object-Based Distributed Systems. X, 317 pages. 2007.

Vol. 4467: A.L. Murphy, J. Vitek (Eds.), Coordination Models and Languages. X, 325 pages. 2007.

Vol. 4466: F.B. Sachse, G. Seemann (Eds.), Functional Imaging and Modeling of the Heart. XV, 486 pages. 2007.

Vol. 4465: T. Chahed, B. Tuffin (Eds.), Network Control and Optimization. XIII, 305 pages. 2007.

Vol. 4464: E. Dawson, D.S. Wong (Eds.), Information Security Practice and Experience. XIII, 361 pages. 2007.

Vol. 4463: I. Măndoiu, A. Zelikovsky (Eds.), Bioinformatics Research and Applications. XV, 653 pages. 2007. (Sublibrary LNBI).

Vol. 4462: D. Sauveron, K. Markantonakis, A. Bilas, J.-J. Quisquater (Eds.), Information Security Theory and Practices. XII, 255 pages. 2007.

Vol. 4459: C. Cérin, K.-C. Li (Eds.), Advances in Grid and Pervasive Computing. XVI, 759 pages. 2007.

Vol. 4453: T. Speed, H. Huang (Eds.), Research in Computational Molecular Biology. XVI, 550 pages. 2007. (Sublibrary LNBI).

Vol. 4452: M. Fasli, O. Shehory (Eds.), Agent-Mediated Electronic Commerce. VIII, 249 pages. 2007. (Sublibrary LNAI).

Vol. 4451: T.S. Huang, A. Nijholt, M. Pantic, A. Pentland (Eds.), Artifical Intelligence for Human Computing. XVI, 359 pages. 2007. (Sublibrary LNAI).